A PURITAN IN BABYLON

THE STORY OF CALVIN COOLIDGE

A PURITAN IN BABYLON

The Story of Calvin Coolidge

BY WILLIAM ALLEN WHITE

CAPRICORN BOOKS
NEW YORK

CAPRICORN BOOKS EDITION 1965

BY WAY OF PREFACE

ONE book about a man should be enough for a writer. Yet this is my second book about Calvin Coolidge. The first was comparatively a short story, a biographical sketch written during the first year of his occupancy of the White House. That book, in the nature of things, could not interpret his life nor relate him to his times.

He lived in the White House during the six most prosperous years and most portentous that the country had seen since the Civil War. They were the years of the great boom. After those years came the deluge. And Calvin Coolidge represented a phase of those times, probably their dominant ideals. This study of the period from 1923 to 1929, with some account of the relation of President Coolidge to his times, is justification for a new story.

The narrative of Calvin Coolidge's preparation for the White House is a necessary preliminary. His whole life from the time he left college until he came down the mountain from Plymouth to take the train for Washington as successor to President Harding, was one continuous political preparation for the task ahead of him. No other President in our day and time has had such close, such continuous and such successful relations with the electorate as Calvin Coolidge had. He knew the people—their tricks and their manners, their strength and their weaknesses. He knew how demagogues could fool them and how honest men could win or lose them: two most important things for a man to know in politics if he retains his faith. Mr. Coolidge retained his faith to the end. Perhaps his faith was futile. It was not based upon a deep knowledge of various environing realities of his country and of Christendom. He may have lived in a dream world. But at least he lived as nobly as any man could live, equipped as Calvin Coolidge was for his task.

When a man starts to write a story, whether his story is buttressed by historical research or by fictional imaginings, he has a hypothesis to prove, a moral to point. My hypothesis is this: That in the strange, turbulent years that brought an era to a close a man lived in the White House and led the American people who was a perfect throwback to the more primitive days of the Republic, a survival of a spiritual race that has almost passed from the earth. The reaction of this obviously limited but honest, shrewd, sentimental, resolute American primitive to those gorgeous and sophisticated times—his White House years—furnished material for a study of American life as reflected in American business and American politics, which I hope may be worth the perusal of readers who would understand their country in one fine, far day of its pomp and glory.

Furthermore, this book is an attempt, nearly a decade after the Coolidgean era, to set down even in that short perspective the story of the man and his times whose administration closed a distinct epoch in American life. President Harding's administration from 1921 to 1923 was an episode, a sort of intermezzo between President Wilson with his illusion of world peace unfairly negotiated and President Coolidge with his speculative boom. Both were calamities that followed in the train of war. The story of the Coolidge period, a stirring drama, hangs on the undramatic and slight figure of the man who dominated the era, and by his qualities rather than by his words or deeds gave it substance and direction.

In listing the various books and contemporary friends and acquaintances who have aided in the composition of this narrative, I must name first of all Calvin Coolidge. I met him in the White House in December, 1924. I was asked by Collier's Weekly to write a series of four or five articles, more or less biographical, about the President, and Collier's made the appointment for me with the President. He sent a telegraphic invitation for Mrs. White and me to join him and Mrs. Coolidge on a party on the Presidential yacht, Mayflower. A December snow had slowed down western railroads. Our two hours leeway for connections at Chicago between the Santa Fe Railroad and the Pennsylvania was clipped from two hours to ten minutes. The Pennsylvania train master held his train while we crossed Chicago. But our trunk could not travel as fast as we. So we landed in Washington with one little black bag. It was easy for me to rent evening clothes. Mrs. White ransacked the wardrobe of Mrs. Victor Murdock and her daughter, Mrs. Harvey Delano, and appeared in rather gorgeous borrowed plumage as we walked up the gangplank of the Mayflower Saturday noon. Being what we were, we could not conceal the joke on ourselves; and the gay party from the White House, including the President and Mrs. Coolidge, Mr. and Mrs. George Harvey, Mr. and Mrs. David Lawrence, Governor and Mrs. Gore, of West Virginia, made festive and ribald remarks about Mrs. White's borrowed clothes when she appeared; and my hired evening clothes were not without puncture from a few slings and arrows. The gaiety thereunto appertaining early broke what might have been the ice of the party. We lived on the Mayflower until Monday morning. It went a few miles down stream and anchored most of the time. My first experience with the President came on the wharf as we embarked. The camera men wanted a picture of the party. One photographer—a moving-picture man—hustled us around for a silent news reel and then cried:

"Look pleasant, and for Heaven's sake say something—anything; good morning or howdy do!"

To which Calvin Coolidge remarked dryly as he assumed his stage face: "That man gets more conversation out of me than all Congress."

At odd times we talked about the *Collier's* articles. He told me whom to see at Plymouth, Vermont, his birthplace, and at Ludlow, and at Northampton. While we were discussing Plymouth, he spoke most beautifully about it and of the hills and the woods, and of his Vermont people. No constriction of language bound him. He named a number of friends and relatives who should be seen in Plymouth and rasped dryly:

"A lot of people in Plymouth can't understand how I got to be President—," he grinned, "least of all my father!" Then after a pause he added: "Now a lot of those people remember some most interesting things that never happened."

I noticed that President Coolidge never grinned after his jokes to punctuate them. This misled people. They sometimes thought his remarks were dumb. Whatever may be said of Calvin Coolidge, he was not dumb. And many a dumb remark of his afterwards recounted merely reflects the dullness of the narrator. The President warned me against the enthusiasm of his school teacher, and perhaps one or two others, saying they were subject to the usual commercial discount. I tried to talk to him after the first interview, rather unsuccessfully. What I wanted was his slant on things, his point of view, the light that glowed in the inner chambers of his heart. He kept it hooded. He tried to be sociable on the yacht, but he had few social graces. At the first dinner I watched him pecking at the nuts before and between courses, eating lightly but with steady competence as course followed course, and venturing few remarks to the lady on his right—I think Mrs. Harvey. But Mrs. Coolidge did the social work of two. She kept the table organized. She headed off cross talk, the abomination of small dinners, and drew the company into a pleasant interested circle. She made the right people say the right things at the right time, and kept such a flutter of gay volubility going that it made a screen before the silent man at the head of the table. Literally we forgot him. After dinner we met in the boat's salon, and the gaiety of the table followed us in. Also the somber figure of the President came. It sat motionless for an hour like a lone fisherman. Talk was good that evening. But at ten o'clock the President rose, walked circumspectly across the rug to his wife, stood there a rather pathetic moment in silence. She laughed up into his face and mocking his Yankee accent asked:

"Time tew go tew bed!"

He nodded, they led the way out, and the party broke into groups scattered around the boat. The next morning Mrs. White and I, being country people, rose fairly early, a little after seven perhaps. When we came out on deck, we found the President faithfully parading up and down and around alone. He stopped when he saw us. We joined him. Small talk followed until Mrs. Coolidge appeared. Then he and I walked behind the women. It developed that he was expecting his son, John, to arrive on a launch from an early morning train bringing him down from his school.

The train was late and the President was nervous and anxious. After breakfast he still walked the deck and every once in a while stopped to inquire
of Mrs. Coolidge if they were holding breakfast for John. Two or three
times he asked that question, and she told him exactly what John would
have, rather an elaborate breakfast including pancakes and sausage, the
heavy New England breakfast that a boy can carry. But the President was
eager for his son's appearance and chose the side of the deck next to the
wharf for his morning walk.

The death of young Calvin still cast a shadow upon his heart. That
death was only five months behind him and his solicitude for the remaining boy was sadly obvious. When John came, the men of the party went
to a smoking-room where the New York, Baltimore, Washington and
Philadelphia papers, great wads of Sunday papers, were waiting. With a
reportorial eye I watched our host read them. First he chose the New York
World, ignored its news pages, threw away its stuffing and turned to the
editorial page which at the time under Walter Lippmann's direction was
the best editorial page in America. Then the President picked up the New
York Times, turned with the practiced swiftness of a seasoned reader to
the foreign news, skimmed the heads, read one or two European dispatches,
laid it down. Then he picked up the Washington Post and looked for the
Congressional news and the local news of Washington, threw it down. To
the Baltimore Sun he gave serious attention. He found Mr. Frank Kent's
political column and read it, looked for the general news of the country
and scanned the headlines, browsed up and down the editorial columns.
He gave scant attention to the Philadelphia papers, merely looking over
their first pages to see what they were playing up in the way of national
news. Probably he spent an hour in this process and seemed thoroughly
to enjoy it. One got the impression of a man who knew his way around, a
competent, handy person which he was. I have never seen a public man
who more quickly, shrewdly, efficiently got through a pile of Sunday newspapers than Calvin Coolidge. I was interested in his skill. It revealed a
sharp mind, a lively set of brains. He rose abruptly after his morning stint
of reading, walked out of the smoking-room without saying a word; indeed
he had passed less than a dozen syllables during the hour and a half while
we sat watching him above the rims of our papers. He knew what we were
up to and got away from the ordeal as quickly and as decently as he could.
He took a header down below. He did not bob up serenely until midmorn-
ing. The next time we saw him, a group of us, three or four women and a
man or two, were seated on a divan at the end of a long salon, talking festively. It was evident that we were having a rather uproarious time. He
appeared through the door leading from his bedroom and crossed to the
other end of the salon. In half a minute perhaps, he crossed again to his
bedroom, looking our way the second time. Again he crossed after a three
or four minute interval, making a diagonal to an exit somewhat nearer to
us, and again came back, glanced at us, and after a time reappeared. On the
third or fourth lap he had screwed up his courage, overcome his bashful-

ness, and by main strength forced himself to pace into our company. We greeted him in something of the gay tempo of our hilarity. He tried to smile, clipped off a few syllables, and stood by quietly. He did not get aboard the rollicking talk. After five minutes, perhaps ten at most, he turned without a word and walked away. Once or twice in the half hour before luncheon, he pendulated between the entrance and the exit at the far end of the room but did not rejoin his guests. That evening, at dinner, Mrs. White sat next to him, perhaps at his right, and she and he talked together, about Calvin, the boy who died. He told Mrs. White many things about the boy, his ways and his charms. He said that the boy took after his people and looked like Mr. Coolidge's mother and reminded him of her. All the while he was doing a slow free-hand movement from the nut bowl to his mouth. At dessert, a large frozen pudding, gorgeously decorated with December strawberries, was passed to him uncut. He glanced down at it over the end of his nose, shook his head and yapped "no." Mrs. White cut out her portion, and looking him squarely in the eye, asked: "Were you afraid to try to cut it?"

He grinned and said: "I thought I'd see how you got along!"

On this trip Mrs. Coolidge told this characteristic story about the President to the women of the party. A few days before, Alice Roosevelt Longworth had called Mrs. Coolidge on the telephone and asked to see her. The call came at half past ten. Mrs. Coolidge was receiving a delegation from somewhere, officially. Immediately after that delegation moved out Mrs. Coolidge was due at some formal function, a convention or something like it, at the Mayflower Hotel. Another formal reception was due at eleven forty-five, and at twelve thirty the President and Mrs. Coolidge were to receive at luncheon the officers of some national woman's organization convening in Washington. With sincere regret Mrs. Coolidge told Alice her predicament. But Alice, being a Roosevelt, refused to take no. So in a few minutes she came bubbling into the White House, and to Mrs. Coolidge's room. She told the lady of the White House about her forthcoming baby. It had just been definitely and finally verified by the doctor. Whereupon she whirled out of the room as Mrs. Coolidge was coming down to receive the delegation, and on the floor below Alice ran into Ike Hoover, the head usher of the White House, and told him the news. She gave him a hug and sent him spinning. Later by way of making the luncheon go off snappily, Mrs. Coolidge told the assembled lady-delegates the news. It was obviously not intended to be a secret. For Alice in breaking her news was properly pleased and proud. But the news made a buzz at the table and kept conversation going for five minutes. No other man than the President was in the room. He sat looking at his plate, saying no word, apparently oblivious of the clatter around him. Finally a woman well down the table called across to Mrs. Coolidge:

"Well, when will it be?"

"Now if that isn't like me," cried Mrs. Coolidge. "In all the excitement I just forgot to ask!"

The women clattered and yammered and laughed, and in a silence the little man picking up a nut quacked:

"Some time in February I understand!" greatly to the consternation of the women at the luncheon and the astounded surprise of Mrs. Coolidge.

Turning to the women of her party on the *Mayflower* that December day she exclaimed gaily:

"Well, how would you like to live with a man like that? He had known it two weeks and never a word said he to me!"

The next morning before we left the boat, I told Mrs. Coolidge I was not getting what I wanted for my *Collier's* story from the President. I talked to her a moment, trying to tell her my needs. She said nothing. That afternoon the President sent for me to come to the White House at a time when I had a most important engagement, one that I could not decently break, and I didn't break it but asked for another hour. He gave it. When I came in, without waiting for more than the merest formalities he said:

"Well, just what is it you want?"

I replied that I wanted a peek at the man behind the mask.

He gazed at me quizzically a moment and said:

"I don't know as I can help you." Then after a sort of prefatory pause, he opened his eyes widely, revealing a man I had not seen in the three days' acquaintance. They were courageous, sentient, purposeful eyes. He said: "Maybe there isn't any; I don't know!"

But anyway we fell to it and I tried then and there to probe for him, and I think I found him—some of him anyway.

One of the significant things he said to me apropos of his family and its proud provincialism, was: "No Coolidge ever went West!" It was not a mere statement of fact but a truth about the family. They were rooted in New England soil, in New England ways, in New England ideals; old country people these Coolidges, and he was proud of it. We played with the idea for a few moments back and forth, and then I left him.

The next time I saw him was at a family dinner in the White House. The meeting happened this way: I had come to Washington at the suggestion of some farmers' organization, or of certain amalgamated organizations, I forget what. Anyway, I came to talk railroad rates to him and ask his help to get freight reductions for trans-Missouri agricultural products to eastern markets. Chiefly at the moment we of the West desired also rates on grain and hay and livestock. I was primed and crammed with figures about the cost per ton of agricultural freight rates, compared with industrial rates and the rates of coal and iron traveling over the same territory. The rates were much higher for farm products than for the products of the mines and the factories. I went to the White House and asked Bascom Slemp, his private secretary, for an appointment. It was made for ten thirty in the morning. And when I came to meet the President, Mr.

Slemp pleaded an emergency meeting with the Congressional committee and said to come back at three. I came back at three and something had gone wrong with that appointment. Mr. Slemp explaining, added: "But the President would like to have you drop in for dinner this evening and he will go over the matter after dinner."

I appeared at the dinner hour, properly garbed. Ike Hoover, the White House usher, met me at the door. I remembered afterwards, he batted his eye and said: "Why, hello!" with a question mark in his voice, and added: "Glad to see you. Come right into the red room."

I had known Ike Hoover since McKinley's day, had seen much of him in Theodore Roosevelt's day. Occasionally we met under the Taft and Wilson regimes, and twice he had ushered me into President Harding's presence. So he and I chatted for a time, and finally the President and Mrs. Coolidge and Mr. and Mrs. Frank Stearns, of Boston appeared in business garb. We took the elevator to the private dining-room. Obviously a simple family dinner was at hand; beefsteak, fried potatoes, lettuce salad and apple pie—good apple pie. I was feeling fit and told them all my Kansas stories; political stories which I am sure were new to the President, and he was more than cordial when we went into his little office on the second floor of the executive residence. We fell to talking railroad rates after he had showed me the picture of Plymouth and a panoramic photograph of the hills that he knew in his boyhood. He grew lyric about those hills before we got down to business. But when we came to talk railroad rates, he easily outsmarted me. For he knew, possibly he had crammed up as I had, not the comparative rates between agricultural and industrial tonnage but detailed rates comparing the freight rates on hay and grain from the trans-Missouri country to the Ohio and Tennessee valleys and the Great Lakes, where our agricultural markets were, with similar agricultural rates covering the same distances in New England and the Atlantic seaboard. The man actually had the definite rate, or at least quoted it, and I could not deny it, on alfalfa, from the heaviest shipping point of alfalfa in my own county to our heaviest customers in the dairy areas of Wisconsin and Tennessee. Also he quoted the rates on corn from my home town to those same dairy points and compared those rates with similar rates from Northampton to the same distance in Pennsylvania. And of course the Eastern rates were higher than the Western rates. But my point was that the Western farm rate was higher than the rate for Western industrial and mining products. I tried to edge in my point of view. He clipped off figures glibly from his point of view. We came to a deadlock. But we had a fine evening and thrashed it all out, and the faithful Frank Stearns sat through most of it, wordless, leisurely smoking his cigar. We fell to talking politics where we got somewhere. I tried to go at nine thirty. He held me until ten. He walked down the corridor with me to the elevator. We paused a moment. He opened his eyes, flashing a warmth and cordiality that ordinarily was veiled. As I stepped into the elevator and the door was about to close, he said, solemnly:

"Bascom made a slip. He didn't tell us you were coming tonight!"

And with that he turned away with a smile in his heart that wasn't on his face. It had been a revealing evening. He knew that I was coming a week before I came, and what I would talk about. I suppose he was prepared for the meeting. But he did not expect it to occur that night.

I saw him in Kansas City when he came to dedicate the Soldiers' Memorial Monument. Our meeting was brief but pleasant. A short chat, and a more than perfunctory greeting from Mrs. Coolidge, and we parted. When the *Collier's* articles appeared and were bound into a small book, I sent him a copy and had a pleasant acknowledgment from Mrs. Coolidge; no word from the President. But a few months later when I had some sort of public business in Washington and my name was on a program, a letter came inviting me to stay at the White House. My public appearance made it wise for me to decline the invitation. Another and another came within a year or two, when the papers announced that I was to be in Washington. I did not accept. A White House guest is handicapped if he expects to see many people or do as he pleases in Washington. And for that reason I felt that my freedom was important, and I said so frankly and he understood.

I suspect that these invitations to be a White House guest over night were instigated by George Harvey. Here are my grounds for suspicion. Mr. Harvey once was a White House guest for several days and wrote me from the White House. Replying, I indicated my pleasure at his being made a baron of the realm. I declared that in the American book of heraldry a White House luncheon guest was a knight. A White House dinner guest was a lord. A White House overnight guest was an earl. And when a man had been around the place long enough to throw his dirty laundry down the chute, he was a Grand Duke of the realm; which, I indicated to George Harvey in a moment of idle persiflage, was my one aim and object in a long and useless life. Probably he tried to pass my persiflage to Coolidge, but fumbled, and the President took what I had written to Harvey seriously and set about earnestly, kindly and insistently to help me realize my life's ideal!

Our last meeting was in the latter days of his administration. An invitation to luncheon came which I was glad to accept. I found that among the other guests was Marion Talley, the young Kansas City soprano, who was to give a concert that afternoon, and Mrs. Coolidge was to sit in a box. I sat at the luncheon near Mrs. Coolidge. It was a small party and I was near enough to the President so that during the luncheon somewhere after the soup and before the salad she leaned toward me and whispered:

"For Heaven's sake stir up those two at the end of the table."

Marion Talley, at the President's right, a bashful child, was sitting there frozen with shyness, and he, eating with sad sincerity, had uttered no word. I began to tell him about western Kansas where Miss Talley had bought a farm. It is in the high, dry plains with an altitude of something over three thousand feet, quite a different state from the lower part of Kansas in the Missouri valley. Naturally a lot of new Kansas stories were available about the western end of the state. And by appealing to Miss Talley's mother, who was a most human person, we got the ice jam out of the conversational floe at the President's end of the table, and he actually made a quip or two, solemnly and with an uncracked face. As a reward of virtue, Mrs. Coolidge asked me to join her box party at the Talley matinee, but again I had promised to go backstage with a Kansas City newspaperman and had to forego the pleasure of the box party.

My last contact with Calvin Coolidge was in the autumn of 1932. I wrote to his secretary for a photograph, enclosing a check and asking the secretary to have the ex-President autograph it. I received a letter from Mr. Coolidge in which he spoke pleasantly of my request and added cordially that a man who had written a biography of him [1] was entitled to more than a photograph. He "rated an etching." So he sent the etching duly autographed. It is pleasant to remember that our last contact revealed him a kindly, grateful, sentimental man. If sometimes in this biography I have slipped my foot on the scales in weighing evidence to shift the balance in his favor, this closing episode of a pleasant acquaintance explains my feeling toward him.

Now these few casual meetings at which I kept my eyes and ears open and my reportorial mind at work gave me some sense of the enigmatical character of Calvin Coolidge. I had gained a number of rather leading and important facts at those interviews, to wit:

First, here was a shy man.

Second, here was a kindly man, grateful and sentimental.

Third, here was a trained mind; a studious, competent man, tenacious of facts, and capable of coördinating them into a hypothesis that made truth as he saw it.

Fourth, here was a man whose social graces were not formal, but were based upon an instinctive desire to be pleasant rather than punctilious.

Fifth, here was a man intuitively suspicious until his intuition was disarmed, then a gentle and a loyal friend.

Sixth, this exterior crust, this curious shell of sardonic silence which frequently overcast him was a protective armor against an encroaching world which until proven not guilty he seemed to believe was made of two kinds of people, fools and self-seekers, neither of which would he tolerate gladly.

This much he contributed to this book. To him I am deeply grateful for the privilege of seeing him at his daily work. He seemed to me one of the most curious human problems that as a reporter I have ever confronted. And because he and his day of glory were so intricately interlocked, when I considered writing of his times, I was forced to write of him as a part of his times. His leadership then seems an inevitable consequence of his times. They and he were mutually self-serving—one and inseparable.

My second obligation must be acknowledged heartily. Of all the books I have read, and of all the persons I have consulted while writing this story,

[1] Referring to the first little book I had published in 1925.

the "Autobiography of Calvin Coolidge," [2] stands out the most revealing and important single document to which I am indebted. When read casually, it seems rather dry and uninforming. But read as an obbligato to the details of his life's story, read with a knowledge of what events inspired certain paragraphs in his story, the autobiography is unbelievably clarifying. In it, Calvin Coolidge, the man, modest, wise, kind and resolute, stands out a clean-cut figure. It needs the interlinear interpretation of passing history to make it shine; but shine it does if one knows the story upon which his document is based.

In pursuing this study of America in the third decade of the twentieth century, I have consulted many books and talked with many people. I suppose my heaviest personal obligation in this book is to the late Dwight Morrow. I went to him when I first thought of writing about Calvin Coolidge in 1924. He promised to give me fifteen minutes. We talked for three hours. He did most of the talking. When I left him, I made full notes of what he said and many things in this book are taken from those notes. I have had help from several invaluable assistants. Miss Lillian Rixey went through for me the files of the New York Times, a most reliable American newspaper. She covered the years of the Coolidge days in the White House. Also she prepared a bibliography of the Coolidge messages and books and documents. Mr. George Homans, of Boston contributed invaluable matter about Coolidge's Boston career and the Massachusetts background. When Mr. Homans went to other employment, his cousin, Mr. Thomas Adams, continued his work and from time to time has made a valuable contribution to the facts hereinafter recorded. For various reports and accounts of Coolidge's career, I am indebted to Washington and New York newspapermen: J. G. Hayden, of the Detroit News, Mark Sullivan, Arthur Sears Henning, of the Chicago Tribune, Leroy Vernon, of the Chicago News, all stationed in Washington. And in Boston, R. M. Washburn, who wrote the first biography of Coolidge, has taken great pains and has been most kind and helpful, quite apart from his illuminating volume. M. E. Hennessy, who wrote a splendid Coolidge biography, and Frank P. Sibley, of the Boston Globe contributed vital items, and Frank W. Buxton, Editor of the Boston Herald has given me the benefit of his lifetime acquaintance with Mr. Coolidge. William Z. Ripley, whose books on American railways and whose studies in modern corporate organizations entitled "Main Street and Wall Street" have furnished much of my material relating to these matters, has gone out of his way to supply me with many facts and has given me from his own experience the color of Massachusetts politics in the Coolidge-Boston days. A. A. Berle's studies in "The Modern Corporation and Private Property" and "Liquid Claims and National Wealth" have made it possible for me to feel fairly sure when I wrote of the growth and tendency of amalgamated and aggrandized capital in the Coolidge era. Also I have read, and profited by reading, George Soule's "The Coming American Revolution"; Louis Brandeis' "Other

---

[2] Farrar & Rinehart.

People's Money"; Lewis Mumford's "Technics and Civilization"; "The Chart of Plenty" by Harold Loeb and Associates; "Recent Social Trends," and the splendid recapitulation of those two fat volumes in Edward Eyre Hunt's "Audit of America." It was necessary to read Frederick Lewis Allen's "Only Yesterday" to get a renewed picture of Coolidge's time; and John T. Flynn's "Security Speculation" furnished me with many useful facts. Mr. Benjamin M. Anderson, Jr., of the Chase National Bank, New York City, sent to me the files of the Chase Economic Bulletin from 1923 to 1930, and from those bulletins I was able to garner factual statistics which gave me a sense of walking on solid ground in dealing with the economics of the market boom of that time. If ever a prophet has seen his prophecies realized, Mr. Anderson is that prophet. I am indebted to him for the economic hypothesis laid down in this book. I have had invaluable help from Perry T. Hitchens in assembling and intelligently rationalizing all of this economic data which, but for Mr. Hitchens and Mr. Anderson, would have made tough sledding for one untrained in economic matters. Mr. Frank B. Kellogg gave me every opportunity to know the exact truth about the Kellogg Pact and its inception. Drew Pearson's book, "The American Diplomatic Game" helped me to write the story of the Kellogg Pact. I saw it signed. To Professor D. L. Patterson and Professor C. B. Realey of the University of Kansas, and to Professor David MacFarlane of the Kansas State Teachers College I acknowledge help in the survey of contemporaneous European history which in the prologues to the four divisions of the book I have made a sort of obbligato to the American story. I was a delegate and reporter to the National Republican conventions of 1920 and 1928 and a reporter at the convention of 1924. My own reportorial files held the accounts of the nomination of Presidents Harding, Coolidge and Hoover which I have incorporated herein. They seem perhaps more vivid than if I had tried to reconstruct from memory those festive scenes and significant events.

The reader will find occasionally in this narrative reference to the apocrypha of politics. That was a considerable secondary source of information in preparing this book. The apocrypha of politics is, of course, the gossip of politics; but a little more than that. It is the folk tales which are formed, somewhat based upon truth, perhaps more truth than fact, the stories that ought to be true, that grow out of men's estimates of their minor gods. I have found these stories of Calvin Coolidge and have taken none that I have not heard many times repeated. It would be impossible to trace these stories to the first narrator. But they have lived because of certain fundamental survival qualities based upon the commonly accepted size-up of their hero. I have tried, in setting down these apocryphal stories, to take none except those that were seasoned, in a way crystallized, by time and usage.

I am greatly indebted to former President Hoover for guidance in conversation and letters about various public attitudes of President Coolidge whom Mr. Hoover greatly admired. Over and over in his letters and conversation, Mr. Hoover reiterated his belief in Coolidge's intellectual rectitude and his moral courage.

In Northampton, I am under great obligations to Judge Henry Field. He is by odds the most intelligent of Coolidge's Northampton main street friends; understanding, sympathetic, appreciative of Coolidge's strength while realizing the oddities which sometimes pass unfairly for his weaknesses. To Judge Harlan F. Stone I am indebted for many recollections of Amherst in the eighteen nineties and for his characteristically careful comments about Coolidge in the White House. Mr. Frank Stearns also gave me vital facts. He was kind beyond words, as was William M. Butler. I talked profitably with James A. Bailey, a prominent organization Republican, of Boston, and with Tom White, Coolidge's political mentor and manager in Massachusetts, and with S. B. Pearmain who added several striking episodes. I had occasion to quote Herbert Parker, who took some trouble to verify the few sentences of his I have used which are of importance. Also H. Parker Willis, M. A. DeWolfe Howe, W. H. Grimes, of the *Wall Street Journal*, Walter H. Newton, former secretary to President Hoover, James Truslow Adams, and James Jackson, of Westwood, Massachusetts, have contributed illuminating facts. Harry Slattery, of Washington, D. C.; Charles G. Dawes, former Ambassador to England; Wm. S. Culbertson, former Ambassador to Chile; Nelson T. Johnson, Ambassador to China; Louis M. Lyons, of the *Boston Globe*; James Derieux, of Summerville, South Carolina; Norman Lombard, of New York City; Adolph Miller, of Washington, D. C.; Dr. Nicholas Murray Butler, of New York City; Robert T. Brady, of the *Boston Post*; Judson C. Welliver, of Philadelphia; Walter Millis, of New York City; Harriet W. DeRose, of the *Hampshire Gazette*, Northampton, Massachusetts; Miss Mary Randolph, author of "Presidents and First Ladies"; and Joseph R. Grundy, of Bristol, Pennsylvania, have furnished significant facts. I am also indebted to Charles A. Andrews of Amherst, who has furnished me much valuable material about Coolidge's student days. Also I am deeply indebted to the family of the late William Howard Taft, who gave me access to his letters now on file in the Library of Congress, and Val R. Lorwin, of Washington, D. C., who went through the files for me.

Contents

BOOK ONE

BOOK TWO

BOOK THREE

BOOK FOUR

BOOK ONE

PREFACE

Calvin Coolidge was fifty years trudging from his birthplace in Vermont to a place in Boston from which he started on his way to fame. Those fifty years in that part of the world known as Christendom—the West, from the early seventies of the last century to the mid years of the second decade in the new century were years of tremendous stress and change. It was for the most part peaceful change, social and economic and so, of course, political change. The story of this book will tell of those changes in America. But to understand the changes in the United States we must glance for a moment at Europe and at the British Empire. In the western world the changes which were transforming America were matched with similar changes more or less synchronous in our so-called civilized neighbors.

The explosion that followed the puff of smoke in the little town of Sarajevo in the Province of Bosnia in Austria-Hungary, in June, 1914, released social forces that had been gathering for fifty years in the world between the Ural Mountains on the east and Shanghai on the west.

The half century before the Great War saw the rise of the middle class. But also the times witnessed the slow decadence of the individual. As industry and politics were democratized, corporate capital, impersonal, anonymous but vested with power, took charge of the various estates of mankind in what by quaint courtesy we were calling in that half century, the Christian world.

In politics the great leaders had begun to pass from the scene—Cavour, Garibaldi, Bismarck, Gladstone and Disraeli. The generals who led great armies were gone. The day of large, loosely articulated national units in continental Europe was waning. Germany had formed under Bismarck. The Austrian Empire had coalesced under liberal policies of respect for racial minorities. Russia had become nationally conscious. The three emperors of Russia, Austria and Germany were trying futilely to make a pact. But the nationalism of small countries was growing. The dreams of power in the hearts of soldiers who would be world conquerors and kings of kings were fading. Patriotism confined in narrow bounds was holding the loyalties of the people rather than kings and empires.

Inside national boundaries patriotism was but one centripetal force. In each land its commodity industries were coming under corporate control. Large-scale production makes the corporation necessary. Individual capital and partnership capital are not big enough for vast industries, great railroads, great ship companies, and the like. The corporation appeared: a device whereby many individual capitalists pool their resources and limit their risks at the same time that a great majority of them surrender their active control of the use of their capital into the hands of the management of the enterprise.

But also this new device, the corporation, was used for something more than large scale production. It was used to merge competitors into monopolies or partial monopolies. A real distinction must be drawn between vast corporations organized to get the technical advantages of large scale production, and still larger corporations designed to eliminate or reduce competition. One is creative, the other instinctively possessive. The far-visioned captain of industry was merging his interest with his competitors. But acquisitive monopolies were forming in the world's commerce.

The corporation was not the only instrumentality for monopoly. The cartel was appearing. It was a more or less loose association of previously competing corporations which tried to maintain prices, or which allocated certain markets to different ones of its members, or which operated under quota agreements by which its members limited production. It was widely known in continental Europe before the World War. Similar organizations were known as "pools" in the United States. In Germany, they were sanctioned and fostered by government. They prospered in good times, and often broke apart in depressed times. Their influence in Germany, as in America, was on the side of high protective tariffs. They were relatively impotent in England, because England had free trade. A country with free trade does not feel an urgent need for a Sherman Law.

The influence of these monopolies was not always on the side of narrow nationalism. International cartels were formed. And capital of many countries shared risks in many competitive enterprises. Big business in the last third of the nineteenth century on the whole tended toward increasing international economic relations, trade building sentiment for peace. International banking and foreign investment especially made for peace. Rumors of war were much more damaging to a stock exchange which had securities of many countries for sale on its floor than to the stock exchange dealing only in national paper.

The corporation owned the machines of industry—the tools of trade. It represented on the whole organized, aggrandized, acquisitive and possessive forces of society. The corporation sought to control and somewhat did control politics. On the other hand, political parties were growing strong. For democratic privileges—the secret ballot, universal suffrage, the rise of parliaments gave the common man his voice in public affairs. With his ballot he bought or bludgeoned economic privileges and advantages. Moreover, even in Germany, businesses of moderate size flourished and fully held their own alongside the great consolidated concerns, as the "Revisionist Socialist" Bernstein long ago pointed out. The proletariat was becoming bourgeoisie. In the fight for the control of politics, the corporation met a growing sense of social and industrial justice in the hearts of the people. The press was being liberated. Freedom of expression was widening everywhere—a little even in Russia under the Czar.

As the old century went out, it was evident that a new order was coming in. Inside the rising and tightening boundaries of every nation, as organized industry was winding its grip on civilization tighter and tighter, hooping its collective grip around the production of necessities, of comforts and of luxuries, the coil of incorporated capital felt the outward pressure of an expanding sense of justice, the urge for a wider participation in the fruits of man's toil. This sense of justice was manifesting itself in politics. The social democrats were rising in the first decade of the century in Germany. Their fellow socialists were coming into power in France. Even in Russia, the Czar let a parliament meet. In England, in the middle of the first decade, the liberals were putting forth their claims for a larger share of the income of industry. Their fellow liberals were active all over the British Empire.

Universal free education which started to march with the rise of democracy in the last half of the old century was moving men at double quick, not to barricade, but with the ballot and with the strike. Sheer sense of justice was growing in the hearts of the liberalized masses. Thus by reason of creative forces in man's heart and by reason of the dynamic expansion of man's possessive forces, deep social changes were moving in the world.

In the first decade of the new century men were living better in Europe and in the English speaking democracies than ever they lived before. The common men, men who had been born in the lower class, emerged in the new century in the middle class. The middle class was multiplying. In its upper reaches, it was absorbing corporate wealth. In its lower range, it was enjoying higher wages, better food, better housing, wearing better clothes and educating its children with a prodigality that no other middle class generation before had known.

Democratization of every human institution was flowering within national boundaries. The democratization was apparent in politics. It was somewhat evident even in the distribution of wealth through corporation shares, and somewhat in public improvements, schools, parks, public hygiene, slum clearance, a rising living-standard. The shame of oppressors at their own greed was establishing a new sense of justice in the ruling classes. This sense of justice made it easier for the oppressed to realize their demands. A new social morality was creating a new economy. Statute laws, business customs, social institutions and the intangible thing called public sentiment which is an agreed statement of the sense of justice—between rancor on the side of labor, shame on the side of the greedy exploiter of labor—all were breaking down the economy of ancient laissez faire.

Two divergent theories manifested themselves in economic policy in

Europe. The theory of the British, American and Austrian economists was on the whole liberal. The dictum of Adam Smith, that "a rich neighbor is a good customer," dominated British economic policy from the time of the repeal of the Corn Laws in 1846 down to the outbreak of the great World War in 1914. The policy of the British, Dutch and Scandinavian countries, and to a considerable degree of other parts of Europe, was to keep trade lines open and let the merchants, seeking profits, buy and sell in international trade.

German economic theory, however, took a different turn under the nationalistic spirit that followed the unification of the German States into the German Empire after the Franco-Prussian War. Economic theory gave way to an astonishing extent to historical and factual studies in German universities, and governmental economic policy became a curious blend of nationalism, politically dominated, and the economic theory of the great socialist, Karl Marx. The German government and German business men, with Marx, believed in the danger of a general overproduction, and felt that their salvation lay in forced expansion of the export trade in the undeveloped countries. They feared the growth of production in other countries. They sought feverishly to expand their export trade by artificial means in the outlying regions of the world—China, Africa and other undeveloped regions. They also sought to build up colonies as markets for their goods

It is an interesting fact that the colonies were not important German markets, nor were the outlying regions important. Russia was not economically of first importance to Germany, though Germany was highly important to Russia. Russia took nine per cent of Germany's exports. Great Britain and Austria-Hungary were the two greatest markets for German goods. Europe as a whole, outside of Russia, took sixty-seven per cent of Germany's exports. A rich neighbor is a good customer.

But Germany did not believe this, and the widespread belief in the myth of the importance of colonies as markets was a very significant factor in creating the international frictions which culminated in the Great War.

The old struggle for colonies, in the seventeenth and eighteenth centuries, renewed its tension following Germany's unification and rise to power, especially after the dismissal of Bismarck by the young Kaiser Wilhelm. The tension was manifested in the Mediterranean basin. Africa continued to be partitioned. England started her last great colonial war in South Africa as the century was closing, and after that, grew ashamed of the policy. She increasingly relaxed Downing Street's control of the British Empire, and allowed the Boers to regain political control of their own country. By 1914 she had so won their friendship that they supported her loyally when the great struggle came. Her shame with respect to old policies manifested itself in Ireland, where, though physically strong enough to control the increasingly rebellious South of Ireland, she temporized, and concerned herself primarily with protecting the loyal North of Ireland. But Germany set a faster and faster pace in seeking to expand her political power in the outlying regions of the world.

Modern industry, bringing great masses together in cities, has usually been accompanied by the growth of democracy. Democratic nations are usually pacific. The common man is not anxious to go to war, sees little chance for economic advantage in war. A democratic government generally responsive to public opinion strives for peace. On the other hand, modern war must have industrialization. So the countries which were strong enough industrially in 1914 to fight effectively were in general democratic countries which were reluctant to fight. The world saw in 1914 two striking exceptions to this rule. In two countries, industrialization had come, with autocratic government remaining. These two were Germany and Japan. The autocratic government of Germany could gain through a successful war immense prestige for the ruler and military leaders who were close to him. It had, moreover, for two decades been increasingly forcing its policies and its theories upon German universities, upon journalists, and upon other makers of public opinion, and it had even engaged in widespread propaganda for the glory and greatness of Germany in foreign lands.

France remembered the dictated treaty that followed the Franco-Prussian War and her lost provinces of Alsace and Lorraine. She dreamed of La Revanche, but, she feared the rising power of Germany. As German armies grew, French armies grew. With the growth of the German Navy, Britain, feeling her naval supremacy threatened, evidently increased her navy, and later announced the policy of building naval ships faster than Germany could build them. France cultivated friendship with Russia. Russian loans were placed in Paris through government encouragement, not because they were good loans for French investors, but so that Russia might build strategic railways and develop her military power, as well as her general economic power as a foundation for military power. Russia and France became allies. Germany and Austria-Hungary were allies and finally drew Italy into their alliance. Britain tried to preserve detachment and to act as a pacifying influence between the two distrustful groups, but ultimately threw in her lot with France and Russia, becoming part of the Triple Entente. International business and international banking hoped for peace and worked for it. International labor hoped for peace and worked for it. International socialists, leaders of labor, proclaimed that the solidarity of labor as against capital transcended national lines, and hoped and proclaimed that German socialists would refuse to fight French socialists, and French workmen would refuse to fight German workmen, if a test should come. The British and French banks continued to lend to German banks, though French banks greatly reduced their loans to German banks after the Agadir incident in 1911.

Social dynamite was forming, and wise men heard the whirling wheels of industry, and saw the brightened faces of the workers, their hovels falling and new houses rising, as education transformed peasants and working men into voters with dreams of social progress. So wise men said: "What a beautiful world! How grace and mercy, justice and peace are rising in the earth!"

CHAPTER I

The Museum That Was Vermont

CALVIN COOLIDGE was born July 4, 1872, at Plymouth, Vermont. In that sentence lies the embryo biography of the man who governed America during a golden age. The time and the place were environing circumstances which as much as his blood, perhaps even more than his blood, determined the kind of a man the thirtieth President of the United States would be between 1923 and 1929. It is, therefore, essential to consider here and now this environment: first, July 4, 1872; second, Plymouth, Vermont.

The American Civil War was seven years past on July 4, 1872. The Union soldiers of that war, who were to wield most of the power in their country and to enjoy all the glory for the next quarter of a century, had turned their swords into plowshares. These soldiers were scattered over the land from Maine to California. The hand of fate luring these young men across the land was sifting them like fruit through sizing meshes. They fell into thousands of communities—into towns, farms, cities, seafaring villages, industrial centers, isolated ranches, proud suburbs, and ugly slums. The place where they fell determined their way of life. If they fell into the Ohio Valley, they became the builders of a strange, new industrial civilization where farms had been when they marched out under Lincoln's call. If they fell beyond the Mississippi on the prairies and the high plains sloping toward the Rockies, they became pioneers struggling with the soil, building a rural civilization where the buffalo and the Indian had roamed while the young soldiers were fighting the war. If these soldiers sifted by fate fell eastward of the Alleghenies, for the most part they fell into an industrial civilization more intensely unified than the new country in the Middle States. But all of them, across the land from the Atlantic to the Pacific, were controlled by a genie that had come newly into the world—

7

the spirit of steam. Steam was making over American civilization.

Rails for the steam engines first had stretched across the continent, two thin gray lines less than ten years before Calvin Coolidge was born. And when he came to Plymouth, that Fourth of July 1872, literally tens of thousands of miles of rails were being pushed like the threads of a great spiderweb across the continent east and west, not once, not twice, but in a dozen places; then north and south and finally diagonally and crisscross. The railroad was dominating politics, engaging finance, distributing the products of industry far and wide, breaking the frontier and making a new age, a new America; a roaring, greedy America, noisily proclaiming its patriotism and denying the ancient gods of the fathers. The pioneer spirit was going mad. The young soldiers were joined by another army, the immigrant army from the north of Europe. The Irish, the Scotch, the Germans, the Scandinavians were hurrying into America and across the continent helping the young soldiers to build their cities and their states, to erect a shiny new terrible and unbelievably ugly civilization upon the fresh-turned sod. Invention rushed after invention to multiply the power of men's arms, but all the inventions were moved by steam. It was the age of the steam troll!

Coal was coming out of the earth. Black smoke was staining the sky of a thousand towns. The coal smoke from the stacks of innumerable engines that roared across the valleys and the plains into the mountain canyons and down the western slope of the Sierras to the Pacific was bringing the pungent, acrid, sulphurous odor of burning coal for the first time to the nostrils of millions of human beings whose forebears for a long millennium had been wood burners. It was a grimy day, and gold was plastered over the high places in a vain effort to beautify the grime. Great fortunes were amassing. Great scoundrels were grabbing power. In Congress, in the legislatures and in the city councils of the towns these scoundrels, honestly abetted by sincere, high purposed men, were working their will with government. Bribery, ill-concealed, became conventional. During Calvin Coolidge's childhood, through his boyhood well into his youth, scandal burst upon scandal in Washington and in the state capitals in the cities, towns, and villages, as the steam giant ravaged the land.

In the South, the political, economic and social reconstruction that followed the Civil War reeked with corruption, cruelty, and violence. That was a black chapter in our history; because in the South the element of race-hatred and the opportunities for revenge upon a prostrate people made man's inhumanity to man more casual than the greed and vanity that moved men in the noisy North and West. America was a babel of strident voices building a tower to some god of materialism. City and country alike far across the continent in those late sixties and early seventies from Boston to San Francisco were under a sombre curse: the shadow of coal smoke and of white steam. Thus the young soldiers of the Civil War, slaves of the steam, were rebuilding a continent, making a new civilization by destroying the wilderness.

So much for the time when Calvin Coolidge came into the world. Now for the place. Consider Vermont. All this turmoil that was the America of the decade that followed the Civil War, all this soot and ashes of corruption within the gilded sepulchre of the era, scarcely touched the soft green mountains of Vermont. There the old America lay peacefully basking in the ideals of its past. Vermont remained calm and sweet and lovely, like the interior of some cool museum preserving colonial life. Of course in the larger towns and cities some smudge of the shiny god of steam touched men here and there. But after the Civil War for a decade or two or three, back in the hills, in the hamlets, in country towns and in little white villages nestling in the valleys, life went on, unspoiled by the devastation of the times. Plymouth, Vermont, was one of the primitive villages, one of half a hundred exhibits in this social game preserve that is Vermont, where men and things retain even today some semblance of their pristine American quality. Here survive the last of the Puritans.

So we may say that Calvin Coolidge, who came into this world July 4, 1872, was born immediately after the Revolutionary War. Water wheels turned the little grist and saw mills that made the articles of sluggish barter which passed for commerce in that place. The New England farm was self-sufficient, a century after other American farms were becoming industrialized. On Vermont farms the food was grown, the clothing made, the logs and lumber hewn and sawed, and from year's end to year's end the outer world sent back little into this quiet Arcady resting under the spell of a day that long had gone. Only loggers and trappers had commerce with the outer world. Here the soldiers returned from the Great War of the Rebellion were repressed even in their boasting. Repression was a way of life. It was not hard nor sterile, this life that men lived in those green mountains. Nor were the silences of men moody. Life flowed sweetly enough. Its outer aspects did not clatter, and tongues did not need to rattle. In the smooth and even processes of existence, each community had its peculiarities. On one side of the Green Mountains farmers wore gray woolen smocks—shirtlike garments, heavy and loose and comfortable, and men said of a farmer on a village street who appeared carrying a hickory ox-goad and wearing a gray smock and a woolen hat:

"He is an over the mountains man."

The Coolidges came from over the mountains, from the Eastern slope, and wore this peasant smock. It had been their customary garment for a hundred and fifty years. They did not change it with the smart fashions that penetrated the fastnesses of the village a few miles to the west. The Coolidges had come up from Massachusetts before the Revolutionary War. Five generations of them lie buried in the little graveyard near the town of Plymouth. They were decent, hard-working, prosperous people, proud to belong to the ruling class. Sometimes they farmed, somtimes they kept store. But they were always well-to-do according to the standards of the place. They went to the legislature, served as justices of the peace. They had been town clerks, deacons of the church, and were leading citizens. John Coolidge, father of this Calvin Coolidge born that July 4, 1872, had married a Moor, and the Moors had come up from the Connecticut Valley where they picked up a drop of Indian blood.[1] They also belonged to the ruling class, quite the equals of the Coolidges. The Moors lived the simple, rather dignified life of landowners, farmers who grew their own food, made their own clothing, built their own houses, tinkered their own machines, lived within their means, went their quiet ways, but bent their knees and pulled their forelocks to no man! The Coolidges and the Moors were aristocrats who knew no betters and patronized no inferiors. Here the town meeting was almost the complete government of the people. Here were few inequalities of wealth. Here one man with one vote, two hands, a horse and a farm and a birthright in the social order was as good as any other. It was this primitive political, economic and social democracy into which Calvin Coolidge was born, with every man literate, with no one rich, no one poor, a survived vestige of a vanished ideal—the Puritan democracy. The offices of the local government were passed around. In his day John Calvin Coolidge, father of the President, had been elected constable, justice of the peace, selectman, tax collector, fence-viewer, pound-keeper, member of the Vermont House of Representatives, and a member of the Vermont State Senate. Excepting the legislative offices, the Moors had been similarly honored. These offices were hardly marks of distinction, but indications of a readiness for duty. The teachers in the village schools boarded around when the teacher's board was not let to the highest bidder. The Moors and the Coolidges took their turn at boarding the teacher as they took their turn at viewing the fences, collecting the taxes, keeping the pound. In the sixties and seventies of the last century every Vermont farmer took a farm paper, possibly the *Rural New Yorker*, and a Weekly from the big world, probably Horace Greeley's *New York Tribune*; also the nearest local newspaper. They were familiar with a few books, notably the Bible, a bound volume of Shakespeare, Bunyan's "Pilgrim's Progress," the diseases of the horse, a family doctor book, Gaskill's "Compendium of Social and Legal Forms," and probably two or three novels.

[1] Conversation of Calvin Coolidge with author. See also *Good Housekeeping*, March and April, 1935.

2

Into such a home, in such a place and at such a time a hundred years back of the calendar, came Calvin Coolidge. The house in which he was born was a five-room story and a half cottage to which were attached the postoffice and a general store; his father's store. Family tradition says that the Fourth of July baby was named John Calvin Coolidge and that his family presumed he would be called John, but the John was dropped early in his childhood and he was known as Calvin. Probably Calvin Coolidge "took after" his mother more than his father. She was a Moor.

His mother was fair-haired, a New England blonde, probably repressed. For the countryside is filled with wise saws of his father, and with stories of the Coolidges who apparently were a more or less garrulous lot. Not a story persists about the fair, gentle woman whom her son idealized whenever he wrote about her. Tradition pictures her as a flower-loving woman. In the yard behind the white picket-fence where grew a mountain ash beside a clump of purple lilacs and a plum tree that flowered beautifully in the spring, she planted her flowerbed. Her father, Hiram Dunlap Moor, was of Scotch and English descent, with a jigger of Welsh in his blood. But Abigail Franklin Moor, the maternal grandmother of the President, was of New England stock for more than a hundred years. His mother's name was Victoria Josephine which places the time of her birth in the middle of the nineteenth century when Victoria was a young queen and Josephine a romantic memory. The President declares that she was "of a very light and fair complexion with a rich growth of brown hair that had a glint of gold in it. Her hands were regular and finely molded." [2] If her words are forgotten in the valley where she lived, there lingers the memory of her beauty. She bore John Coolidge a daughter three years after the birth of Calvin and was an invalid for seven years. It is only as a sick lady that her son remembers her. Yet he recalls that "there was a touch of mysticism and poetry in her nature which made her love to gaze at the purple sunsets and watch the evening stars." [3] Like most repressed people she was sentimental and her son recalls her love of beauty in form and color and writes: "It seemed as though the rich, green tints of the foliage and the blossoms of the flowers came for her in the springtime. And in the autumn it was for her that the mountain sides were struck with crimson and with gold." [4] The depths of this man's feeling fifty years after she was gone may be indicated by the lyric rhythm in which he puts down his

[2] "Autobiography of Calvin Coolidge."
[3] *Ibid.*
[4] *Ibid.*

memory of her. He remembers her deathbed; that she called the two children, the little boy of twelve and the girl of nine, to her and the two children knelt down at her bedside and received her "final parting blessing. In an hour she was gone. . . . We laid her away in the blustering snows of March. The greatest grief that can come to a boy came to me. Life was never to me the same again." [5]

Heartbreak in childhood and in youth makes deep changes in the channel of one's life and destiny, and those who knew Calvin Coolidge best found in him a tenderness and an unusual capacity for sentimentality, the more powerful because it was repressed. It came to him somewhat out of the memory of his mother whom he reverenced with pious devotion. All through his life we shall observe a repressed sentimentality flickering like a secondary personality, chained in the prison of a smooth, flinty New England exterior. For more than half a century Calvin Coolidge clung to his mother's memory. Like an ikon, he cherished her picture. He died with it in his watch next to his heart. [6]

As a little fellow young Calvin played with his grandfather, Calvin Galusha Coolidge, a tall, gaunt Yankee who stood better than six feet high and was always running for office. His grandson, writing of the elder man fifty years after his death, remembers him as immoderately taciturn. But Bruce Barton who went to Vermont to garner the truth about President Coolidge, gathered from the neighbors there that Old Galusha was a chatterbox! The boy was once greatly impressed at seeing his father's ancient Aunt sitting in the chimney corner smoking a clay pipe after the fashion of old women a hundred years before. The boy remembers that he was shocked. Modern women in his day had dropped the pipe.

This grandfather must have been a delight to the boy. He was a horse-breaker and kept colts and puppies around, and even had a peacock. He lived within a little fenced garden filled with scarlet flowers. And the sadist tendency which sometimes cropped out in the President came from the Coolidge strain, for we are told that the old grandfather was never happier than when playing practical jokes. His most gorgeous moments came when he was able, inadvertently, to lure a man into a nest of bees. It was the native humor. The old gentleman taught young Calvin to ride and also to ride standing on the horse. He was a farmer and the family tradition says that Calvin Galusha was disgraced when Calvin's father began keeping store. The gaffer was of the aristocracy that believes in the soil as the only respectable means of support. Yet he was something of a wastrel. He and his wife brought up two orphan children of an only sister and charged nothing to their estate for their keep. More than that, Galusha actually turned over to the orphans when they were eighteen years old, eight hundred dollars of the grandfather's own money as he did to Calvin's father. The prodigality of that act still lives in story at the Notch.

The old gentleman, hoping to bind young Calvin to the soil, left him, with a deed for life, forty acres of his farm and a mare colt and a heifer calf which came of a stock strain from Calvin's grandfather's grandfather's farm. It was an impressive gift. And on top of that, the old gentleman entailed the forty acres to young Calvin's lineal descendants. And that when the boy was six years old! The iron purpose of the New England will was not to be idly thwarted. Yet he had his sentimental moments, this ancient of days, who was the grandfather of a President. For when his son John, the President's father, was elected to the legislature, the grandfather took the baby Calvin and his mother up to Montpelier to visit the legislator in his grandeur. Later he even carried the child to the State House and set him in the governor's chair. But the stuffed animals in the museum in the State House impressed the little boy, and he knew that he sat in the governor's chair only because they told him. He really remembered to the end the stuffed catamount that was ready to lunge at him. The boy had learned to read the Bible at an early age, and the grandfather probably shared with the mother the credit for teaching Calvin. And when the old man was on his deathbed he asked the child to read the first chapter of the Gospel of St. John,[7] which Galusha explained he in his childhood had read to his grandfather. Little Calvin began:

"In the beginning was the Word, and the Word was with God, and the word was God.

"The same was in the beginning with God. All things were made by him; and without him was not any thing made that was made. In him was life; and the life was the light of men. And the light shineth in darkness; and the darkness—"

And here the boy stumbled. Here was a new word, a long, four-syllabled word for a little boy of six. He could not make it. And the grandfather helped him over it and he read:

"—and the darkness comprehended it not."

And so through the darkness that comprehended it not, Calvin Galusha Coolidge was gathered to his fathers.

[7] We shall meet this passage again in the grandest day of Calvin Coolidge's life. He remembered it sentimentally and turned to it in pride.

CHAPTER II

The Museum Piece

IT IS no wonder that Calvin Coolidge, though never pious, a church member only in his fifties, all his life was deeply religious. Religion was a part of the spiritual air men breathed in that New England, as it was in the earlier New England two generations before. The church life in Plymouth was not well organized. Apparently itinerant preachers of various creeds came to Plymouth Notch. For the most part they were Congregationalists, but occasionally a Baptist or Methodist appeared; and Sunday school at the Notch was held every Sunday. The little fellow went to Sunday school through his childhood and well into his youth. It was an orderly, rather intellectual religion, the folk religion of that time and place. The children heard the Bible read in school every morning and they sang the gospel hymns which celebrated the homely virtues of a thrifty and diligent people. "Pull for the Shore," "Work for the Night Is Coming," "Bringing in the Sheaves," "What Shall the Harvest Be" were favorites. These songs taught the children to work and to save, and influenced their lives probably more than the theology which indoctrinated the jingles. Indeed formal religion did not seem to influence the childhood and youth of Calvin Coolidge. No tradition of the revival nor of any emotional phase of religion survives in the Coolidge story. Doctrinal controversy had no place in Plymouth and particularly in the Notch where the Coolidge store was the core of the social crystal. But nevertheless John Calvin Coolidge and Victoria Josephine did not rear their children like heathen. Children who learn their letters from blocks and begin to read out of the Bible generally pursue their pious way, and when they are old they do not depart from it. And what with the mystery of death and bereavement and the mysticism of the Bible, with all the beauty of the mountains and their ever-changing moods,

14

naturally this New England child became set in his ways, a Yankee mystic to the end.

On the other hand he was trained in the use of things. No course in manual training set down by a wise pedagogue could have done more to make a handy youth out of a New England visionary than the training Calvin Coolidge, as a child and boy, had in the Notch at Plymouth. His grandfather initiated him in the lore of horses and dogs, chickens and sheep, cattle and pigs. Domestic animals were about him from his infancy and he had chores to do almost after his babyhood.

He liked the circus. But his father said of him that as a child Calvin did not care for play, and he himself repeatedly admitted the same thing with perfect frankness. If in later years he took to fishing, it seemed to be more for the gesture than for the pleasure. Certainly there was none of the passion that Grover Cleveland threw into it.[1]

He made kites, cut out arrows from the pine and the ash, made bows from the hickory, had a pocket-knife before he could read, and used the adz before he knew the multiplication tables. After his grandfather's death he had two playhouses, the barn and the blacksmith shop which his father had installed a few hundred feet from the Coolidge home. The blacksmith, who worked for Esquire John Coolidge for a dollar a day, replaced the grandfather as the boy's hero. He was a swarthy giant, this blacksmith, with a tender heart, who loved the child. And when the boy was hoeing his row in the field and got behind, his great sooty playfellow would throw off his apron at times and run out to spell the boy for a few moments and bring his row up with the others. The child loved to see the sparks fly from the anvil and hear the swish of the hot iron in the tub, and see the horses and oxen shod. The smith could throw a horse who became restive and shoe him lying down. Such a hero easily becomes a boy's god.[2]

When the little boy became President they exhumed his school records and found that he could read when he started to school at five. But two things are remarkable in the record which covers the years from his childhood to his early youth. One was his promptness. In those years he was tardy less than a dozen times in seven years and absent not at all save for the illnesses that visited childhood. Incidentally his deportment was good. When he was thirteen years old he was able to take the town teacher's examination and pass. His sister at thirteen took the teacher's examination and taught school. The Coolidges had quick minds. In school Calvin was a quiet child, addicted to mild devilments but never boisterous. He romped

---

[1] "The Quick and the Dead" by Gamaliel Bradford, p. 235.

[2] These notes and recollections of the Coolidge childhood come largely from the "Autobiography" and from "The Boyhood Days of President Coolidge" by Ernest C. Carpenter.

with a race of Brobdingnagians. The Vermonters of that neighborhood, full grown, ranged well into six feet and weighed from one hundred and eighty to two hundred and forty pounds. As boys they were lubberly and rough. But by some miracle, though his father and grandfather were lanky, Calvin was not much more than five feet eight; as a man and as a boy, compared with his kin he was about two sizes above a runt. His teachers testified that he did not join the rough and tumble but stood about doing the things he could do and making dry, cackling comments on the playground battles. Thus caution was developed. But at home and in his father's field he made a hand. Oxen helped with the farm work in Vermont two decades after they had passed from the American farms of the West and Southwest, and the boy could plow with oxen in his early teens. He worked in the hay field where he used old-fashioned reapers. Men with scythes sometimes mowed the stony fields. When times changed he drove a sulky plow and helped with the mowing machine. There is a story that as a little boy he tried to harness the horse to the family buggy all alone. His father came along and saw that the boy could not quite make out what was wrong with his harnessing. His father said: "Perhaps if you would put on the tugs and the breast plate you might do better," and turned away.

All sorts of handwork the boy learned. In the attic of John Coolidge's house, after the squire's death in 1927, neighbors going up to list the old junk that had lain there for a generation, found a ship's crowbar, a flintlock rifle, a canteen, a stonemason's hammer, a wooden fan for winnowing chaff from wheat, a homemade hoe, a mattock, a bush scythe, a loom, a flax wheel, a spinning wheel, a reel, hatchels used in preparing flax, a spool rack, bobbins, a shingle horse, a frow, two mallets, a handmade trough for catching sap, a jack plane, a log beehive, a jointer; and in a neighbor's attic was found a sap yoke made by young Calvin when he was a boy. He saw his grandmother carding, spinning and making the wool smocks that his father wore and that he wore as a boy at his field work. And to his death the ex-President cherished the homemade bed linen and towels from his grandmother's flax wheel which she made for him. He was a part of it, of that vast labyrinth of daily handicraft in the home where bored log pipes brought the water from the well to the kitchen.[3] He was jack of all trades, in the field, in the barnyard, in the blacksmith shop. In the sugar camp where the family made from a thousand to two thousand pounds of sugar every spring young Calvin, according to his father, could get more sap out of a tree than any hand on the place, and that in his early teens.[4] He had the knack of it. It was his gift.

[3] Classically in that day a pump and a privy were known in New England as Vermont plumbing!

[4] Reported by Bruce Barton.

He was not particularly a good boy, nor did he work harder than other boys, nor was he brighter. He was just an average country boy in a New England farm village who skated on the ponds in winter and swam in them in the summer. He hunted, trapped in the woods and brush, fished in streams, and learned to love the gifts the seasons brought—the field flowers, the distant velvety hills all green with spruce in winter and aglow with the new foliage, the hickory, the ash, the maple, the elm in spring. He had a calico horse which was his delight. This gave him distinction. It was marked like a circus horse and young Calvin could ride standing. And when at the end of a hot day they took the horses to the ponds for a swim, Calvin could wrestle with the men as the horses frisked and played there. His agility gave him equality with the biggest and the lustiest of the boys and the hired men of the village. But he learned this important lesson from the hired men and the hired girls, that they were as good as anyone and when they wanted to go to town with the family young Calvin was left at home that the hired girl or the hired man might have a seat in the wagon. That was a right that went with the job.

They were frugal people, these Vermonters. In the seventies their land was on a decline. It had fewer people at the last quarter of the nineteenth century than it nurtured in the eighteenth century. The boy Calvin saw pock marks on the hills—the cellars of abandoned homes, of little factories and of farm buildings. The last of the forges was passing. The workshops that made wagons, buckets, barrels, tubs and clothespins, the little foundries that made stoves, were gradually going down under the competition of the larger factories and foundries and forges in Pennsylvania, Ohio and New York. Capital was consolidating in corporations. The individual owner in industry was passing in bewildered defeat. Land in the Vermont hills often went for taxes, and a farm improved with the house and barns sometimes went without bidders at tax sales for a debt of ten dollars an acre. The struggle was hard. If the Vermonters were parsimonious, if they bit every penny that came into the village to mark it and hold it, they must not be blamed. George Harvey is author of a story that when Calvin was eight years old his father gave him a silver three cent piece, the wealth of Croesus, with which to buy a ticket of admission to some local entertainment. But the day before the entertainment the father came to the boy and said:

"Now son, we are liable to need all the cash we can get and I may need that three cent piece. We must save every penny. If the election goes wrong and General Hancock wins we are going to have hard times."

The boy gave up the silver-piece. The election passed. General Garfield won. The father appeared the day after election, returned the three cent piece, admonishing: "Now don't spend it foolishly!"

His father was running the country store successfully. The annual rent on the whole place, store and farm, was only forty dollars,[5] and Colonel Coolidge's merchandising bills were about ten thousand dollars yearly. He had no other expense. He made about one hundred dollars a month and saved it all. When he retired he had about twenty-five thousand dollars. When a child left his playthings in the road to be run over, he heard from it. And as a man the President used to remember that he was often routed out of bed to fill the woodbox as a penance for his wastrel ways.

Naturally a people who wrested from the soil their food, their clothing, their shelter by handmade devices, relied upon homemade diversions. The literary society in the schoolhouse was the week's high point in Plymouth, and the Sunday school picnic in the summer was an oasis of delight in the work-a-day week. Pictures in the village are extant showing scores of children picnicking in the woods around the cabinet organ brought from the schoolhouse for their delight. As he went into his early teens, and boys and girls began to couple off and go to parties, he took his girl and when they played "Copenhagen," he kissed her if he could, and when they played "postoffice" he kissed her anyway if she called for him. It was the way of the time and the place. The people danced. But Grandma Coolidge set no store by dancing. She gave the boy a dollar, such a price as only the rich could bestow upon their children, if he would not go to dances. He accepted the bribe. But he was no bound boy at the huskings. The red ear meant as much to him as to anyone. And tucked under the buffalo robe with a red scallop, he cuddled up with the little girls and boys as lustily as anyone. There is no tradition of a pharisaical aloofness haloing Calvin Coolidge in the village. He was up to tricks, to practical jokes, sometimes a bit mean and cruel, but after the prevailing sense of humor of youth the world over. Evidently he was not cruel to animals nor unkind to those whom the Coolidges held to be their inferiors. For remember that the Coolidges were folks—meaning the nobility. Sometimes in the village the election was Coolidge and anti-Coolidge, which means they were the accredited gentry, the storekeeper, the squire, representatives of the ruling class.

2

So the boy in his way was a young prince. Yet, for some reason, he was unbelievably shy. He remembered his shyness in his later years and to a friend who was puzzled at this shyness when he was the ruler of a great

[5] Many of the detailed facts, including the one in this chapter upon which this narrative is based will be found in the early biographies of Coolidge by R. M. Washburn, M. E. Hennessy, Roland Sawyer, and Robert A. Woods. Many of these facts are found in all of these biographies, some in only one or two, but nothing is set down here which has not been printed elsewhere, or except as I have indicated otherwise.

people, he recalled those days of his childhood. He remembered himself as a bashful, old-fashioned boy who ran away and hid when visitors came.[6]

"Most of the visitors would sit in the kitchen with father and mother, and the hardest thing in the world for me was to have to go through that kitchen door and greet the visitors. By fighting hard I used to manage to get through that door. I am all right now with old friends. But every time I meet a stranger I have to stand by that old kitchen door a minute. It's hard!"

A fairly authentic description of this little boy has been preserved by Ernest C. Carpenter, former teacher in the Plymouth schools: [7]

"He had red hair, blue eyes and freckled face, and a nasal twang that left no doubt as to his New England ancestry and caused children a good deal of amusement when they tried to imitate him in speaking his abbreviated first name."

So obviously this Coolidge quack was a hangover from Puritan days, becoming a bit antiquated even in the eighties so that the Vermont children noticed it. And the children at that school, according to the same teacher, were all of New England stock. He lists them: Browns, Chamberlains, Moors, Wards, Wilders, Pinneys, Willises, Smiths. Two or three Kavanaghs are the only children whose names might not have come out of seventeenth century New England. Yet these natives delighted in mocking the nasal twang of their schoolmate.

"He was never of large stature, though healthy and able to do his share of manual work. He was never a youthful prodigy. In a school of about thirty he was among the first half dozen. He was methodical, faithful, honest and punctual, never tardy, never much ahead of time." [8]

His teacher says that this story was repeated about the boy: that he got up late at night to fill his woodbox and when the noise he was making attracted the family, he explained himself:

"Well, I forgot it and I got up to do it."

Probably the twang of the sentence more than any virtue inherent in it preserved the incident, but it does not show the Sunday school-book child so much as it shows the rigor of paternal discipline. It was easier to get up and fill the woodbox than to face trouble in the morning. Teacher Carpenter recounts that the boy was not hilarious, nor a goody-goody; just an average boy distinguished for those leftover traits of old New England, caution, reticence, reliance, alertness; Indian traits, the ways of trapper

[6] Conversation with W. A. W. in the White House, 1924. He told this story to several other biographers. Frank Stearns refers to it.

[7] "The Boyhood Days of President Calvin Coolidge" by Ernest C. Carpenter, Marble City Press, Rutland, Vt.

[8] "The Boyhood Days of President Calvin Coolidge" by Ernest C. Carpenter, Marble City Press, Rutland, Vt.

people who live near the Indians; pioneers who had to emulate Indian virtues to thrive in an Indian land. He was quick-witted enough. When the hired girl [9] upbraided him for his dirty face and asked him if he was afraid of water, he said:

"I don't know about that. I have known people to get drowned in water." An ancient tag!

When the teacher, shaking him until his teeth rattled for some impudence or misdemeanor, managed to shake two or three buttons from his clothes, "to get even with her he left the buttons conspicuously on his desk for several days and came to school (reproachfully) without the buttons on his clothes." [10]

A human boy who was father of a most human man. And only as the father of the man are we interested in this boy.

In this same teacher's book is another significant entry. It is a report of the expenses of Plymouth Town for 1884. The total expenses of the whole town, or as Westerners would say, township, including three or four villages, half a dozen schools, and the mountain roads in a rather large wild area, the police, and the care of the poor for the year 1884 were $3,182.82. Here are these items:

Paid John C. Coolidge for superintending the schools for year 1884 $11.40
A. N. Earle for recording 17 births and 11 deaths for year 1884   1.40

Expenses of Selectmen

Paid for one day dividing school money	1.00
Paid for three days making highway taxes	3.00
Paid for one day to see Bidgood road	1.00
Paid for one day labor near A. W. Brown's place	1.00
Paid for one-half day fixing winter road	.50

Plymouth Expenses for Town's Poor, 1884

Paid May A. Sawyer for keeping C. J. one year	104.00
Paid Geo. Chamberlain for keeping H. H. one year	50.00
Paid for one pair of shoes for child	1.00
Paid L. J. Green for coffin, robe and funeral services	16.00
Paid for two cords of wood fitted to stove length	6.00
Paid L. J. Green for coffin, robe and services	15.50
Paid A. F. Sanders for digging grave	2.00

The significance of these entries is their parsimony. Not a penny is wasted beyond decencies, possibly not up to the decencies. The school

[9] Never to be confused with a servant.
[10] "The Boyhood Days of President Calvin Coolidge" by Ernest C. Carpenter, Marble City Press, Rutland, Vt.

teacher had $3.85 a week wages and his board cost him from a dollar to a dollar and a half a week. Now these people were not poor, neither were they stingy according to their lights. They were just close—tight. They called it "near." Whatever prosperity they had, and they were well-to-do after their fashion, was bred of thrift. To a bragging Iowan who returned to Plymouth and the Notch, telling about the wide acres of Iowa, its twenty foot soil, its six foot corn, its fat herds, and ended his eulogy by asking:

"How in the world do you people manage to get along here on these barren hillsides?"

The classic Plymouth reply was: "Waal, you're right. 'Twould be a hard life if 'twa'n't for our Iowa six per cent mortgages which help some!"

But it was a civilization based on thrift, hard thrift, but not ever to the point of penury, in speech, in emotion, in money, in comforts and in delights; not hard, not mean, adequate—no more.

Now this civilization, peculiar almost to Vermont, certainly to Vermont and New Hampshire, rural Maine, and western Massachusetts, persisted through the sixties, seventies and eighties and into the nineties when all the world outside those areas was running amuck; squandering, bragging, scattering taxes, voting bonds, carousing spiritually in revivals, and sousing in saloons, dreaming nightmares and realizing them by day in vast schemes to populate a continent and build a new land in its own exultant image: Corruption in politics, vulgarity in popular art, vainglory in business, cruel contrasts between the rich and the poor, the industrialists and the exploited immigrants, all the opium dreams of the age of steam hardening in stone and steel and lumber were waiting beyond the hills of Plymouth for this boy. And the boy, entirely unconscious of his destiny, was begetting the man who should go out into that bedlam to fight with it to the end of his life.

And what did he bring from childhood? Largely what every child brings into his teens as his inheritance: habits and morals. He brought the habits which were inevitable in a meager but happy civilization: thrift, energy, punctuality, self-reliance, honesty and caution. These marked the man to the end. His morals were Hebraic. Plymouth Town, with its villages nestling in the hills, was governed something as the ancient Hebrews were: the town meeting was supreme, the voice of the people was the voice of God. There was no doubt about God. In the morals of this boy God was the force outside himself that governed that inexplicable chance in the game of life which makes in the end for righteousness. There was no doubt about righteousness. Righteousness was neighborly consideration, justice between man and man, kindness institutionalized in various social forms and political works and ways. The barn-raising was one of the social forms of righteousness. The common school which guaranteed educational equality was a political work of righteousness. And the public care of the poor to eliminate begging was an economic way of the righteous life. And these morals which were founded upon the hypothesis of an orderly cosmos were ingrained in Calvin Coolidge in those years of his childhood. He never forsook them. Always he had faith in the moral government of the universe. He felt that God keeps books and balances them with men and with nations. He believed obviously that it pays to be decent and kindly and just. He had the Bible for it—that righteousness exalteth a nation and sin is a reproach to any people. When he came out into the world to fight his dragons he was never unsure. He never faltered, i. or halted in a complex world; a cynical world, a world full of a thousand doubts, an order rocking in a flood of disconcerting and terrible facts. His truth was unshaken. He looked back to Plymouth and Vermont to renew his faith. At the climax of his powers he wrote:

"Vermont is my birthright. People there are happy and content. They belong to themselves, live within their income, and fear no man."

At the end, after his life's struggle with the strange gods of the plains, his heart turned back to the gods of the mountains and he wrote:

"As I look back on it, I constantly think how clean it was. There was little about it that was artificial. It was close to nature and in accordance with the ways of nature. The streams ran clear, the roads, the woods, the fields, the people, all were clean. Even when I try to divest it of the halo which I know always surrounds the past, I am unable to create any other impression than that it was fresh and clean." [11]

At the close of his childhood, Calvin Coolidge came like a waxwork figure of a Puritan boy, out of the social museum that is rural Vermont. There time had slowed down a century. When he left Plymouth Notch for Ludlow to attend high school, he traversed fifty years in two hours' journey to the railroad. Even at the rail head he was in the corridors of the museum. He could see the world outside, still fifty years away. But young Calvin walked circumspectly into youth—a museum piece.

[11] "Autobiography."

CHAPTER III

Wherein the Museum Piece Peers Through His Enclosing Walls

THE eighteen eighties were going well when Calvin Coolidge came out from Plymouth Notch and took his first real peek into the wide world. He came down the mountain twelve miles to Ludlow. Not that he had never journeyed so far from home before. He had been to Woodstock once or twice, fifteen miles, to see a circus, and with his lawmaking father the lad had gone to Montpelier and to Rutland, decent Vermont towns which to the child seemed like busy cities. But this trip in the mid-eighties was no casual day's journey. Boys and girls a year or so older than he were teaching school in those parts in those days. Colonel John Coolidge, country squire and local statesman, his own groom and footman, drove his son, Calvin, to Ludlow to enroll him in Black River Academy. It was a momentous journey. Important preparations preceded it. In a motherless home these preparations were difficult. New clothing was provided, as also were school texts and such impedimenta as young boys take on their school journeys. Miss Chamberlain, the hired girl, had done her motherly best to outfit the boy. And the father, being proud of him, contributed what he could without appearing prodigal. The two rode down from Plymouth to Ludlow in the Coolidge farm wagon.

More than forty years after, the boy wrote that the preparations and packing required more time and attention than he spent collecting his belongings when leaving the White House. His whole outfit went into two small handbags. And he remembers the winter snow lying on the ground in October.[1]

"I was casting off what I thought was the drudgery of farm life sym-

[1] "Autobiography," p. 32.

bolized by cowhide boots and everyday clothing, not realizing what a relief it would be to return to them in future years."

He wore his Sunday suit and instead of boots his shoes with rubbers, because "the village had sidewalks. . . . No one could have made me believe that I should never be so innocent or so happy again. . . . As we rounded the brow of the hills, the first rays of the morning sun streamed over our backs and lighted up the glistening snow ahead. I was perfectly certain that I was travelling out of the darkness into the light." [2]

Tradition [3] declares that they took to the Ludlow market a calf in the back of the wagon and that in parting with his son after the crabbed New England fashion, the father suppressed his emotion in a joke:

"Waal, goodbye John.[4] You may some day, if you work hard, get to Boston. But this calf's going to beat you there."

And in lieu of a parting handclasp and a tear, the father, probably realizing the truth, that it was goodbye forever to the little boy he loved, turned away casually and went to the stock buyer and began bargaining to sell the calf.

And then he turned back to the motherless home where his little daughter Abbie and the hired girl furnished the only solace to the widower. It was not entirely the money that John Coolidge sacrificed when he sent his son away to the Black River Academy. Surely loneliness came upon him when the boy was gone. Twelve miles and back is a long, hard journey with a team on a winter's day. Winter begins in the New England hills in early November and runs through until late March. Yet for four years, save for the few times when Calvin walked home or stayed over Sunday with his aunt, John Coolidge made that week-end journey to have the companionship of his son. Probably neither of them ever spoke of what it cost in time and weary hours, plowing through the snow and slush on the mountain roads, but the boy never forgot.

Ludlow in that day was the nearest railroad point to Plymouth. It still is. It always will be. Change and decay do not come to Vermont. Ludlow in that day and in this was and is a small manufacturing village. Water power was its reason for being. Its millwheels turn by the water from the hills. As a mill village it had its quality who lived on the heights and its mill hands who lived in their quarters. But even so caste lines were flexible. An upward movement steadily pushed one generation of mill hands into the next higher industrial class and out into the ownership group, so that even in the purlieus of Black River Academy Young Calvin met children

---

[2] *Ibid.*

[3] Preserved in early biographies of Coolidge and later newspaper stories.

[4] Named John Calvin for his father. The name John dropped off in high school. Grover Cleveland lost a Stephen, and Woodrow Wilson sloughed off a Tommy in boyhood.

of other bloods and breeds than young Yankees whom he knew in Plymouth. He was beginning to touch the seething world beyond his horizon.

Outside that periphery in the middle eighties, America was boiling her melting pot. The exploited millions of Europe were pouring into the United States, splashing over the rim of the cauldron into economic oppression—if not social—as harsh and cruel as that they left in Europe save for this: the opportunity to rise, to climb out of the hell into which they were dumped. These foreigners, crowding into the drab industrial towns of New England and the Middle States, were manning the mills and factories. They were building the railroads of the West, digging the ditches for the pipes and stringing the wires on the poles and cutting the poles to bind the distant states as neighbors into one compact family of states. And while young Calvin, in Black River Academy was worming his way through the curriculum, there following what today would be called a high school course in our public schools, slowly the industrial boom that rose from the Civil War was nearing its climax. Soon it would explode into the seven hard years of the nineties. Grover Cleveland was President in those Ludlow years of Calvin Coolidge, the first Democratic President since Buchanan. He was cleaning the Augean stables that had become befouled after twenty-four years of Republican rule. He had entered a grand battle with the veterans of the Civil War who were debauching the pension system of the government. He was assailing the protective tariff and declaring for a revenue tariff. He was establishing a civil service and at the same time trying to "turn the rascals out"—a difficult task. It would have been impossible for a strait-laced Yankee boy down at the rail head, the son of a blue-bellied Vermont Republican, not to hear some echo of the vast political caterwauling that was reverberating across the land rising from Grover Cleveland's battle with the Republican dragons.

In Ludlow, young Calvin Coolidge remained as he had been in childhood in Plymouth Notch, slightly aloof, a sensitive, inquiring youth who eschewed athletics not because he was unsocial but because his muscles trained behind the plow or in the hayfield or in the woodlot or in the sugar camp did not ache for exercise. When he went home occasionally he had plenty to do. And on week days when he had an idle hour there was much to see and learn in Ludlow, and he saw it all and learned much. His school work took him well into the earlier courses in Latin, through Caesar into Cicero, with a forward look into Livy. In mathematics he finished algebra and went into plane geometry. Studying English he completed the later grammar texts and went into rhetoric, not unlikely Hill's rhetoric, which was standard in those days. He had some experience in writing essays. He joined the school debates and became one of the roisterers in the school literary society. Friday afternoons occasionally were high days and holidays when there was speaking and recitation, dialogues and music and a round of revelry by night in Black River Academy. All of which were good for the boy.

Physically he was changing from a child to a youth. His red hair was toning down. In Plymouth he was nicknamed "Red" and sometimes "Red Cal." There is no tradition in Ludlow of those nicknames. He was fair, and the sun did not tan his skin. As he grew older his freckles faded and left a rather pale but healthy blond skin where they had been. He was slight but not skinny. His muscles were hard and his jaw was well set, his features clear cut. The photographs of that period indicate that he did not pass through the lumpy age that sometimes accompanies adolescence. He was spare, trim, agile, with the quick, free, feline movements of a cat hurrying along a wet walk. Coming home over Saturday and Sunday he was the young prince of Plymouth. He took the lead among the boys and girls of "The Notch." A story persists that when the boys were raising the devil in the assembly room over the postoffice and the quiet villagers around the stove rapped on the ceiling for order, Young Calvin brought up a board; standing some other boys on one end of it Calvin bent it as far up as he could and let it slam kerbang on the floor. Listening for the response and hearing none, he quacked: "Guess that must have put them to sleep," and went on with the ruckamuctions that had caused the original protest.[5]

No Sunday school boy this. He liked to tease, was not above a bit of meanness which most boys enjoy, and which his type often chooses in lieu of more robust exercise. So naturally he was a practical joker. But even at home over Saturday and Sunday and during his vacation he was not of Plymouth Notch. His face was turned permanently toward the world and its pleasures and palaces. He was following the calf to Boston "out of the darkness into the light"—out of the darkness that "comprehended it not."

His four years in Ludlow in Black River Academy gave him something more than his textbooks told him. The Cleveland administration was set upon one thing, the change in our tariff policy from a tariff made under the protective theory to a tariff made for revenue only. One of the points of President Cleveland's attack was upon wool. It was a famous saying in the late nineties that no Democrat could look a sheep in the face, and when Cleveland turned his guns on wool the echoes stirred the hills around Ludlow. Textile mills in Vermont were affected. One of the periodic depressions struck the land, and Ludlow felt it. Young Calvin saw men and women out of work and hungry, children and youth denied education, and lives blighted by the denial. Being what he was, a sensitive boy, surely the impression became a part of his conviction that the protective tariff was

---

[5] "The Boyhood Days of President Calvin Coolidge" by Ernest C. Carpenter, Marble City Press, Rutland, Vt.

sacrosanct, and a part of God's established order for America. The old soldiers sitting around the stove in the store at Plymouth Notch to whom the boy listened at odd times were furious in their repressed New England way at Grover Cleveland's free trade and the revenue theory of the tariff. Calvin recalls those soldiers:

"They were not talkative but took their military service in a matter of fact way. They drew no class distinctions except for those who assumed superior airs. Those they held in contempt." [6]

Grover Cleveland drew out their scorn as a poultice draws the virus from a wound, as they spat into the sawdust boxes in the postoffice where Squire Coolidge reigned. The youth in his middle teens should not be too deeply blamed if he confused his Republicanism with his patriotism and merged them both into that religious faith that he held unshaken all his life. Other Vermont boys of his day and time felt as he did. "The Coolidges never went West." Vermont never went Democratic. For fifty years, its Republicanism was rooted in its granite and stood unshaken by the breezes. The drum stove in a hundred Vermont postoffices was the burning altar around which a hundred thousand boys like Calvin Coolidge in the eighties, acolytes in their protectionist faith, served and fed the sacred fire.

Another extra-curricular influence upon the youth in Black River Academy was reading—books at home. The Vermont farm house contained its quota. Of course there was the campaign biography of Abraham Lincoln, but the small type in it discouraged the boy. At least he remembered the small type as an alibi for his listless interest. But probably the biography was unutterably dull. Also there was a dull biography of Garfield, hurriedly thrown together to sell on subscription after Garfield was assassinated. Two novels which his grandmother loved, the boy read, novels written in the grand Walter Scott manner against a Vermont revolutionary background. They were "The Rangers or the Tory's Daughter" and "The Green Mountain Boys," by D. P. Thompson. The lack of imaginative literature of that time is an illuminating reflection upon the home, upon the blood that was in him. Other children in the mid-eighties elsewhere were reading "Little Women," "Tom Sawyer," and the Cooper novels, and Scott's tales of love and glory. Dickens was at the height of his American vogue, and "David Copperfield" was not too old a book for the boy at Black River Academy, nor were George Eliot's romances. But he writes that in that day he encountered Headley's "Washington and His Generals," Harriet Beecher Stowe's "Men of Our Times." It is odd that he does not list "Uncle Tom's Cabin," a hangover from ante bellum days, which even then was selling by the millions in the eighties. Instead he read "Livingston Lost and Found" by the Rev. Josiah Tyler, and Redpath's "John Brown," along

[6] "Autobiography."

with Trumbull's "History of the Indian Wars." In the Academy he was passing from the "United States Reader" by John J. Anderson into Hilliard's "Sixth Reader." But he does remember a volume called "Choice Poems and Lyrics," probably a volume issued by John B. Alden, or possibly this was one of the "Redline Poet" series. In school there was the "Young People's Bible History," and in Latin he remembers the orations of Cicero, showing the marked political trend of his mind in his teens. Virgil must have been a chore to him, and Caesar's story of the struggle with the Helvetians apparently fed no military microbes in his blood. An early biographer [7] says of the young man who spent his summer vacations making a hand on the farm that "in due time the father came to feel that his son could manage the farm as efficiently as he," and calls attention to the notable fact that "such cumulative experience in meeting and overcoming all the farmer's problems is sufficient to develop capacity for statesmanship."

This biographer may be right. But the fact is that Calvin also often worked as a piece-worker many Saturdays in Ludlow in a toy factory where they made wagons for other boys. And this boy himself remembered through a generation that he put his earned money in a savings bank and declared that his father advised it "because he wished me to be informed of the value of money at interest. He thought money invested in that way led to self-respecting independence that was one of the foundations of good character." [8]

That lesson stayed with him all his life. He remembered that he painted and repaired around the farm during one vacation. Work and savings and the high morality of accumulated savings in property he cherished as the lesson he learned as a boy on the farm. That was his chief preparation for statesmanship when he met the farm problem in a generation that was to come. Work and save was all that he was ever sure about!

Yet he had learned from his grandfather, who had them from his grandfather, the traditions of the battles for liberty which Americans, his own kith and kin, had won. He was deeply patriotic. The hard mold of environment at Plymouth Notch, in the books he read and in the widening horizon of the manufacturing village of Ludlow, closed about the plastic mind of this farm boy and predestined him to politics, made him a foreordained Republican. His blood kept him constant, incapable of change.

In those Black River Academy days, young Coolidge was a part of the boy life of the place. An old teacher used to like to tell [9] about a prank the boys played in the Academy. They tied a rope to the leg of a stove in a classroom after dusk. And later in the night at the end of the rope, they

[7] "Preparation of Calvin Coolidge" by Robert A. Woods.
[8] "Autobiography."
[9] Interview in Boston Globe, January 7, 1933.

dragged the old stove clappity-bang down the stairs to the lower hall. The leaders were caught. Fishing for accessories before the fact or deed, the Professor examined Calvin who was obviously sweetly sleeping in his dormitory bed far from suspicion.

"But Calvin, you must have heard the noise when the stove rolled down the stairs?"

Calvin admitted it. "Then why didn't you do something, give the alarm?"

"It wa'n't my stove!" quoth the boy who was father of the man. Whenever he faced temptation to "rise and spread the alarm" he always remembered that "it wa'n't" his "stove."

2

A year before he left Ludlow, his sister Abbie died. She had been a student with him at Black River Academy. They were an affectionate pair. She was stricken with appendicitis which was not defined in those days. He sat by her bedside a day and a night, dumbly, helplessly to the end. Young Calvin seems to have lavished upon her the love which his mother might have had, and his sister's death was a blow from which he never entirely recovered. It is curious to note how death clouded him. As a child his grandfather's passing left him lonely, though the resilience of childhood removed all marks of sorrow. But the death of his mother six years later and of his sister six years following his mother's death did affect his life deeply. That his emotions were hidden did not lessen their power. In Ludlow old people live who still remember the pale, prim young fellow of that day and how his face seemed permanently saddened when his sister passed out of his life. He was a little more aloof, a little less mischievous, a little more serious for even a serious boy when he came back to Black River Academy after his sister's funeral.

He was graduated that summer. The picture of the graduating class shows him a slender, lean-faced youth in his gawky teens. The other members of the class scattered to the four ends of the earth, but "no Coolidge ever went West." The only class picture extant turned up in Nevada.

Young Calvin expected to go to Amherst, but owing to an illness he failed in his entrance examinations there and went to an Academy at St. Johnsbury. During the spring following his graduation and in 1891 before entering Amherst, a pageant crossed his path which impressed him tremendously. It was the dedication of the Bennington Monument at Bennington, Vermont. He heard and saw President Harrison there. The boy remembered that in 1888 in Ludlow when Harrison defeated Cleveland the faithful Republicans of the town spent two nights parading the streets of the village with drums and trumpets celebrating the victory. So that the hijinks in Ludlow put the boy in a proper mood to enjoy the appearance of President Harrison, the first President he had ever seen.

Now that day at Bennington was one of more than passing interest in the boy's life. His father, a Republican war horse of Plymouth Town, took him to this Bennington dedication—a pious pilgrimage! For here were assembled to dedicate this monument all the high priests of the Republican temple. The governor of the state was there, and all the living ex-governors. The secretary of war was there, and the attorney general of the United States, William H. Miller, General Alger of the Civil War, General Alexander Wendt, General O. O. Howard, John King, President of the Erie Railroad, W. Seward Webb, President-General of the Sons of the American Revolution—all were there. Senators Dawes, Edmunds, Chandler, Gallinger, and ex-Governor Rice, of Massachusetts, all Republican high priests, were among the notables. The train bearing the President had come up the Hudson valley where local statesmen met it and boarded it; among others, two young men who were to be important leaders of the Republican party in the next decade, young Ben Odell, of Newburgh, and William Barnes, of Albany, New York. The Rev. Dr. Charles Parkhurst, of New York, opened the exercises with prayer.

It was an impressive ceremonial. Here Young Calvin first heard the ram's horn blare and the cymbals clash and the populace lift their voices to the tribal gods—the gods of the mountain, Vermont's fireside idols, the leaders of Republicanism who fought the war of the sixties and cherished its holy cause! John Coolidge led his son there as Jesse took little David up to the court at Hebron, or as the pious Mary and Joseph led their son to the temple at Jerusalem.

The boy saw for the first time in his life a huge triumphal arch under which the parade of the priests and the pharisees passed. The turrets and embrasures for the arch were filled with damsels clad in red, white and blue, and on the top of the arch were more than a hundred white-clad maidens singing patriotic songs. In the loftiest turret was a gorgeous throne of gold occupied by a handsome young woman representing the Goddess of Liberty. Fifteen thousand people gathered around that battle monument at Bennington which still stands, a square shaft of limestone more than three hundred feet high.

Young Calvin heard Edward J. Phelps, one of the golden-tongued orators of that time, deliver the dedicatory oration, and he heard President Harrison's response. Those were the days when it was proper, even necessary for Republicans in the most nonpartisan gatherings of the North to speak of the participation "of this State in the War of the Rebellion." [10] "Vermont troops," quoth President Harrison, "took to the fields of the South that high consecration to liberty which had characterized their fathers in the Revolutionary struggles. They did not forget on the hot savannahs of the South the green tops of these hills ever in their vision lifting up their hearts in faith that God would again bring the good cause of freedom to a just issue."

A dinner (they called it a banquet) was served under a tent, where Calvin and his father sat down with the great Republicans of the land. There President Harrison again spoke and declared:

"If we seek to find the institutions of New England that have formed the character of its own people and have exercised a stronger molding influence than that of any other section upon our people, we shall find them, I think, in their temples, in their schools, in their town meetings, and in their God-fearing homes. The courage of those who fought at Bennington was born of high trust in God. They were men who fearing God had naught else to fear. That devotion to local self-government which originated and for so long maintained the town meeting, establishing and perpetuating a true democracy, an equal, full participation and responsibility in all public affairs on the part of every citizen, was the cause of the development of the love of social order and respect for laws which has characterized your communities, has made them safe and commemorable abodes for young people." [11]

Now these remarks are quoted because they are typical of the pabulum that was sustaining the New England youth of the nineties, and because Calvin Coolidge, feeding upon it, became what he remained to the very end, an old-fashioned, God-fearing primitive Puritan democrat. It was his strength and his weakness when the test of his life should come. Curiously enough, looking back through forty years upon that day of triumph in Bennington, when he had laid down his scepter as the ruler of his country forty years later, he writes wistfully of President Harrison:

"As I looked on him and realized that he personally represented the glory and dignity of the United States, I wondered how it felt to bear so much responsibility, and little thought that I should ever know."

At the close of young Calvin's academy days his father married Carrie A. Brown. The boy grew fond of her. He responded to kindness. She was warm-hearted and the step-mother complex never developed in the family. For he was too well coördinated to cherish that kind of neurosis.

---

[10] Only mugwumps and softies called it the Civil War and he who in Vermont would refer to the late unpleasantness as "the War Between the States" was a marked traitor!

[11] These words and the description of the dedication are from "Speeches of Benjamin Harrison," compiled by Lovell, Coryell and Company, New York, 1892.

CHAPTER IV

Our Hero Stages Himself as a Cinderella in the Academic Ashes

In 1890 Amherst College took Calvin Coolidge mentally cast in the granite walls that circumscribe Vermont. Of course he was callow, no prodigy here; an average high school boy with an average high school record and an average capacity for acquiring more or less useful information in a college course. Amherst merely developed him, filled him with its fire, made him come alive and turned him from youth into manhood, finished and ready for life.

Now for Amherst and its influence upon the youth from Plymouth Notch: Amherst in the early nineties was less than a hundred years old, one of a dozen New England colleges of its type with a student body of four hundred when he entered and nearly five hundred when he left. It was fairly well endowed, paying faculty members from two thousand to four thousand a year, having one member of the faculty for every ten students. Amherst College was founded by the Congregationalists, and a succession of Congregational doctors of divinity had been its presidents. The college was set in a small town of between four and five thousand people. It was a hundred miles from Plymouth, and less than seven miles from Northampton where Calvin Coolidge was destined to live thirty-five years and where he died. Any of a dozen small New England colleges which then lay one or two cuts educationally under Harvard, Yale, Princeton and Columbia was distinguished for three or four faculty members outstanding and dominant, who left their impress on the students of that day. One of the things which gave Amherst distinction was its library. Its three outstanding members of the faculty were Professor Charles E. Garman, who taught philosophy, Professor Anson D. Morse, who taught history, and

Professor George D. Olds, who taught mathematics. The latter was one of the younger members of the faculty. Garman and Morse, Calvin Coolidge haloed in his memory.

In that day socially, the Amherst campus was like a hundred fresh-water colleges throughout New England, the Middle States and the Missouri valley, except that as one went west one found coeducation. The boys at Amherst for their diversion went down to Smith, a girls' college at Northampton, an hour's drive with a horse and buggy and fifteen minutes by train. Eight or nine college fraternities were organized in Amherst, comprising in their membership seventy-five or eighty per cent of the student body. They gathered into their ranks the well-to-do, the socially gracious and often the distinguished among the students who stood out either as athletes, as debaters, as orators or as exceptional men. But primarily the fraternities were for the well born—the good dancers.

Those who were not invited into the fraternities were known in Amherst as "oudens." Naturally Calvin Coolidge was an "ouden." He had no distinction. In Plymouth he was the son of a merchant prince, a scion of the ruling class. He ranked well as a student at Ludlow in Black River Academy. And with prudent foresight, which was inbred in his kind, he wrote to his father when he was still in preparatory school that he "had better go down to Amherst some time to see about getting into a society there,[1] the societies are a great factor at Amherst and of course I want to join one if I can. It means something to get into a good society at Amherst. They don't take in everybody, but Dick thinks I can if we scheme enough." [2] Apparently he did not "scheme enough," but when he was not taken into a fraternity certainly he kept his disappointment to himself. He entered Amherst handicapped by his lack of social charm, by his innate shyness and by the meager sum he had to spend over and above his needs for tuition, board, room, and textbooks. He lived in a small cheap bedroom in a rather remote private house. He was—in short—a sort of male Cinderella. Curiously all his life he entered each expanding phase of it a Cinderella, and quit each phase dancing with the prince.

He paid $3.50 a week for his board at a large boarding-house. The fraternity men whose houses served no meals mingled democratically with the rest of the student body in boarding-houses. It was a democratic college. All the freshmen were herded into the gymnasium together and wrestled with dumbbells and Indian clubs. Incidentally, they vaulted, jumped and had other setting-up exercises. Also they marched. And note this well, for it reveals an interesting mental habit of Calvin Coolidge: he records that the marching about the floor of the gymnasium produced "a military flavor

---

[1] Meaning a college fraternity.
[2] Letter reprinted in *Good Housekeeping*, February, 1935.

which I found very useful in later life when I came in contact with military affairs during my public career." Like most of us, he enjoyed rationalizing his conduct after the fact.

Two or three other students roomed in the house with Young Calvin, men like him who were pinching and scrimping their way through college. He had no time or talent to sing May Irwin's ditties in the glee club or play in the banjo club, or to flit around nights with a guitar or mandolin serenading the girls at Smith, or to play baseball in the spring, or in the autumn to romp with the rangy, rough-and-tumble football team. He was utterly undistinguished—a typical "ouden" during his freshman year at college. But characteristically he appeared at the first meeting of the freshman class and for the four years he was present at every meeting of his class. For two years he was just one of a hundred and fifty freshman and sophomore boys. When his class met the second or third time, the boys were ganged off into cliques, societies and fraternities. Coolidge did not gang. No one thought of pulling the lone, silent freshman into any of the many self-crystallizing social units of the school. But he kept on attending class meetings, patiently sitting through each session, lifting no voice, making no motion but growing wise about his fellows, learning much about the class—Cinderella devoted to the pots and pans of the hearth. When it became a sophomore class, they say he quacked a few short sentences at one of the meetings and what he said had weight, not much weight but a little; even though what he said was not backed up by any group. So he went his way alone through his sophomore year.

In his sophomore year Young Coolidge took a pair of shoes to Northampton to be half-soled and met there "Jim" Lucey, shoemaker, born in the South of Ireland of whom we shall read more as the story progresses.

Young Coolidge's classmates remember that he was given to long walks through the woods and hills around the town of Amherst.[3] The hills, softer than those around Plymouth, were of the same kind, and it is more than likely that the boy harbored a love of beauty which he rarely voiced. This love for beauty—a by-product of his sentimentality—appears throughout his life obliquely in many of his written works. As a freshman and sophomore he was known to but few boys beyond his rooming-house or his classes. Utterly outcast was he from the merry, noisy life in the fraternity houses. But he saw this life. He understood it. He desired it. His letter to his father from Black River Academy attests his shrewd appraisal of it before he became a part of it. He loitered casually around the edges of the crowd at the ball games, attending the public functions punctiliously, the lectures, the concerts, the intramural debating events, the oratorical contests, too shy at

[3] See early biographies of Coolidge by Washburn, Woods, Sawyer, Whiting, Hennessy for many of the data for this chapter.

first to participate more than it was absolutely necessary under the college rules.

Twenty-five years afterward at a dinner in Boston, where he sat at the head of the table, the lady on his right trying to make him talk, asked: "Did you participate in the athletic events when you were in college, Governor?"

"Yes," he answered, munching a nut. And to the question in her eyes he extended his remarks: "I held stakes," and lapsed into silence, thinking perhaps of the youth on the fringe of the crowd, or thinking back to the boy before the kitchen door at Plymouth, battling bashfully for courage to meet the company around the stove.

Of course that kind of boy is not invited to become a fraternity member unless he has a rich or distinguished father. Despite the fact of his faithful attendance at every class function and college gathering, he remained an outsider. In 1892 he joined the college Republican club, supporting President Harrison and solemnly followed the drums of his Yankee gods in a torchlight parade, his first appearance in American politics. Harrison was defeated that year by Cleveland,[4] running for his second term—his third Presidential candidacy.

In his junior year he had his first distinction. The class traditionally used to put on a plug hat race in which all members donned high silk hats and ran a foot race. Calvin, from some remote fastness, begged or borrowed or stole a plug hat and entered the race conscientiously with characteristic pains by way of doing his full duty by the class. He knew he was slow. He was the last man in the race. That kind of boy would be the last man, handicapped by a plug hat. The penalty to the loser—and as a loser he won his first distinction—was that he should make a speech. Whereupon through the speech he gained his first distinction. It was a funny speech. It ended in an ovation,[5] the first call of his fairy godmother!

His classmates remember his concluding words, the modest and hopeful words of a loser who had faith in himself. He said, 'Remember boys, there are firsts that shall be last and the last first.'" During his junior year Young Calvin began to attract attention as a debater. Public speaking and debating were compulsory parts of the curriculum. His classmate, Charles A. Andrews, remembers that "as a reciter of other men's speeches Calvin was not especially noted. But when we got into the debating, he began to emerge and become one of the best debaters in his class, distinguished for the directness and brevity of his style, the soundness of his logic, the aptness of his humor. By the time we got to be juniors we all knew that Cal-

[4] "Autobiography."
[5] "The Preparation of Calvin Coolidge" by Robert A. Woods, a classmate.

vin was a real fellow and that the quality of his college course was excellent." [6]

2

So it was that in his senior year young Calvin "made" a fraternity. It was a new fraternity established in Amherst the year before, Phi Gamma Delta; a young fraternity at that time just edging into eastern colleges after western triumphs. The Amherst chapter needed members. The plug hat race with the oration that followed had distinguished Young Coolidge. Two fraternity men from the brand new Amherst chapter of Phi Gamma Delta sought him in his room. He knew who they were and what they came for. He knew all about Phi Gamma Delta as a national fraternity. No senior with as keen a mind as Coolidge's and as sensitive a spirit could have escaped full and bitter knowledge of what was going on in the new fraternity. The boys who came to ask him rather elaborately prepared their program of invitation. One took the national chapter. The other discussed the local chapter. He let them talk. Never a word quoth he until they said flatly and finally:

"Will you join us?"

He replied: "Yes," and lapsed into silence.

His initiation into a fraternity furnishes a striking example of this capacity for covering the hurt in his heart. For three years he saw the gilded youths of the college strutting about in little cliques and clans wearing their fraternity pins, eating and drinking and making merry with the girls, mostly with good little girls, but sometimes with naughty little girls. These gilded youths had more money to spend than he, more social graces to recommend them, more personality with which to win friends than he. But he had more patience than they. He waited. In those years he sedulously refrained from attacking the fraternity system—as many an ouden had done. He knew the faults of the system. But he needed it in his business. So he waited. When the time came to serve the fraternity he revealed how much he knew of the chapter and its membership. The boys put him on the house committee, turned over largely the business of the chapter to him. They sought to make a butterfly out of him; took him down to Smith College, tried to make him dance. But Grandmother Coolidge's dollar bribe which had kept him off his feet at the dances in Plymouth, left him awkward at the dances at Smith College, and he would never do a thing that he didn't have to do unless he could do it well. That was one of his vanities that marked him to the end. He disliked being a wallflower at parties and dropped out. But inside the fraternity he developed a certain volubility, and in a shy way took business leadership. The

[6] Letter to the author, April, 1933.

boys liked him but never knew him intimately. He drank nothing, smoked a little, talked only when he felt mellow, avoided the girl proposition, did more than his share of work in the chapter and let it go at that—forecasting his whole life in little!

In the meantime, as a student he was faithfully attending classes. He was an average student,[7] no more, no less. He often was seen in the college library where he seems to have read more than was usual for an undergraduate. His classmates remember him best in the library—a pale, silent, long-necked, freckled, fair-haired youth, poring over a book assigned by a teacher as supplemental reading.

Obviously he was not eligible to Phi Beta Kappa. He won no prizes for scholarship nor for any scholarly attainments. But nevertheless his spirit awoke in Amherst, and the man who aroused him was Professor Charles E. Garman, head of the Department of Philosophy. Everyone who writes of Coolidge in that period notes the influence of Professor Garman and his philosophy upon the young man. He studied mathematics into calculus. He studied history and found it absorbing under Professor Anson D. Morse who lectured on medieval and modern Europe. But he writes: [8]

"It was when he turned to United States history that Professor Morse became most impressive. . . . Washington was treated with the greatest reverence. A high estimate was placed on the statesmanlike qualities and financial capacity of Hamilton. But Jefferson was not neglected."

A fellow student [9] remembers Coolidge in the library with his nose in the "Federalist," [10] and there can be no doubt that American history under Morse was one of the subjects which Young Coolidge tackled with feverish interest. His class was the last Amherst class to read in the original Greek, Demosthenes' "Essay on the Crown." But Demosthenes did not set him afire until Garman, interpreting life and knowledge in the terms of a purposeful moral government of the universe, gave the boy from Plymouth Notch the key which then and until he died, unlocked for him the philosophic mysteries of life. Calvin Coolidge came into Professor Garman's classes, a thorough-bred Yankee Puritan. There for the first time, he heard

---

[7] Scholastic record of Calvin Coolidge during his senior year at Amherst College:

		Grade
Debates	3 terms	B
English Lit.	1 term	B
Italian	3 terms	B, C, B
Philosophy	3 terms	B
History	2 terms	C, B

[8] "Autobiography."
[9] Robert A. Woods.
[10] Where he learned from the text to associate "the rich and wise and good!"

the philosophy of modern Yankee Puritanism expounded. It was the philosophy of Vermont illuminated, oriented, convincingly expounded. Young Calvin got what he brought to Garman's classroom—but polished up.

3

So, it is important to consider Professor Charles E. Garman. He was in his early forties when Calvin Coolidge knew him, a scholar at the height of his power. He was a religious democrat, a Congregational mystic, a Neo-Hegelian in his philosophic attitude. He believed and taught that back of the thing we know as matter, as material creation, there is mind; a conscious, purposive force which directs all processes of universal life. He reasoned because human consciousness is a part of universal life, that the part may not be greater than the whole. Therefore he concluded that the purposive force which directs life, which guides the stars in their courses and spurs and speeds the energies inside the atom, is of itself a consciousness. This cosmic consciousness Garman logically presumed to be greater and wider than human consciousness is; and it follows in sequence that the human mind cannot know or understand the larger consciousness. That cosmic consciousness, that force that was the law, "the way, the truth and the life" was Garman's God.

He was a charming man but with great personal power in his charm. Justice Harlan F. Stone, of the United States Supreme Court, who was an Amherst student of those days, remembers Garman "as a tall, spare man, clean shaven, hair and eyes coal black, the latter deep, cavernous, glowing. He reminded one of the pictures of John C. Calhoun. His personality was impressive. Casual acquaintance with him gave the impression that he was a mystic. Fuller acquaintance with him revealed that he was a great teacher, a man of intellectual power. His influence was not limited to the classroom. He had close personal contacts with most of his students who visited him at his house and corresponded with him for many years after leaving college. In no home could one be more charming or more attractive, especially to young men. Everything he said or did stirred their higher aspirations and their desire for worthy accomplishment." [11]

It is easy to imagine what happened when this tall, spare man with coal black, glowing, cavernous eyes and with a charm and grace of manner which goes with a sophisticated personality, turned the electric current of its force upon the shy young student from Plymouth Notch with bran-flake freckles peppered on his innocent face. Calvin Coolidge marched in Garman's train.

"We looked upon Garman," wrote Coolidge, nearly forty years afterward, "as a man who walked with God. His course was a demonstration

[11] Letter to the author, April, 1933.

of the existence of a personal God, of our power to know Him, of the Divine immanence and of the complete dependence upon Him as Creator and Father. Every reaction in the universe is a manifestation of His presence. Man was revealed as His son, and nature as the hem of His garment. . . . The spiritual appeal of music, sculpture, painting and all other art lies in the revelation it affords of Divine beauty." [12]

Let us look at him in the classroom: this tall, gaunt man with glowing eyes. His was what has come to be known as the "Socratic method" of teaching interspersed with occasional lectures, informal talks. He posted mottoes around the classroom: mystic slogans, and inspirational aphorisms which were advanced theology in their day. He printed in pamphlet form excerpts from a wide range of writers dealing with subjects suitable for philosophic discussions. These, selected by him, were placed in the hands of students who were required to read and digest these pamphlets. The students' critical faculties were stimulated. It is easy to understand how this reticent youth from the Vermont hills, seeing his universe mapped out and for the first time related and made known to him through his own processes of thought under Garman's stimulation, was immersed under Garman's spell, baptized for life!

The boy was prompted to weigh evidence, to draw his own conclusions, to defend them. In other words, Garman was nurturing Calvin's spiritual self-respect, making a man of him, commanding him to rise out of the environing confusion of his adolescence, helping him to shake off his inner baffled sense of inferiority and helplessness before the forces of life. This teacher and his method brought Calvin Coolidge to himself. He became a disciple. Long years afterward he testified to the faith that Garman put in him. We must know this reborn spirit whom Garman begot if we understand the President who sat calmly in the White House for six years, even complacently, and let fate take her course, while tumult and the roar of the doomsday rapids raged about him. Calvin Coolidge in those crowded years when the country was rushing to the precipice was fortified by an invincible faith in Garman's conclusion:

"It sets man off in a separate kingdom; makes him a true son of God, a partaker of a Divine nature. It is the warrant for his freedom, the demonstration of his equality. It does not assume that all are equal in degree but all are equal in kind. On that precept rests a foundation for democracy that cannot be shaken. It justifies faith in the people. No doubt there are those who think they can demonstrate this teaching was not correct. I know in experience it has worked. In time of crisis my belief that people can know the truth, that when it is presented to them they must accept it, has saved me from many counsels of expediency. The spiritual nature of men has a

[12] "Autobiography," pp. 65 and 66.

power of its own that is manifest in every great emergency from Runnymede to Marston Moor, from the Declaration of Independence to the abolition of slavery." [13]

What a faith Charles Garman the spiritual progenitor of the President gave Calvin Coolidge! Bred in the heart of youth there in that Amherst classroom appears the creed which the President of the United States put into his works. In Amherst College with its compulsory calculus, with its slight course in the natural sciences, with its stress upon history, literature and the humanities, one man shone out for Calvin Coolidge—this lank, almost cadaverous leader with the glowing hypnotic eyes, who was at once evangelist of the boy's spirit, the awakener of his self-respect, the recipient of his unqualified youthful adoration.

We must take one more look at the Coolidge philosophy as it came through Garman, to understand the passionate conservatism of this disciple. He writes of the Garman ethics that "there is a standard of righteousness that might does not make right, that the end does not justify the means, and that expediency as a working principle is bound to fail. The only hope of perfecting human relationships is in accordance with the law of service under which men are not so solicitous about what they shall get as they are about what they shall give. Yet people are entitled to the rewards of their industry. What they earn is theirs, no matter how small or how great. But the possession of property carries the obligation to use it in a larger service. For a man not to recognize the truth . . . is for him to be at war with his own nature, to commit suicide. That is why 'the wages of sin is death.' Unless we live rationally, we perish psychologically, mentally, spiritually." [14] There speaks the Puritan out of a philosophy that had gathered itself into form for two hundred years.

4

Now we may take a final glance at Calvin Coolidge in those closing Amherst days. He was slightly above medium height and slim. His blond skin was still splotched with big freckles that came from his outdoor childhood. His features were clearcut and his eyes even in youth had a flashing spark of intelligence when he opened them suddenly and gave an upward glance. His shoulders sloped a little. As a freshman he was not well dressed. His clothes had come from Plymouth and Ludlow, and he was outgrowing his trousers and his coatsleeves perceptibly. But as an upper classman he was conventionally clad. For Sunday he wore the derby hat of the period and a dark suit which the merchant prince of Plymouth had probably ordered ready-made by number and measure from Boston. But it was a

---

[13] "Autobiography," pp. 65–7.
[14] "Autobiography," pp. 67–8.

good, warm, well-made suit, and was his everyday suit the next year even if he had grown a bit. Harlan F. Stone, who was a class or two ahead of Coolidge, remembers the young freshman's red hair and freckles. That red hair darkened only a little for a dozen years after he left college.

"I doubt," writes Justice Stone, looking back on that period after forty years, "if many of his fellow students were intimate with him. His extreme reticence made that difficult. To those who knew him casually he seemed odd or queer, but it only required a slight acquaintance to appreciate his quiet dignity and the self-respect which commanded the respect of others.[15]

Harlan Stone, his fellow student, did not realize how much there was in Calvin Coolidge until after Stone had graduated and had come back to commencement where he heard Coolidge, a senior, delivering what was known as the Grove Oration. It was supposed to be the humorous number on the class day program. But not even then did Stone realize what was going on under the quiet exterior which he glimpsed that first year after returning to the campus, a graduate. He was "impressed by the humor, the quiet dignity and the penetrating philosophy of his oration."

The oration made a hit. Calvin Coolidge was beginning to stand out in the college as a personality; but he must have emerged in his class before that, for he was chosen by a popular vote of the class to deliver this Grove Oration, a distinction itself. Only three seniors were elected to speak. He was one of those. His comment upon his success is characteristic:

"While my effort was not without success, I very soon learned that making fun of people in a public way was not a good method of securing friends or likely to lead to much advancement, and I have scrupulously avoided it." [16]

This comment, of course, was made in the perspective of more than forty years after the event, but it is singularly important first, to note that according to his own testimony in his callow youth he was anxious for success, ultimate success, not immediate popularity; and second, to note that the man deliberately suppressed his talent for humor and his tendency to be sarcastic in public. Calvin Coolidge at eighteen definitely was laying the foundations for a life career, pruning his dangerous talents, interesting himself in the long view of life—something young men rarely do. Ambition defines itself commonly in the late twenties. It is curious—indeed rather portentous—to see a boy reveal in his teens a prudence that would suppress his natural clownish delight in the exhibitionism of adolescent humor. But here in his callow youth, Calvin Coolidge seems deliberately to have stunted his frolicking elf; but often as the years passed, the puckish imp would peek from his life, squint-eyed and abashed, to spit out some

[15] Letter to the author, April, 1933.
[16] "Autobiography," p. 71.

dry hard quip—the clinkers of a hidden fire, so bitter that men often thought it was dumb. Another view is tenable: that these words written by the ex-President forty-five years after the Grove Oration are what the man thought he did in his youth. It is quite possible for a man to rationalize his conduct after the fact. Rather habitually in his autobiography one sees the ex-President looking back over the past and doing that very thing. That also is a human trait which helps to run the boundaries of a man's vanity.

At any rate, those intimate with him testify that he was given to mild tantrums, and that he liked to bedevil his inferiors, not seriously, but by way of diversion. A boarding-house tradition says that they served hash one morning for breakfast. Quoth Calvin to the landlady:

"Where is the dog?"

"Why, he's in the kitchen."

"Well, call him in!"

The landlady's cook called in the dog. He came wagging his tail.

"All right then, pass the hash!" quacked young Calvin while the table roared at the cook. He would have his little joke.[17]

One finds him writing near the close of his life: "I have never been inconsiderate of my superiors." [18]

Heredity in Plymouth and environment in Ludlow and Amherst had made their man. Let us look at him for a moment as he stands on the day of his graduation, which he accomplished "cum laude" but not "summa cum laude." He is one of a goodly company of three score young men who were destined to go out into the world and achieve the thing that Americans call success. They were to live without physical toil—live by their wits and brains. They were to belong to the ruling class, some of them; a few to the governing class. From this group of boys, assembled chiefly from environing New England out of middle-class homes, this young Yankee from the museum that was and still is Vermont, had emerged. He was known as a debater rather than as an orator. His fellow students knew that he could crack a joke. They realized that he was methodical, patient, long-suffering, and cannily ambitious, that he would sacrifice his own immediate gain for his larger good. In a rather consciously profitable silence he scorned the social butterflies. To the end of his days he held them in low esteem. When he could, he released his contempt for their riotous ways. We find him writing of the playboys a generation later: [19]

"There were few triflers. A small number became what we called sports,

[17] "Coolidge Wit and Wisdom" by J. H. McKee. Frederick Stokes, N. Y.
[18] "Autobiography."
[19] "Autobiography," p. 51.

but they were not looked on with favor and they have not survived. While the class has lost many excellent men yet it seems to be true that unless men live right, they die. Things are so ordered in this world that those who violate its laws can't escape the penalties. Nature is inexorable. If men do not follow the truth they cannot live."

Saith the Lord! Here across a generation speaks Garman's Jehovah. Remember, this was written thirty-five years after the festive youths of Amherst had neglected the "ouden" from Plymouth. He could bide his day for an invitation to join a fraternity. He could snap up the invitation with a monosyllabic "yes" and stop there. Also he could hold in his heart the unction that comes to him who knows that the "ways of the Lord are good and righteous altogether."

On his graduation day he had attained his full stature, five feet nine. He was slight, lithe but not scrawny, thin featured but not sallow, with a carroty head of hair that in a few years would darken into chestnut. He had kindly and, certainly for the most part, gentle ways and most interesting eyes wherein were reflected the prophecy of the coming years. Intelligence was there, determination and strength, strength that begot confidence in him when others looked into those eyes.

He had heard the throbbing of the drums of the minor gods who called to him all his life. In Ludlow when he was sixteen, for two long nights and days the Republicans celebrated their victory over Grover Cleveland. In Bennington he had attended the gathering of the tribes when the high priest came. The boy had listened to President Harrison exorcising the tribal gods who prevailed over the South in the "war of the rebellion." In 1892 he had marched in the Republican torchlight procession and had heard the drums rattle for the gods of prosperity, the gods which taught men to work, to earn, and to save. At the end of his career, looking back over this period, he extolled savings. For savings might also work and earn, and so he reverenced "money invested," which "led to a self-respecting independence that was one of the foundations of good character."

He had learned to cherish the works of the great prophet of the gods of prosperity, Alexander Hamilton. Set out on his upward journey was this Vermont youth. Body and mind and spirit were cast into the iron mold of a fate which guided him through life.

After the graduation exercises at Amherst were over, he and a stocky, brown-eyed youth, who had come to understand and love young Calvin, were wandering across the campus. They sat down under the trees and looked over the hills. Then, lying on their bellies in the shade, with their chins in their hands, they kicked the bluegrass sod and talked of the future. Both were going into the law. Dwight Morrow felt he was headed for Pittsburgh, which was the West. But "no Coolidge ever went West." Coolidges took the line of least resistance.

"Well, where do you think you will go?" asked young Morrow.

After a silence young Calvin answered: "Northampton is the nearest court house." [20]

It was seven miles away, and there he went.

[20] Story told the author by Morrow in 1924.

CHAPTER V

In Which Our Hero Takes His First Faltering Forward Steps Toward the Big Wide World

WHEN Calvin Coolidge came to Northampton in September, 1895, he made the final break with his past. His father and grandfather had been farmers and peace officers in and around Plymouth Notch for seventy-five years. His grandfather had entailed a piece of land for the youth, and the neighbors remember that his father was saddened by the boy's departure. Yet the father must have known when he took Calvin down to Ludlow with the calf that autumn day ten years before, that his little boy was lost to Plymouth and to Vermont. He hoped that the boy would "keep store"; preferably a drug store. Calvin demurred: "Well, you wouldn't want me to sell rum, would you, father?" And the slick answer turned away the elder man's wrath.[1]

Calvin Coolidge came to Northampton after spending a summer vacation on the farm following graduation. He has recalled that he read Walter Scott's poems during one of his vacations. He was an industrious boy. His education did not stop when he passed out of the front gate of Amherst. And it is well to note that in those post-college days he was reading "Shakespeare and Milton and found delight in the shorter poems of Kipling, Field and Riley."[2]

But when he set out definitely to make his fame and fortune in the big world, he gave up childish things. First of all he records that he had his college haircut shorn. The law was the dignified Delilah that took his carroty locks. Then, accompanied by an Amherst friend, he walked into the

---

[1] Story printed in the newspapers at the father's death; revived when the son died; still current in Plymouth.

[2] "Autobiography," p. 73.

law office of Hammond & Field, of Northampton. His friend introduced young Calvin who stood grim and speechless while the young man from Amherst extolled the boy's qualities to the elder lawyers. His would-be employers in return addressed not Young Coolidge but his friend. In less than twenty minutes the deed was done. He had a place in that law office to study law. At the end of the interview Calvin Coolidge had contributed two words as he turned away with his friend:

"Good morning!" [3]

The law office of Hammond & Field in the middle nineties did not differ from any country law office in any country town from California to Maine: an outside reception room lined with law books, carpeted with brown hemp runners from desk to desk. Out of the room opened two small consultation rooms also filled with law books, and in each room a desk and two chairs, a swivel chair behind the desk, and an old-fashioned office chair with round arms at the side of the desk. Calvin Coolidge had a small table in one corner of the outer room. He sat there like a mouse for months, ran errands for the two elder lawyers, copied legal papers, took over collections and read. Always he was reading—or writing. He was improving his English style, also writing on a thesis which he submitted in an essay contest promoted by the Sons of the American Revolution. He won the prize. Neither partner knew he was seeking the prize. The partners read the announcement of his success in the *Springfield Republican* which identified the winner as a young law student in their office. The prize was a hundred and fifty dollar gold medal. The essay discussed the principles of the American Revolution. Judge Henry Field, with the *Springfield Republican* in hand walked to the youngster's desk and asked:

"Is that prize yours?"

"Yes sir," answered the clerk.

"When did you get it?"

"About six weeks ago," replied the lad.

"Why didn't you tell us?" asked the curious Judge.

"Didn't know you would be interested," explained the apprentice.

The Judge in amazement exclaimed: "Have you told your father?"

"No! Sh'u'd I?"

A month later he brought the medal into the office and asked the lawyers to put it in their safety deposit vault. A year or so later he laconically asked for it. No one in the office ever knew what became of it. [4]

It is illuminating to know the Coolidge side of this episode. He explains it, and in his explanation reveals his sentimental side:

"I had a little vanity in wishing my father first to learn of it from the

---

[3] Conversation of Henry P. Field with the author in 1924.
[4] Conversation of Judge Henry Field with the author, 1924.

press, which he did. He had questioned some whether I was really making anything of my education. . . . He wished to impress me with the desirability of demonstrating it." [5]

So, like the Little Red Hen, he did. Colonel John first read of it in the newspapers. But never a word said Calvin to his employers when they asked if he had told his father; "No! Sh'u'd I?" was all they got of the little sentimental play he was staging in his heart for the family at Plymouth.

The thing that impressed Judge Field about the youth after his indefatigable energy was his seriousness. Coolidge explains this by saying that in that first year in the law office, in addition to his law books he was spending his evenings with the masters of English composition, reading the speeches of Lord Erskine, of Webster, of Choate, the essays of Macaulay, the writings of Carlyle, and, of course, John Fiske. In the middle nineties John Fiske was a prophet of "manifest destiny." In that day, the evolution of the town meeting of New England democracy unto the land of pure delight with an assured six per cent interest, was a bauble always shimmering in the vision of the American academic world north of Mason and Dixon's line.

Young Calvin, reading John Fiske, was translating the orations of Cicero,[6] by way of practice for the ascension! Is it then any wonder that he resolved, and for forty years remembered his resolve, not to go back to his college for intercollegiate games, but only to attend the alumni dinners scrupulously? His father was sending him thirty dollars a month, and he lived within his income.

"Not much," he wrote forty years after, "was left for any unnecessary pleasantries of life."

Though he did, probably because it was good for him, join a canoe club and went rowing on the Connecticut a mile from town on Sunday. The Puritan Sunday was fading, but the Puritan conviction of virtue in discarding the "unnecessary pleasantries of life" was deeply imbedded in his heart.

In his new abode, the young lawyer had acquaintance with a few storekeepers, and he renewed his acquaintance with "Jim" Lucey, the shoemaker who lived on Gothic Street, not far from his law office.

Coolidge sat at his desk for twenty months, and then having been duly admitted to practice law before the courts of Massachusetts, after he had finally got his certificate, formally notified his father. He had done three years' work in twenty months, but not a word said he to his father about his hope to clip off sixteen months. He explained:

[5] "Autobiography."
[6] So he told John Buchan in the White House.

"I wanted to surprise him if I succeeded and not disappoint him if I failed; I did not fail. I was just twenty-five years old and very happy."

One begins to see what the Coolidge silence concealed; pride and sentiment among other things.

"No! Sh'u'd I?" was not the quaint dumb response of the dullard. It was the acridly humorous defense habit—peeking Puck spitting ashes!—of a sensitive spirit, proud but fearful of failure; the drab cover of an acute inferiority complex bred of Puritan repression. One hundred and fifty years before, Jonathan Edwards thundered in the Congregational church at Northampton and Cotton Mather came there to preach his dour doctrine. But Calvin Coolidge, moving about town, going with unbroken regularity to every session of the court,[7] sitting silently watching the proceedings that he might gain months in his law course to show his father he was not dumb, found at Northampton an atmosphere that would cherish and preserve his virtues.

2

Northampton in the middle nineties of the nineteenth century was developing economically and politically the Puritan thesis which Jonathan Edwards and Cotton Mather in their day and time stated as a religious creed. All that continence did by way of thrift and prudent savings, all that godly conduct vouchsafed in the way of material prosperity, all that saintly living in meekness, frugality and piety returned to men who kept their eye on the main chance, were coming through flower to fruit in western Massachusetts in those days. Northampton, an industrial town, was realizing that the philosophy of John Calvin, Edwards and Mather would produce property. Even though the increase in property was for the moment denied by the panic and hard times of the early nineties, yet was property held in reverent esteem; and Northampton knew for a truth that to work and to save were the highest human virtues. The Hamiltonian doctrine of the right of the propertied class to rule in those days was translated in the shibboleth of the times: "Brains win." By brains one meant avarice, cunning, forestalling, cheating in its politer legalized forms, foresight as it affected money-making, and greed as it goaded men to build up fortunes, piling dollar on dollar for the sake of the hoarded pile. Northampton, however, cherished no town secret. All America recognized these prescient truths. The world was approaching the new century aflush with this evangel!

Of course it was tempered, this creed of the accumulator. Philanthropy was supposed to soften the wrath of God. Pleasant living in spacious homes underneath New England's tall wine-glass elms surrounded by

[7] Conversation with Henry Field and the author.

velvet grass and garden flowers, gave the successful ones who kept the Puritan faith in its economic phase some sense of Heavenly peace and beauty. Moreover, colleges were endowed. The lame and the halt and the blind were cared for and cured at public expense and by private beneficence, while prayers were said in lovely white-painted, tall-steepled churches. Music rose from the organ lofts to anoint the ears of the faithful, so that Greed might be decently clad and not shock men by his indecorous nakedness.

Into that world with short, quick steps, with a slightly rigid body giving him a rather prissy gait, walked Calvin Coolidge circumspectly before the Lord, following the law. It was foreordained that he should find his first lodging house in Ward Two, the soundest Republican ward in Northampton. And when he had acquired the right to vote there it was predestined that he should appear at the Republican precinct caucus. He took no part. He just appeared and voted. The next spring he appeared and voted again. And the next fall he appeared and voted and helped to nominate a court house candidate for something, not much of an office but he had a toehold in the party organization. The year was 1896. That year in America saw the first gigantic struggle between the haves and the have-nots; the challenge to Puritanism in its political and economic phase made squarely by William J. Bryan, its first national critic of prosperity and the rule of prosperity since Jackson's day. Into that struggle Young Coolidge, law clerk in the office of Hammond & Field threw himself on the side of the haves. All that he knew and thought, all that he had lived and all that he had been taught from Grandpa Galusha Coolidge, through Ludlow to Garman in Amherst, revolted that year at the doctrine of the Democratic National platform and Bryan, its candidate. The young debater from Amherst took the stump for McKinley.

The echoes of the drums of Ludlow, the bugles of Bennington and the Republican applause at Amherst were throbbing in his ears. He was a Republican by blood and breed and birth. To understand how deeply moved he was in that campaign, we must consider the issue before America in 1896, which arose from a major depression. For thirty years we had been settling the West and rebuilding the South. A continental commerce sped on rails, through pipes, upon rivers, canals and coastal waterways, along wires. The ox and the horse for the first time since the Caucasian left the Aryan lands of the distant East were no longer his first friends and dependents. Steam, beast of burden whose flesh knew no weariness, to whom rest was cruel rust, drove men into shops who had prodded horses and oxen for thousands of years. Machines rewarded their owners and no one gainsaid them, gouging into the forests for lumber, delving into the earth to bring up stone, coal and iron, and fashioning cities—all with borrowed money. The mad, unhampered debt-scramble of the sixties, seventies, and eighties suddenly stopped with the bang of a panic. The rails were all laid, the pipes dug into the earth and the wires strung over it, the cities built and the states hewed out of the desert. It was done—finished. For the moment no more material things were needed. Billions of capital or what seemed to be capital—but which was mostly debt—the unearned and accumulated increment of commerce, suddenly became paralyzed in idleness. Factories closed. Banks began to fail. Farm prices slumped. Riots broke out in the great cities. The hungry, the miserable, the exploited who had poured into America from Europe were left deserted, dazed and despised by those who had brought them here to batten on their sweat. We had borrowed of our tomorrow too deeply to pay the day's wages. The industrial machine stopped. In 1893, and '94, and '95 strikes and civil disorder became so widespread that the President called out the federal troops to quiet disorders. Prices of commodities slumped tragically. Mortgages began to default. Unpaid interest upon bonds of states, cities, railroads, and public service corporations shrivelled their values. Savings banks tottered and the gods of New England were mocked. Out of the new country across the Mississippi and below the Ohio came the cry of the debtor. The Western borrower tried vainly to cut down the dollar, to inflate his resources that he might put them at par with the dollar.

In those years from 1893 to 1896, Congress was thrown into turmoil by the demand for "the free coinage of silver at sixteen to one" when the commercial ratio was about thirty to one. President Cleveland, deserting his party, battled alone for the gold standard and the big dollar. The Democrats of the debtor West left President Cleveland, and the Democrats of the South who were deeply in debt because they had been reconstructed with borrowed money joined the West. It was a mad world.

CHAPTER VI

His Baptism of Fire

REMEMBER that Calvin Coolidge was a Republican; congenitally through his physical father, the Colonel, and spiritually through the Puritan mystic, Charles Garman. He had his baptism of fire in the fight against Bryan and fiat money. The differences between the parties were fundamental! The Democratic convention nominated as its Presidential candidate, William J. Bryan, handsome young Congressman, from Nebraska, who roused the convention by a speech dynamic with emotion and swept the delegates off their feet by his fiery eloquence. Surrounding him in power were the Dantons, Robespierres, and Murats of troubled times. Wall Street became an Octopus. The East and New England were rising to face the menace of political rebellion. It is easy to see how Young Calvin Coolidge out of the green hills of Vermont, where the farmer's thrift had bought for him a few Iowa and Kansas mortgages and where the whole civilization basked in a pre-revolutionary calm, should inevitably set his flinty New England face against the tenets of the Democratic party and its rabble-rousing candidate. Calvin Coolidge was instinctively attracted by the Republican position. Mark Hanna, the kingmaker, had contrived the nomination of William McKinley in the Republican National Convention and made "the gold standard" the issue of the campaign. Hanna's gold standard platform forced a bolt of the Republicans from the silver-mining Rocky Mountain states. Moreover, McKinley, who was first of all a politician but often at incidental moments a statesman, forgot that he had once advocated free silver and casually accepted the gold standard plank in the Republican platform. Young Calvin Coolidge, a Republican by blood and in the birthright, would have accepted the Republican platform, with or without the gold standard.

So the budding Yankee statesman painstakingly went into the issue presented by the challenge to the gold standard. He studied the currency question carefully, if with partisan bias, went out speech-making to the little towns of his district, even went to Vermont and had his say in Plymouth. He became identified with the Republican party in his precinct and in Hampshire County, his adopted home. Local leaders marked the youth and when he thought he could be elected precinct committeeman of his party, his party chose him.

At the end of the fray he wrote in his life story: "Of course Northampton went for McKinley." Nothing in that laconic phrase recalls the turmoil of the hour, the processions, the bands, the old soldiers riding in shotgun cavalcades down the streets of Northampton, the factory workers and mill hands cheering for Bryan, the angry Democratic leaders charging Republicans with corruption, and the slow boa constrictor grip of Mark Hanna's Republican money on the aspirations of the rebelling Democrats who would change the existing order and exalt the have-nots into power. Mark Hanna's money bought few men directly and corruptly in 1896. But Mark Hanna's money and the power of money did penetrate the electorate consciously and with terrible force. For money was respectable and the respectables rallied together. It is always hard to be an outcast. And so during September and October, in Northampton, and in every town, city and township throughout the land, the respectable forces, the governing and the ruling classes rather irrespective of party alignment, used their influence on those next under them who passed their conversion down the line without bribery but through the force of a well-greased [1] organization. In Northampton the banker talked to the mill owner, the mill owner converted the superintendent, the superintendent labored with the foreman, the foreman with his would-be successors, and so in the end "of course Northampton went for McKinley" and Calvin Coolidge put his foot on the first rung of the ladder to Republican leadership. He was a little leader but authentic, tried and true, who marched circumspectly in the parades, minced his way after the election from his rooming-house in Ward Two, the staunch Republican ward he had chosen, to the law office of Hammond & Field on the main street of Northampton, an anointed and respectable Republican even before he had been admitted to the bar. It was inevitable that such a young man should within a few years after he began to practice law become vice president and counsel of the new Nonotuck Savings Bank of Northampton. Such things are written in the stars.

The national election of 1896 marked the peak in the control of the United States by organized capital. Over that peak control continued for

[1] Meaning that it had plenty of money for all legitimate needs.

ten years on a slightly declining slant. Yet politics and business united frankly with the election of William McKinley. Mark Hanna assessed business with something like scientific fairness to support government. McKinley was honest, gentle-spoken, kindly; but Hanna's supergovernment recognized organized wealth as an estate in politics, and politics as the hand-maiden of organized wealth. The alliance was no longer disreputable. Steam had welded the interests of the country's aggrandized corporate wealth into a huge mass of sacrosanct property—largely incorporated—a new social machine. And when Coolidge, a rising young attorney in Northampton, entered upon his political career, automatically he became one of the small cogs in the Republican unit of that machine, a precinct worker, later a precinct leader in Ward Two.

Party membership meant something in those days. Only partisans could lead in the ward and precinct caucuses. Only partisans could sit in the county conventions. And generally speaking, across the land in every hamlet and village north of the Potomac and of the Ohio, to the western ocean, energetic Republicans like Calvin Coolidge who were wise enough to live in the silk-stocking wards that were safely Republican, dominated county conventions. They controlled the Republican nominations for county offices, for district offices, which is to say members of the legislature, the state senate and the judiciary. Rising in the scale of power these silk-stocking Republicans from the wards where lived the well-to-do dominated delegations to district conventions where congressmen were nominated and to state conventions where governors and state officers were named.

Ward Two into which Coolidge moved when he came to Northampton was typically an upper-class section of the Republican party. Here lived and ruled the major gods in the party organization. Did Calvin Coolidge know this or sense it or did he just happen to locate in Ward Two? In Northampton they say it was Coolidge luck. Luck often is a lazy man's explanation for the result of intelligent diligence.

At any rate there he was well placed for a political career. It so happened that in the last half of the last decade of the old century Massachusetts politics, compared with politics of many a western state in that period were reasonably clean. The Massachusetts tradition of political service coming down through the century from the Adamses to Webster, to Sumner, to Hoar, who was contemporary in his declining years with Young Coolidge entering Massachusetts politics, had been a high tradition. Its governors almost invariably had been men with some sense of obligation of their nobility. Its congressmen, as for instance Young Henry Cabot Lodge, were erudite. And while railroads, insurance companies, manufacturers and shipping companies had their way and their say in matters which affected them, still the use of money in Massachusetts politics and particularly in Republican politics was not in those days a public scandal. The Republican state chairman had funds. He sent money down to the various senatorial and legislative districts to help the nominees of his party. These funds came from business men who had interests before the legislature. They were buying the party support, not bribing individual legislators. The transaction was conventional and highly respectable. Power along with money came from the top in the Republican organization. Young Coolidge, working in Ward Two, attending every party meeting with scrupulous regularity—as he attended his class-meetings at Amherst—seemed never to have questioned or deeply considered in those days the plan and scheme of things by which his party was shackled with plutocratic control. He sacrificed nothing either of honor or freedom in being a Republican committeeman. Moreover his small prominence in the ward gave him a little power, a certain prominence in the town, and naturally brought him a few small clients. He became a ward leader, known to organization men in Hampshire County before he was thirty. He ran errands for his political superiors, county Republican chairmen, the county officers, the scribes, the high priests in the court house, the local temple of Republicanism. He was not asked to do compromising things. He learned to love his party as a source of power, power for him, power for what he regarded as good government, the rule of the well-to-do; brains in short. God was moving in his mysterious but appointed way through the Republican party "his wonders to perform"— establishing an orderly expanding civilization among his chosen people, the Americans. John Fiske's "Manifest Destiny" was doing its daily miracle. What Calvin Coolidge had learned in Vermont, singing in school "Work for the Night Is Coming," and "Bringing in the Sheaves"—all about the virtue of thrift, the nobility of saving as it affected human character—what he had learned from the mottoes on the walls of Garman's classroom, made him feel comfortable in serving his party. Its ideals of incorporated capital as a means of social, economic and political happiness and salvation hardened the cement of his loyalty. So he stepped without qualms upon the escalator in Ward Two as precinct committeeman in the days of McKinley, and started his long upward journey through hard ways to the stars.[2] He told his friends, and wrote it later in life, that he always expected to be the kind of country lawyer he saw all about him. But he said it was decreed otherwise. "Some power that I little suspected

[2] *Ad Astra Per Aspera*, the motto of the state of Kansas, which is a state settled by abolitionist Yankees in the fifties, was written on the Kansas state seal by a young Massachusetts Yankee, John J. Ingalls, United States Senator for eighteen years and active vice president, who had the Puritan theory that no good comes easily, that success through hard ways is a proof of virtue.

in my student days, took me in charge and carried me on from the obscure neighborhood at Plymouth Notch to the occupancy of the White House."

2

The nineteenth century, swinging into the twentieth found the thirtieth President of the United States in a law office of his own in Northampton, in the Masonic Building. He had two rooms. He remained in those two rooms for twenty-one years. He put the money he had inherited from his Grandfather Moor into the office furniture and a good working library— in all eight hundred dollars. He paid two hundred dollars a year rent. A few months after his admission to the bar he was paying his way. "I was alone," he often said to his friends. "Of course I had acquaintances that I might call friends, but I had no influential supporters who were pushing me along, sending me business." Always in that period, and indeed through his whole life, he spoke of himself as alone. Indeed he was an isolated man in many ways.

He became a member of the Republican state committee in Ward Two. There again he attended meetings of the county organization regularly— silent, eager for work. He worked and worked well. He asked nothing, thus saving his political capital. Reward came as the result of work. He had no other formula in his life than the mystical truth of the Garman mottoes, shibboleths which exalted the Puritan virtues. When he became a member of the state Republican committee, in Northampton, he became a vital cog in the Republican machine of his county, of his city, of his state.

Three years after he came to town, he was elected one of the members of the common council of the town of Northampton, from Ward Two. It was in that year that the Northampton City Council was faced with the first serious traffic problem that had come up in a hundred years. Fred Jager brought an automobile to town. It was called a Locomobile. It went chugging up and down the streets and scared the horses. The Council resolved that something ought to be done. But nothing was done. Councilman Coolidge was invited to ride in it but refused. As a councilman that year it was Coolidge's particular job to solidify himself with the heroes of the Spanish American war by getting behind their efforts to build a memorial armory which later he explained was the beginning of his interest in military preparation; again rationalizing his conduct thirty years after. Four years after he came to Northampton he became vice president of a savings bank and by reason of his official position the attorney therefor. He was making fourteen hundred dollars a year [3] and saving most of it.

[3] He spent two hundred dollars for office rent, one hundred and fifty dollars for board, one hundred and twenty dollars for room, fifty dollars for clothes, fifty dollars for dalliance—cigars, an occasional beer, amusements and nonsense, twenty-five dollars for laundry—on the estimate of his friends remembering his habits after his death.

He served two terms as member of the city council. And by the way, no biographer who ever talked to him as much as an hour fails to note that the first public act he performed as a city councilman was to offer a resolution of respect for a dead Democratic councilman in a Democratic ward. Obviously he regarded it then and at the end of his life as good politics. Always he played good politics. Which does not mean that he was dishonest, nor that he was a demagogue. For he was never that. But he always kept his eye on the main chance at the next election, and the next, also the next, and even the next. He had a sense of the electorate, and its whims and fancies. After retiring from the Council, Calvin Coolidge was appointed city solicitor and he has declared many times that he wanted to be city solicitor because it would make him a better lawyer. He was elected city solicitor and held the job until March, 1902, at the salary of six hundred dollars a year. But he had a sort of foothold in law which was worth many times more than the salary in the end. He won a lawsuit from his old preceptor, Mr. Hammond, of Hammond & Field, and recalled it in his autobiography with a chuckling delight after thirty years. He had the legal work of the bank, managed a little real estate, settled a number of estates, did some collection business. Being a competent lawyer, he settled more cases out of court than in it, but he was proud to brag that "very few of my clients ever had to pay a bill of costs." During those first years he worked so hard that for three years he did not find time to go back to Plymouth. But when he went, he went as city solicitor of Northampton, and was a somebody. A year after his last election as city solicitor the justice of the Supreme Judicial Court appointed the young man to the office of clerk of the court for Hampshire County, a job that paid twenty-three hundred dollars a year, more money than he was making at that time. But he was too smart a politician to play the money game. He got out of the office by refusing to stand at the election, and by that time had taken the second round of the ladder. He had been elected chairman for his county of the Republican state committee. He was beginning to have some local fame as a speaker. People in the smaller towns around Northampton often invited him to be "the speaker of the evening." It was in the autumn of 1904. The young clerk of the court was to make the principal speech at the country club of Blandford in the Berkshires at the dinner after "the Fox Hunt." But the notable thing was not the speech. The memorable thing was that Calvin Coolidge took his first automobile ride that night to and from the dinner. His neighbor, Fred Jager, being the only man in Northampton with a car, carried him over. It was a noisy, rattlety-bang car, going on a dark and stormy night over the hill roads of the Berkshires and Jager always remembered that Young Cal was scared. Climbing into the car to go back he said he didn't care much to ride in an automobile, but he had to get home. That banging, clattering, pounding, hiccoughing, hesitating, tooth-shaking first gasoline car in Northampton was the precursor of an industry which twenty years later was to be the symbol of a civilization that went pell mell dashing headlong through a new economic frontier to its fall while Coolidge was in the White House. He might have been vastly more frightened than he was that night on the Berkshire hills if he could have sensed the spiritual and economic danger that would come in his early fifties to the established order of his ideals as he sensed physical danger in his early thirties. As it was, getting out of the car that night Coolidge said to Jager: "It's wonderful to ride in a horseless wagon." Then a pause: "But it won't amount to much!" [4]

Always one finds him moving up, getting a little more influence even if a little less money, but always depositing service in the bank of party gratitude, service that would in the mysterious alchemy of his philosophy bring him power. In 1904 when he was nine years out of college, he was at the head of the Republican party in a town of thirty thousand, a man to be reckoned with politically, a rising young Republican, firm in his faith; a party man consecrated to the party cause. And all this because he had faith in the mystic puissance of friendly service to bring its ultimate rewards. He was that prudent from the start.

[4] Letter from Fred Jager to the author.

CHAPTER VII

Our Hero Does the Best Day's Work in His Life

EVEN before Calvin Coolidge became chairman of the Republican central committee for Northampton, the worm was in the bud for his party. The triumph of the militant plutocracy led by Mark Hanna, which elected William McKinley President in 1896, enjoyed a short, sharp climax. The gold of the Klondike and from the African mines inflating the currency, the Spanish American war stimulating trade, brought back prosperity with a rush during the last three years of the old century. With prosperity came its handmaidens—stock gambling, chicane and corruption. It was a boiling mess of pottage with the lid off. Naturally, this broad statement must quickly be qualified. Gambling, chicane and corruption blighted only a small area of American life. But the blight colored the picture far beyond that small area. The Democratic party as an opposition party was divided under Bryan's leadership. Republicans controlled government unquestioned at Washington, and in most of the state capitals and in the larger cities of the North. Swindlers and cheats were inadequately checked in small business, and the higher realms of business ran with a loose rein. Still put together large and small they made but a fractional percent of the population. Small grafters and corruptionists, if not directly aided, were at least comforted by the respectables. Mark Hanna's alliance between business and politics in Washington, without buying and selling men, soon gave Wall Street leaders political control in many states through state bosses who named federal Representatives and Senators. When they appeared in Washington these Congressmen were bought and paid for—not all, not even a majority, but many.

It was a great day at the court of Belshazzar. Then suddenly death played a trump card. McKinley was assassinated. Theodore Roosevelt appeared. The scene changed. Roosevelt took the leadership of youth in the Republican party. In the first four years of the century when Calvin Coolidge was county chairman of his party in the town of Northampton, he heard strange voices in the political air. Roosevelt had begun preaching a thing he called "righteousness"—a pious sounding word, offensive to his party elders and betters. In those days Uncle Joe Cannon, Speaker of the House, the high priest of conservative reaction jeered that Theodore Roosevelt had discovered the ten commandments. Which was not quite true. However, Theodore Roosevelt's contribution to American life was like that of Moses. Roosevelt was an agitator. He used the front porch of the White House as a sounding board which sent his preachments and moral proclamations booming through the land. He denounced conventional corruption. He drove the thieves out of Cuba, sent them scurrying from the postoffice departments before the monument had been set over McKinley's bones. First of all we must remember that Theodore Roosevelt was young, a President in his early forties. His appeal was directly to young Republicans. He awakened hope in the colleges. It was not strange that Calvin Coolidge heard him. Calvin Coolidge had much to unlearn from Theodore Roosevelt, much that Vermont had taught him. All the smug conservatism that had anointed Coolidge from the dripping torches in the parades for Harrison and McKinley had to be reconsidered, restated, oriented anew to be available for a young Massachusetts Republican in the days of Roosevelt. And such a stirring figure was Roosevelt—the Teddy! So glamorous was his leadership that thousands of young men like Calvin Coolidge, out of high schools and colleges, filled with ideals of service that they had learned from books and in classrooms like the various Garman sanctuaries, gathered around the Roosevelt banner. They accepted his leadership even without realizing the implications of his cause, but certainly not denying those implications. Calvin Coolidge ex officio became one of these acolytes of the new temple. Possibly Roosevelt, who thrilled Coolidge, also frightened him. For Calvin Coolidge was apparently never deeply convinced that Roosevelt's creed was sound. Indeed the Rooseveltian seed fell on stony ground in Coolidge's heart, and when the sower passed, the thorns and brambles of his Yankee conservatism choked the Rooseveltian crop. But in the first decade of the century, Theodore Roosevelt was a winner. The people were with him; yet Young Coolidge was a winner, too, with a keenly developed sense of what voters will do in elections. As chairman of the city Republican committee, he wrote of one unsuccessful campaign:

"We made the mistake of talking too much about the deficiencies of our opponents and not enough about the merits of our own candidate. I have never again fallen into that error." [1]

[1] "Autobiography."

Also the young man in Northampton had realized that power in politics as an organization leader outside the office takes much time. An honest, poor man cannot afford it. So he refused reelection as chairman of the committee, but stayed on the Republican central committee where he could have a finger in the pie. He was committed irrevocably to politics—politics for the sake of politics! He had saved a little money in his first ten years out of college. He had been elected twice to office—councilman, and city solicitor—and had served as clerk of the court. He was established as a rising young attorney, vice president of a savings bank with a wide acquaintance in his town and county. He was a poor mixer, using the word mixer to mean one who rather deliberately associates with men who have votes to give, in order to gull the voters by graces and charms of manner. Yet he was a vote-getter. In the evening it was his habit occasionally to loaf downtown. He took his monthly glass of beer at the town beer gardens, stiffly and stodgily without joy in it. He sat almost silently in James Lucey's shoeshop while the loafers' parliament babbled through an evening. The Lucey friendship was one of those illogical attachments that men sometimes make; purely sentimental. Coolidge was given to these odd but genuine displays of affection. Lucey's case was classical. Lucey became attached to the youth in Coolidge's early days, gave him sententious advice, which Coolidge seemed to consider but did not. Lucey probably did teach Coolidge something about the technique of that democratic process known as "handshaking"—personal, direct vote-getting. Moreover, the shoemaker's shop was a good place in which to meet and mix with men whom Coolidge rarely met elsewhere. His barber maintained another such shop and Coolidge frequented one barbershop for nearly forty years. He had his favorite bootblack and barkeeper, he never forgot his father's old hired girl in Plymouth, nor his childhood school teacher. And through all the years while he rose to fame and power, he warmed their friendships in his heart. For after all only his face was frozen! It was said of him at the end that he was as thrifty with his friends as he was with his money. He had Jim Lucey and his first nickel at the end of his life.

Coolidge in those callow days at times might be seen sadly smoking a five-cent cigar in the drug store with the young men of his age for a few minutes after supper before hurrying to his office. Even then old settlers [2] recalled that he used to purse his mouth as though his heart was bursting with some secret sin or sorrow or as though his lips were swelling with an unrelieved chew of tobacco. But he never chewed. Neither was his heart

[2] Conversations with the author of Coolidge's Northampton friends—Lucey, Field, Jager, et al—in 1924 and again in 1933 are authority for statements made in this chapter about Coolidge and his Northampton environment. Something of the same material is found in early biographies by Washburn, Woods, Hennessy and Sawyer.

broken. It was his way, his sad visaged demeanor. Yet withal men trusted him; probably because in a gabbling world he did not babble. He learned about men in the shoeshop, at the barbershop, around the soda fountain, before the bar of the tavern. For he was a shrewd Yankee, smarter than chain-lightning about the ways of men. He could tinker with certain cogs and levers in the human heart better than most, and not in cold blood either. For he loved his kind.

In those days his body filled out. Its gaunt gawkiness went. His face grew full but never plump. His hair was still carroty, though darkening into chestnut, and his eyes, always the beacons of his intelligence, were expressive. When he opened them they revealed a man that men could trust even if they did not return his repressed, undemonstrative love. So knocking about Northampton with his heart set on two things, his account in the savings bank and his popular strength at any possible election, this Yankee tinkerer with humanity must have realized deeply and definitely how completely Theodore Roosevelt was capturing the American people. But the realization did not make him a little Roosevelt. Little Roosevelts were popping up all over the land in cities, in counties, and in states, noisy young county court-house candidates who thought they were idealists, eagerly trumpeting at the local dragons of sloth and corruption. But Calvin Coolidge was never one of those. He bided his time. He saved his money, played politics with his cards close to his vest, said little and watched.

While he was watching he met Grace Goodhue. She was a teacher in the deaf and dumb school located on the outskirts of Northampton. The teachers in the school were a few years older than the girl students in Smith College which dominated the social life of the town. The women members of the faculty of Smith College wore their stockings too blue for a young man who was as proud of his erudition as young Calvin Coolidge. His kind like to think they prefer their females dumb. Also the women in the Smith College faculty were making more money than Calvin Coolidge when he was ranging the woods looking for a mate. So he picked a teacher at the deaf and dumb school making $75.00 a month and keep, who would not put him in his scholarly place. She was a lovely creature, Grace Goodhue, then in her twenties; a graduate of the University of Vermont, but with no indigo in her hosiery. And when she, having the social experience of a sorority girl in a co-educational college, took a look at his obvious specifications, Calvin had no chance. He was foredoomed. Grace and Calvin met casually, went with the same boating, picnicking, dancing, whistclubbing set—not the royal families of Northampton but the young nobility and gentry on the edges of the Congregational church. Her family lived in Vermont and the awkward young statesman of Northampton did not fall under the appraising eye of his future mother-in-law until it was too late for her to stop the love affair. The mother-in-law tried vainly to postpone the wedding from June to November. Those two were never friends. But Grace Goodhue was right about Calvin Coolidge, her mother was wrong. It was love at first sight. She, however, probably saw him first. Not only was her social experience wider than his, her emotional intelligence was keener. So when her lover and her mother clashed, she followed her lover.

2

It is characteristic that in his "Autobiography," the page which begins with this sentence: "My earnings were such that I was able to make some small savings," is the page on which he records his first love affair. And on the next page, writing of Grace Goodhue, his first-acknowledged sweetheart, he declares:

"We became engaged in the early summer of 1905 and were married in October, 1905. . . . I have seen so much fiction written on this subject that I may be pardoned for relating the plain facts. We thought we were made for each other. For almost a quarter of a century she has borne with my infirmities and I have rejoiced in her graces." [3]

It required some emotion to pen those last three sentences nearly a third of a century after the fact! For in those three sentences are chronicled the essential details of the story of a happy, successful marriage. It was the story indeed of the woman facing one reality—the minor infirmities of an exceptional man. She understood him and still loved him. She knew he kept a mistress—politics—and that as his wife she had the deep, dependable loyalty of the man who trifles rather idly, but always turns homeward for courage, if not for wisdom, at least for solace. He was a moody man who had a cruel streak of which he was ashamed, given to tantrums when crossed, not hot tantrums but cold, peevish ones. He was the kind of man who quarrels or is irritated by one person or circumstance and "takes it out on the next person." [4] He was full of mischievous deviltries—meaningless, unimportant, which never affected his career or scarred his life. One sees them in his boyhood. They were to show up in his last days. And as a husband there was that mistress with whom he divided his talents and to an extent his income, his seven-devil lust for politics. Next to his pride in Grace Goodhue, her children, her home, his love for politics was the dominating influence in his private life.

The courtship of Calvin Coolidge and Grace Goodhue naturally was queer. Years afterward, when he was gone and her mother had been dead

---

[3] "Autobiography."
[4] Frank Stearns told me this; also Frank W. Buxton; and Mrs. Coolidge reveals it plainly in her articles in Good Housekeeping, 1935.

nearly a decade, Mrs. Coolidge, writing for a magazine [5] frankly declared that his mother-in-law and he never got on. She opposed the marriage. In the family wrangle he defeated the Goodhues and speeded up the marriage from June, 1906, to November, 1905. Before the marriage, Grace Goodhue took her gaunt, taciturn, repressed suitor up to Vermont to visit kin and friends. She remembers that he wore "a dark blue serge suit, new and perfectly tailored, and a black derby hat." And that he put a whisk broom in the back of the buggy in which they were riding so that he could spruce up at the end of the journey. When they entered the house, and after the introductions were over, he sat on one end of a long parlor sofa, she on the other, while their friends chattered, and Calvin said nothing. "Not one word did he utter." [6] And when at last he could bear it no longer, he rose and said simply, with one of his best smiles:

"We'll be going now!"

Naturally, she was put out. And naturally, they jangled about it a little when they were riding home, and naturally, the jowering was futile. He did the same thing over and over again when he cared to. He could not help it.[7] Being what he was, profoundly loyal, he was a devoted husband; but being also self-centered, undemonstrative, reserved and like most shy men crusty and, when irritated, grumpy, he did not make her a full partner in his life. "He was not favorably impressed with my education," she wrote for Good Housekeeping after his death. Also she recalled that he never talked politics with her nor took her into his business confidence. If he came home moody and irritable, that was his affair. She learned to keep quiet and let him stew in his own juice. "If I had had any particular interest," she wrote about his political affairs, "I am sure I should have been properly put in my place." [8] Some men are just that way in their homes. Introverts and can't help it. Naturally Grace Coolidge, looking back at the close of a happy life which she lived gorgeously, beautifully, would write that she was always suspicious of those wives who said of their husbands, "He never gave me a cross word!" She had received her quota. Yet she never had to lie awake nights worrying over his whereabouts!

But she knew what he was before her marriage. In those days when he went to a picnic he brought more than his share of the food, and then before clearing up the debris, counted the left over macaroons and wondered what had become of one of them.[9] The other picnickers snickered. And probably Grace Goodhue realized then that she never could break his parsimonious habits. They were inbred out of a Vermont type that she

---

[5] Good Housekeeping, February, 1935.
[6] Ibid.
[7] Ibid.
[8] Ibid.
[9] Mrs. Coolidge's Home Companion Confessions.

understood. She accepted him as "a character," a primitive. Modern slang would call him "a sketch."

Yet he was vastly more than "a character." And shrewd Grace Goodhue sensed it. Back of his obvious gaucheries he had a mind. She recalled thirty years later that when they were married he had a small golden-oak bookcase in which five shelves were filled with standard books. The sateen curtain in front of it concealed a set of Shakespeare, a history of England, a row of Kipling's poems and stories, George Ade's "Fables in Slang," "The Prince of India," a set of Hawthorne, "The Rubaiyat," Whittier's, Longfellow's, and Tennyson's poems, and a row of lexicons and grammars of the Latin, Greek, French, German, and Italian languages. And if he had a book she might be sure he had got his money's worth of it. Never was he a showoff. He told Clarence S. Brigham, of the American Antiquarian Society, that he had begun translating Dante's "Inferno" before he was married and he liked it so well that he kept right on with it and finished it afterward. A young man who will interrupt his honeymoon to translate Dante really should be allowed to count the macaroons after a picnic. His wife wrote that from the beginning he liked to read in bed, to "improve his mind"—his phrase for it. And at the head of his bed always he kept his Bible and "The Life and Letters of Charles E. Garman." She also remembered for a generation those two small paper-covered volumes of the "Inferno" on his bedside table—his escape from Northampton reality. She must have seen that bedside table and must have lifted the sateen curtain that hid those books before she married him. It is an abiding testimonial to her perspicacity that she let him count the macaroons and took him while the crowd snickered.

They had intended to take a two weeks' honeymoon. After a week in Montreal, he decided they had seen everything there was to see and she, being tactful, suggested that they come home, which they did and saved a week's spending.

Coming home they stopped in Boston and visited the State House and the governor's office in the Capitol. In one of the rooms of state was a huge chair, the governor's chair. The happy couple thought they would try it. As one of them was about to sit down, the guards came and chased them away. They remembered the abashed confusion of that incident many years.[10]

When the Coolidges came home from their economical honeymoon in Montreal, a few hundred miles from their wedding place, they went to housekeeping as soon as they could. But before they had settled down, he appeared one day with fifty pairs of undarned socks and told her when she

[10] This story is found in nearly all of Coolidge's early biographies. Of course he told it in pride.

got those done he would have some others. He had been saving them up apparently against the day of matrimony. His doctrine of work and save began early. He saved and she worked.[11] Theirs was a modest menage. It was within their means. Here again was the shadow of his mistress, for he declared twenty-five years later: "We lived where we did that I might better serve the people." Characteristically, and this was one of his infirmities which she had to bear. After nearly a year in a boarding-house he attended to the furnishing of their home himself and took Mrs. Coolidge into it ready furnished! Two weeks later their first baby was born, eleven months after their marriage. Of this event he wrote:

"The fragrance of the clematis which covered the bay window filled the room like a benediction where the mother lay with her baby. It was all very wonderful to us." [12]

Now for the graces which upbore those infirmities: Above all, Mrs. Coolidge was amiable. She had the tolerance of an understanding heart. She could mimic her husband and did for the delectation of the family and her friends without mocking him meanly. She knew his strength and understood without blinking the facts his many minor weaknesses. Several months before their first baby came, Grace Coolidge let a fast talking frontdoor salesman sell to her "Our Family Physician" for eight dollars—a lot of money in those days for that family. When the book came she was shy about telling her husband of the purchase. The Northampton story declares that she decided to say nothing, but left it on the center table. He said nothing. But one day several weeks later she picked up the book and glancing inside found written on the flyleaf:

"Don't see any receipt here for curing suckers! Calvin Coolidge." [13]

Funny? But under the circumstances not entirely kind. Husbands have been poisoned for less. Probably she managed to get along, as most wives do who have to live with spoiled children who are essentially sound but have little soft specks here and there in the wholesomeness of their characters. She was as quick as he. She supplemented his natural shyness with a lovely candor, offset his occasional lapses into taciturnity with a gay loquacity that kept a table going when he would have let the conversation sag unwittingly perhaps if a mood was on him not caring what happened. And then above all she had faith in him. She was never jealous of his mistress. In the first years of her marriage when her children were little, she ran the house with one maid and did the washing.

Then and for years thereafter, Fred Jager, the automobile dealer, re-

[11] *Good Housekeeping*, March, 1935.
[12] "Autobiography."
[13] This story also is printed in "Coolidge Wit and Wisdom" by John Hiram McKee, Frederick Stokes Company, New York City.

garded the Coolidges as a good prospect. But he never could land them. Jager tried to get Coolidge to ride many times but Coolidge always shook his head and said he wasn't interested, "but Grace was. She would be glad to go." And Grace had no authority to buy. Which Jager full well knew! Jager gave them up.[14] A car's cost was beyond the Coolidge cosmos. If the budding statesman had gone up the other route in politics, if he had not rejected reappointment to the profitable office of clerk of the court, he could have had his car and Grace would not have appeared back of the little house on Massasoit Street, Monday mornings, along the clothesline holding a clothespin in her teeth, hanging out the wash. But Coolidge knew that the money end of politics was not his vocation. Money which he venerated as a source of social and political power, he never seriously cared to acquire. But when it came his way, he saved it. Instead of taking the road to wealth, always he took the road to honors, small honors, but not financially profitable. So Grace Coolidge budgeted her household expenses well within the Coolidge income, the income of a struggling young lawyer, still a second-rate lawyer in a fourth class town. Probably his income never exceeded fifteen hundred dollars a year until he was elected lieutenant governor. He told his friends [15] that he could live in a better home if he cared to but that he wanted to be free to serve the public so that he would not be bothered with the worry of having to give up his office and come down in the world if the luck of the game of politics turned against him. He told James Lucey, the shoemaker of Northampton, that he could always sleep nights without worrying about the rent. For which the credit was not entirely his. For living within her income, running the house by keeping the family out of debt, Grace Coolidge had her full and proper share. She worked hard, kept up appearances, did her part in the town's activities, took her place alone in the Congregational church, played the game of the politician's wife, smiling the while, and knew that this man of hers with five talents of which the most shining coin was a sort of diligent honesty had to travel fast if he went up the political road. She realized that he could not climb far under the burden of debt. On the ever upward journey her shoulder was in the yoke as well as his. If there was nagging to be done she let him do it. So they were happy. For she had the priceless jewel of an understanding heart which to the end maintained for him his valiant self-respect.

3

Three months after his wedding, Calvin Coolidge received his first and only defeat at the polls. He was beaten in the race for member of the

---

[14] Letter from Jager to author, 1934.
[15] Conversation with Northampton neighbors by the author.

school board. Men who were his neighbors at the time [16] declared that he was defeated for the school board because he had no children. He, himself, always held that he was defeated because he was a Republican politician and the tradition of Northampton, as of most American small towns in the North, runs strongly against allowing politics to get into the school board. Part of his political success was because he could get such a perspective on himself. He saw himself clearly with his strength and with his shortcomings. The New England habit of introspection was ingrained in him. The next year he supported his successful opponent as a good politician always does. Apparently he took his one political licking with a smile, which proves he was the soul mate of his mistress, all but wedded to politics. The next year, in 1906, he was elected to the Massachusetts House of Representatives.

There Calvin Coolidge first touched the big world beyond Plymouth, far beyond Ludlow, outside the environs of mystical Amherst, over and above Northampton. Yet his journey to Boston was straight as the course of an arrow. Chance never seems to have deflected his purpose. He had no ups and downs. He could not have thought it all out, envisioned it far in advance; yet his destiny was implanted in his character. Behind the New England mask dwelt always the mystic who created like Dante his own dream world, an inferno peopled with his own pet philosophical ghosts. In his Northampton career, definitely and yet with perhaps subconscious forethought, he chose offices that carried no salary; offices requiring service. But these offices gave him more than a courthouse fame. They carried him along in a career. There again he exemplified his faith in a moral government of the universe. For on the whole he did not seek nor accept attorney's fees that came as sugar-coated bribes from persons or corporations whom he might favor or oppose, as a public servant.

He went to the legislature a gaunt young man with two suits of clothes, an everyday suit and a Sunday suit, a firm jaw, a rather mean mouth, strong nose, most eager, intelligent, searching eyes—generally squinted and veiled, and a good brow behind which was a tough-fibered brain. He was reading constantly in those days, law, contemporary literature, magazines and occasionally an old novel, one of the classics, and chiefly books on politics and history, preferably American history. The lives of statesmen in the old American Statesmen series cheered him and helped him to understand himself and his country.[17] But for all his reading his intelligence did not show on his face except when he opened his eyes. His casual aspect was that of a dumb, starved, suppressed young Yankee, struggling against an inferiority complex. Richard Irwin, ex-Senator from Northamp-

---

[16] Conversation of Northampton neighbors with the author.
[17] Conversation with author, 1924-5.

ton, gave to young Coolidge a letter to the Speaker-elect in which the elder statesman of Northampton declared of Coolidge: "Like a singed cat he is better than he looks." Coolidge enjoyed that phrase for a quarter of a century and often quoted it. It was his idea of himself in perspective.

He never visualized himself as a Galahad. Rather he liked to cast himself as Cinderella.[18] He was pleased to clown quietly about his conspicuous defects; though in his inner heart most of the time he had a fairly good opinion of himself. Men like Calvin Coolidge only strut in the deep dark cellar of their subliminal minds.

Armed with this "singed cat" note, Calvin Coolidge left Northampton in the morning twilight on an early train in January, 1907, bound for Boston. It was his first real trek from the middle of the nineteenth century into the new day and time.

4

Coolidge generally took that early morning train for Boston while he was in the legislature, both as member of the lower House and as state senator. The habit did not leave him while he was lieutenant governor nor did he acquire luxurious tastes while he was governor. Thus from 1907 until 1921, fourteen years, with two years out as mayor of Northampton, Calvin Coolidge rode a day coach twice a week, at least three months in the year [19] through his beloved Massachusetts. Over and over the same panoramic scene of one hundred miles went flicking out like a mighty film through his brain. It impressed itself upon him, became a part of him, and he a part of the picture.

Slowly the train pulling out of Northampton, crossed the Connecticut River and sped by the long Common of Hadley. Here Coolidge saw the Connecticut Valley, the best farm land in Massachusetts, a lovely ordered rural picture of peace and contentment. To the southward his eye, always greedy for the beauties of New England, browsed along the wavy ridge of Mount Tom, and so across the fields to Amherst set among the trees. There was his alma mater, a dingy, respectable, solid, prosperous campus. There his youth lay buried, and being a sentimental man at the core, the picture of the Amherst campus could not but quicken his heart as he passed from Amherst to Ware through a wild, rough country, save for a few wide open fields near Belchertown; fair, green fields, soft, gentle fields like a merry little scherzo tune in a stately symphony leading into the wild, rough country around Ware. There in the middle of the first decade of the century gray granite dominated the town. It was a farm-factory town. In the mills they made awnings and underwear. Hundreds of men and women

---

[18] Conversation with author, 1924-5.
[19] All the year around for the last four years.

were at work, happy for all his eyes could see and contented. A typical mill town was Ware with its dam, its main street paved with brick, where cars and farmers' wagons stood parked on both sides of the narrow street. It was all busy, trim, prosperous.

Out of Ware the train ran to Gilbertville, a factory town with a worsted yarn mill; men and women at work, thrifty, busy, and to his eyes always seeing that prospect as God's noblest work, a pleasant prospect. Between Gilbertville and Coldbrook Springs, below Hardwick, the walls of the Ware River Valley open out into neglected fields, a bottom land. It has not been used for farming for a hundred years, if it was ever broken with a plow. Upon the first ledge of the bottom one finds that the rails lead past, but not through, Hardwick, an unspoiled, lovely old town. He knew it well. It was the New England town of his dreams, with wooden buildings glistening white, with here and there a perfect example of that once common type of wooden house, square with pillars running up to the gable to make a porch on the first and a balcony on the second floor, respectability petrified in architecture, not without its beauty. And then, beyond the town, perhaps in the township, ranged a row of old-fashioned, one-story two-family wooden tenements, the social footing stones upon which the architectural crystallization of Hardwick rests, the humble and the loyal in their appointed lot, the industrial system typified in root, branch and flower. At Coldbrook Springs, where the Ware River runs away from the rails, Calvin Coolidge saw an ancient factory painted white with a belfry almost colonial, also the quintessence of New England and particularly of the Massachusetts he loved. On goes the railroad to Rutland, across a swampy trail through the woods, over a willow causeway, into another valley. Rutland is almost purely and typically a farming town. Here the house of General Rufus Putnam struck and kept the Old New England note, the note that Coolidge loved because he understood it, because it was ingrained in all he saw and knew, believed and felt. Passing a modern mill the railroad entered Holden with its white church and its peaceful cemetery beside the road. Rutland and Holden were in Coolidge's day straggling hill towns, full of open country, passable farm land even now. The barns looked big and neat, some were painted dark red, but more were white; or they stood gray and unpainted, blending into the mauve of the winter hills. Along a plateau the traveller passes the reservoir, the Wachusett Reservoir, its bank more or less covered with pines, to West Boylston, a village by the lakeside. "Mount" Wachusett looks over the wilderness from Clinton northward. At the bottom of the valley, flooded for miles, lies the mill dam and far below the gray face of the dam lies the mill pond and power dam of Clinton, and the brick factory building of the Colonial Press, with long, parallel lines of ventilating windows on its roof. To the north of the valley, on the high ground opposite to the reservoir dam, lies the town of Clinton with its usual New England main street; a tram line down the middle, asphalt paving, and portentous solid brick store fronts. Here the French Canadians and the Italians mingle their names with the Yankee names on the sign boards. Here were the carpet mills where the foreign mill hands worked, square, dingy brick buildings with endless rows of blank windows and a squat tower, supposed to be ornamental, carrying the date of the mill construction on it. The treeless slope of the western side of the valley is covered with small, harshly angular wooden tenements and the railroad crosses the valley on a high iron trestle just below the dam. The huge gray stone face of the Wachusett Dam seems to overhang and dominate all the city like a menace, the threat of what would happen if the protecting feudal hand of capital were withdrawn. Beyond that is Hudson, a small factory town that made men's hose, and then leaving the town one passes the houses of the well-to-do citizens, the industrial rulers living in roomy nondescript houses well above the dam and the small square brick mill. Thence down hill pell-mell through second-growth timber the train rushes to Sudbury. We are in mid-Massachusetts now and the high tableland gashed by deep valleys runs across the state north and south, a country of hill towns [20]—mill towns where the Lord's shepherds shed the beneficence of generations of accumulated wealth, giving jobs to those whom Providence has entrusted to the care of the industrial nobility and gentry. Sudbury is almost on the western boundary of the Middlesex country and in Middlesex the scene changes. Leaving the station with its piebald slate roof, the railroad plunges across a marsh from which arises a hill, purple red in winter, gorgeous and beautiful.

In winter scenes always a man seems to be hunting with his shotgun over his shoulder, and along the highways other men, as Coolidge rode through Massachusetts, were forever digging holes for electric poles and linemen were stringing wires. Tufts of yellow grass stick through the marsh ice, save for one glittering steel blue ribbon where the river current runs. At Wayland an old white church appears and flashes by with an open belfry.

So comes Middlesex. Small uneven hills and valleys roll away from the tracks; open fields and orchards separated by thousands of stone walls, worming their way over the hillsides, and through marshes with willow causeways running straight across them. Over the hilltops hardwoods spread their carpet red with the unfallen autumn leaves, the color broken here and there by green splashes of pine. By the highway underneath the spreading

[20] Hill towns are not confined to mid-Massachusetts. Any town in the highland counties west of Middlesex is a hill town. Coolidge uses the term in his report on street railways.

elms a few finely proportioned Federalist farm houses stand each with a barn attached and the wood for the winter piled prudently in the yard. Their lovely doorways too often are hidden by nineteenth century porches in vile taste, and beside these aristocrats of architecture are scores of houses in the swollen, lumpy post-Civil War style of our architecturally gilded age, the homes of the new masters of industry who came in with the seventies and eighties. Here again Calvin Coolidge saw men stringing wires as he rode through Massachusetts. But he saw them merely as men stringing wires. Little did he realize that they were the pioneer scouts and sappers of a new age.

There across these Middlesex meadows and marshes, until he came to the Boston suburbs as he looked from the window at the trim, prim civilization of his state, he saw its prosperity. Coolidge, the average middle-class American of his generation, saw its ordered life with the rich rewarded for their virtue and the poor provided with opportunities for diligence. With the eyes of his class in that day, he saw the improvident bubbling gently in their proper purgatory. But he was sustained by the fact that the poor were able always to climb out if they were thrifty, persevering—and lucky! Otherwise he knew, and all Americans understood, that the poor could simmer in their purgatory or sink to some lower hell if they had not the acquisitive virtues. It all seemed good as he rode for a decade and a half through the middletowns, through the marshes, across the hills blazoned with purples, reds, and browns through the long hard winter, checkered with farms, dotted with decent, dignified white houses; all set, all completed, all wrapped up, tied and ready for delivery into a Massachusetts millennium which became a part of his ideals as he finally became inextricably a part of it. So rose his faith in Massachusetts as he went forth through the years across this glamorous countryside. It was Coolidge done into landscape, and Coolidge was Massachusetts, molded into a political image.

Massachusetts politics in the early days of 1907 was in the midst of the turmoil that Theodore Roosevelt in the White House was stirring up, and Young Coolidge was feeling the swelling in his heart of the seed that Roosevelt was sowing to the four winds. Indeed Representative Coolidge was more than half persuaded to be a Rooseveltian. Theodore Roosevelt, as he "cometh forth to sow," in those days was yammering away on what were known as "the Roosevelt policies." They were all builded upon a new theory in American politics, the theory that we should establish a more equitable distribution of wealth than the financial, industrial and political institutions of the first decade of the new century were giving to the American people. The movement was world wide. All the world realized that we had fairly well solved or were on the way to solve the problem of production. The machine age was reaching its heyday.

CHAPTER VIII

Our Hero Meets a Powerful Patron

IN America, in the first decade of the new century, President Theodore Roosevelt's demand for railroad regulation was regarded as socialistic. Also he was enforcing the anti-trust laws. A campaign for pure foods and drugs was closing victoriously in Washington. The direct primary was coming into vogue in the West. The scandal of the legislative election of United States Senators was stirring up a clamor for their direct election. In the West state after state was adopting the initiative and the referendum. Nothing was sacred in business or politics. On the news counters, those newspapers, magazines and books were piled high which dramatized the story of malfeasance, even corruption, in business and politics. New England was shaken by the charges against the control and manipulation of the New York, New Haven and Hartford Railroad and the Boston and Maine. Business and political altar boys were busy picking mudballs off their idols. The tintinnabulation of the Roosevelt tomtoms had reached the grass roots, and in Northampton Calvin Coolidge knew what was in the heart of the voter. Coolidge, honest and forthright, desiring to represent the voters of his district, was ever a Calvinist—a Puritan, born with a conviction about the sanctity of property. In the passing political conflict Young Coolidge gave an ear to the voice of change clamoring on Main Street, Northampton. But he had his doubts.

Northampton, in those days, a typical unit of American life, was a town of thirty thousand or so, dependent largely upon the agriculture of its countryside—the American norm. Also Northampton was a college town. Smith College was on its outskirts; Amherst College a few miles away. Factories were there and a labor vote was organized and potent. Also capital was represented in Northampton and Calvin Coolidge was vice president of his savings bank. Smith College was growing radical. Labor was radical—for that day! The farmers around about Northampton were liberal. The conservatives were in a minority but were powerful.

So out of this background, into the turmoil of Boston, the political and industrial capital of New England, went this queer, silent, earnest, conscientious, shrewd little Yankee with an eye out always for the main chance —but not for money—deeply infected with the philosophy of his college days. He wrote, looking backward into that period of his political youth:

"I have always remembered how Garman told his class in philosophy that if they would go along with events and have the courage and industry to hold to the main stream without being washed ashore by the immaterial cross currents, they would some day be men of power. That may sound like mysticism but it is only the mysticism that envelops every great truth. One of the greatest mysteries of the world is the success that lies in conscientious work."

And so with this quirk in his brain, this almost fanatic belief in the moral government of the universe, stepping out of a smoke-stinking day-coach in the North Station in Boston "a child went forth."

In Boston, getting off the morning train from Northampton, he took his valise and walked to the Adams House, a second-rate caravansary near the Capitol where unknown rural legislators from western Massachusetts were wont to congregate. It is said he engaged a room for a dollar a day. Probably this is fifty cents under the truth. But he kept the room through his entire legislative career, a matter of ten years or so. He was living within his legislative income, which with stamps and privileges all told amounted to less than five dollars a day. He even may have saved a little as was his wont. The Speaker paid little attention to Calvin Coolidge when he presented the "singed cat" note. Coolidge's first committee assignments were humble though he was a Republican from a Democratic district. But with the habitual industry which had marked him in Plymouth, in Ludlow, in Amherst, in Northampton and in love, he went to work. He is recorded in every session and voted upon every measure when he was not sitting in committee working hard upon some other measure. His colleagues found they could load routine work upon him. He espoused a bill for a Northampton oil-dealer that would have disturbed the Standard Oil monopoly and the beef trust in Massachusetts. He helped to get an anti-monopoly law on the statute books. Being a banker by brevet he helped to codify the banking laws, and by way of balancing his record was careful to see that labor had his vote. He could point to votes for measures restricting work to six days in seven, a bill restricting the work hours of women and children, a bill providing pensions for families of firemen, a bill to provide half fares on the street railways for school children, a bill to equip factories with surgical equipment, and he could prove that he was for these measures. A colleague [1] remembers that Coolidge saw a senator who he thought was lobbying for the beef trust in the House and said: "What's he mousing around the House for?" and he pronounced house and mouse with a Vermont accent. In his first term he accepted few favors from the lobbyists of the sort that always infest legislatures. He was also chary of accepting any favors from anyone; probably because he kept social books. R. M. Washburn, who was his first biographer, recalls that he invited Coolidge to lunch at the Tremont, then a resort of wastrels of the State House, and Coolidge shook his head. Later, thinking it over, Coolidge looked Washburn up and said in sotto voce: "Mrs. Coolidge and I lunch at the Touraine at one. Come if you want!"

He was always extending, rather than accepting favors. It was his thrifty way of being parsimonious in politics where what you take, however small, is your liability, and what you give is your asset.

A year later the Speaker, who had taken one look at the "singed cat" and had dumped him on the ash heap of unimportant committees, visited Northampton, pursuing a candidacy for lieutenant governor, and Coolidge took him around town, staked him for a lunch and took him home. It was a good investment. Coolidge's next term in the legislature found him emerging into some prominence—the old story, the story of Ludlow, Amherst, Northampton, the Cinderella story repeated every time he moved— over and over again—in his life. He always began among the pots and pans wherever he landed but rode away in the fairy coach.

His fairy godfather in the legislature was Senator Murray Crane. Few persons influenced Coolidge's life more than Crane; his father, and Dwight Morrow perhaps, then Crane. So we must pause here a moment and consider Winthrop Murray Crane, twenty years older than Calvin Coolidge, a papermaker, who having been dead a decade and a half as these lines are penned Crane may well be called a statesman. Yet in life he was an astute, active and most practical politician. He entered politics in Massachusetts at forty, became national committeeman, was elected lieutenant governor in 1896, served three terms as governor of Massachusetts, was at first nondescript in factional Republican politics, was accounted on the whole conservative; but he was a dry and he broke a teamsters' strike in Boston by bringing both sides undignifiedly by the scruff of their necks into his office in the State House and jamming them into a settlement. Then he urged the same conduct upon President Roosevelt in the anthracite strike and Roosevelt, following his example, had great acclaim. Crane's first relations with Coolidge came through the Congregational church. Crane was high in the councils of the Congregational church in Massachusetts where he

[1] R. M. Washburn.

was a sort of secular archbishop, and in the nation where he was well placed in the Congregational hierarchy. Northampton was a Congregational town; Mrs. Coolidge, a member of the church; and Coolidge a sort of super-brother-in-law of the Northampton church without a voting membership. Crane was a business man—successful, indefatigable. Coolidge naturally accepted him as a mentor. When Coolidge went to the legislature, Crane was a Senator in Washington. His vicegerent in Boston was William M. Butler, a Massachusetts textile manufacturer, former member of the legislature, former state senator, and holding political distinction chiefly as "Crane's Boston man." Early in his Boston career Coolidge met, admired and followed Butler who was also conservative, more so than Crane. But Coolidge never was entirely subservient. In politics he knew where his bread was buttered, but no boss could depend upon his vote. So Coolidge's meager liberalism, a little of which was the instinctive challenge of every countryman who visits the great city, was also tempered and strengthened by the comforting knowledge that the Rooseveltian liberalism was strong in Northampton. Also Crane was an easy boss. He desired adherents but not sycophants. In those days, and until his death in 1920, Crane controlled Massachusetts Republican politics at a time when Massachusetts was generally Republican. He had a Republican machine. He maintained it by rewarding his Republican friends and punishing Republican treachery. He did not pursue his personal enemies. Pursuit is always expensive. Massachusetts politics in the days when Calvin Coolidge was growing from a "singed cat" in the legislature to a full grown sphinx in the governor's chair in Boston, was of course, like the politics of any other northern American state, supporting the capitalistic American democracy. It was Republican to the core and dominated by a keen Republican sense of property. But those who dominated Massachusetts politics in Coolidge's day had the nice caution of blue-blooded respectability. For two hundred years adventurers had been going out of Massachusetts to the West. For two hundred years caution had become inbred; emotions carefully repressed. The politics of Colorado, a child of Massachusetts, in those days from 1907 to 1917 were full of riot, stratagems and sudden death. But Massachusetts Republican politics, while Coolidge basked there, sublimated its corruption in the tacit understanding of allied respectabilities. Raw money and raw whiskey were not potent in the legislature. There were bunds and pacts. When the Catholic church and the Harvard corporation made common cause as they sometimes did, they were unbeatable. Textiles, railroads, savings banks, all Massachusetts interests had their scouts around every session of the legislature. The Methodist church was there, and organized labor. These scouts Coolidge knew. He helped them —surreptitiously. They helped him—when they could. When Charles Mellen, the railroad magnate rose, outside capital, specifically New York capital, was becoming interested in New England. It brought in a more imaginative and less scrupulous type of financier than the native type and was more competently organized than the public utilities of the Bay State. Coolidge in his day saw this invasion. In so far as it was crass probably it shocked him by its crassness rather than by its objectives.

2

To understand Massachusetts politics, to explain Senator Murray Crane, and to realize Calvin Coolidge as Crane's satellite, we must not forget that Massachusetts politically was two states. The eastern part included the cities and towns near the seaboard centering upon Boston, and the western part held the cities and towns of the Connecticut Valley and the Berkshires where Coolidge lived. This division was more than geography. Eastern Massachusetts a hundred years ago became serenely Unitarian. The Connecticut Valley sternly followed John Calvin and remained Congregationalist. The Catholics came powerfully into eastern Massachusetts. They remained in minority in western Massachusetts. It is interesting to note that Senator Lodge, Crane's rival, was invincible in Essex County, the most northeasterly in the state, and Senator Crane was moated in Berkshire, the most southwesterly. Harvard and the Catholics and an urban civilization dominated the seaboard. A sophisticated Congregational industrialism— farms, fields, and workshops—gave color to the Republican cast of thought of western Massachusetts.

Politics are more social in Massachusetts than in any other American state except perhaps Virginia. Massachusetts society, particularly western Massachusetts society, was, and to an extent is, in 1938, dominated by a set of small capitalists. Their ancestors came into the country, set up their little mills, textile mills and others, wherever a river or a brook falling down the hills might be harnessed to produce waterpower. These mills grew, but in many cases they remained in the families of the founders. Even in the first decade of the twentieth century, these mills were not amalgamated. The larger towns were dominated by their cliques of mill-owning families. In the smaller towns a single family ruled. They were the Drapers, of Hopedale, the Crockers, of Fitchburg, the Wellses, of Southbridge, and a typical family in western Massachusetts was the Cranes, of Dalton, papermakers. Once they had the contract for making the paper upon which American currency was stamped. Murray Crane to his last days coming into a stranger's office would pick up a sheet of writing paper and peer through it for the watermark. These ruling families of Massachusetts, and particularly of western Massachusetts, formed a Republican squirearchy not unlike the English landed gentry. Boston, dominating eastern Massachusetts, had bankers, merchants, traders for its overlords. But the Connecticut Valley squirearchy with its petty enterprisers and their following formed the backbone of the Republican party in Massachusetts which dominated the state during the whole of Murray Crane's political career. It ruled behind its riparian portcullis during the rise of Calvin Coolidge from Boston to Washington. Squires and their followers contributed heavily to Republican majorities. Murray Crane and Eben Draper became governors. Their end and aim in politics was a Republican high tariff. Incidentally the squires scrambled in Boston for the state funds which might be distributed among their banks as patronage. They asked little else. Postmasters and the court house officials were of the lower orders. But the squires supported them and required a sort of political military service from them. Thus the squirearchy reigned feudally in a capitalistic democracy. In that particular region which Coolidge knew best and where he was a shining fugleman from 1910 to 1915 from the Connecticut west over the Berkshires, Murray Crane was chief baron of the domain. He was honest according to his light and leading, a little more honest than the times demanded. And he ruled not with an iron hand but by reason of his kindly heart. As a squire he helped his neighbors with money and advice. Most of the money he lent to Republican politicians was returned. For they had a money sense.

Murray Crane who made few speeches, became Governor and a good one. He fought the Boston elevated railway and the control of Boston by corporations. In Boston, another baronetcy than his own, he was free. So he preached restraint and moderation as a means of grace to the public service corporations. When Senator Hoar died in 1904, Crane was appointed to his seat and later elected. Coolidge therefore met Crane as his United States Senator—the Senator from the west. Lodge was Senator from the east. Naturally Coolidge became Crane's liege man. Crane, also being of sentimental and inwardly gentle nature, accepted this liege man and bound him deeply in his heart.

Crane, succeeding Senator Hoar in the Senate, would be congenitally the foe of Lodge. Lodge was of the Brahmin caste in Massachusetts, and Harvardian of the deepest die, a Cabot by inheritance who liked to be known as the scholar in politics; the author of several books, the friend of the literati in America and England, a travelled man, meticulous of dress, of deportment and of speech, intellectually mediocre, emotional, highly charged but heavily suppressed, capable of hatreds and friendly loyalties that were deeply affectionate. He gave one of these to Theodore Roosevelt. Lodge had erudition without understanding. "Thin soil like Massachusetts highly cultivated" was the estimate of Congressman Lodge made by Thomas B. Reed, Speaker of the national House of Representatives. Lodge, returning from Washington, visited Massachusetts surrounded by the aura of his Brahminism, addressed the legislature in joint session at times, delivered scholarly addresses, made his contempt of the ways of politics obvious if not audible and profited by those ways when he must. His cynicism was fashionable in the high intellectual latitudes wherein he roamed as a god remote and austere. Coolidge, full of Garman's maxims, distrusted and disliked Lodge's kind.

But Crane whose highest moment in politics came when he controlled the machine and tendered Lodge a nomination with a sort of disdain that a bulldog might manifest to a hungry high-bred, ginger-cake dachshund to whom he vouchsafed a bone, Crane was Coolidge's ideal. A slim, trim little man was Crane with a cloudy complexion, thin tousled hair, a ratty moustache, disdainful of the accordion wrinkles in his vest and trousers and the crease across the back of his sack coat which he wore except on formal occasions. Crane's supporters were real to him. He knew them, their families, how they lived and thrived. He fathered them, and William M. Butler was his shepherd who saw that they did not want. In Boston, Coolidge, in his whole career from the legislature to the White House, was unknown on Beacon Street. But while Murray Crane lived, Boston, outside of Back Bay and Beacon Street, was to Coolidge a land of milk and honey, the Canaan of his political dreams. The two men, Coolidge and Crane, had much in common. Crane left school in his late teens. Coolidge went to a school where they taught him a simple faith long since outmoded among the Boston Brahmins. Coolidge was a lawyer and a bank director. Crane made white paper and was a manufacturer by heredity. It was early in Coolidge's legislative career that Crane saw the gaunt, natty, soft-spoken vote-getter of Northampton and marked Coolidge for the Crane sheepfold. A business acquaintance of Crane's was going to Northampton to visit his daughter in Smith College. Crane, learning of it, said:

"Find out all you can about a young man named Coolidge there. You will save trouble in looking him up now! He is one of the coming men of this country." [2]

Crane had the Congregationalist's faith in democratic processes as Coolidge had. They were both mystics, idealists of a sort. In his second term in the legislature, Coolidge became an influential member of the judiciary committee, a place which "I wanted," he wrote, "because it would assist me in my profession." He helped to frame a bill which failed: "a bill to modify the law so that injunctions could not be issued in labor disputes to prevent picketing." Behold the Rooseveltian influence. But in his later years and when Theodore Roosevelt was dead, Coolidge came to feel "that what was of real importance to wage-earners was not how they might conduct a quarrel with their employers but how the business of the country

[2] "The Preparation of Calvin Coolidge" by Robert A. Woods.

might be so organized as to insure steady employment at a fair rate of pay. If that were done there would be no occasion for a quarrel and if it were not done a quarrel would do no good." [3] In this sentence is embodied the whole Coolidgean philosophy, indeed the philosophy of the Republican party from 1912 to 1932, when it was the palladium of the American liberties under a benevolent plutocracy. This feudal reverence for the dignity of wealth amounted to belief in the existence of a wise overlordship of the affluent. Under the kindly guidance of an intelligent government, justice and mercy and peace would bless its subject classes. Calvin Coolidge's Republican creed of course required faith in a moral government of the universe. This guided his course like the star of his destiny, and gave him a sense of power, a sense that he was the pampered child of fate. Looking back on those early days in the Massachusetts legislature, viewing his career across the decades, he wrote:

"I did not plan for it, but it came. By my studies and my course of life I meant to be ready to take advantage of opportunities. I was ready—from the time the justice named me clerk of the courts until my party nominated me for President."

It is curious, and in this tale important, to look back from the viewpoint of twenty years and with the knowledge of a later day to gather a perspective upon the first three decades of this century that we may realize how utterly unprepared actually Calvin Coolidge was for really noble, wise and valiant service when his great time came. When the evils of a privileged plutocracy threatened to wreck his world Calvin Coolidge at the helm consistently, determinedly, honestly, with a fanatic's faith denied the existence of such evils, in our national life. To him wealth in those Boston days and always was sacrosanct because thrift, diligence, purposeful industry, were high virtues, and because the piling up of penny upon penny and dollar upon dollar was a temple ritual in his mystic religion. His courage was undoubted, his honesty was never seriously questioned. He saw everything before him clearly except the truth about the worm in the bud of his democratic faith. While he was walking circumspectly, Crane's meek and loving disciple through the Boston State House in 1906 and 1907, predatory but highly respectable capitalists were grabbing corporate privileges which controlled railroads. With these privileges the railroad managers were wresting millions in unearned pennies from innocent citizens in unfair charges. Calvin Coolidge, in the Massachusetts legislature, apparently did not dream of this raid on the economic liberties of Massachusetts and his country. His was the common view of ordinarily intelligent men. Even Theodore Roosevelt, who was beginning to see the truth, saw through a glass darkly. Calvin Coolidge, who was only a sort of neb-

[3] "Autobiography."

ulous satellite of the great sun in the White House, could shine only with a broken reflected glory in the Massachusetts legislature, one of myriads of his kind, twinkling cloudily, in a sort of progressive Milky Way across the land.

After two years in the legislature, taking his per diem for five days in the week for nearly six months in the year, he found the legislature unprofitable. He had no distinction. He had gained little useful experience. He came out as he went in, a nobody in particular, going nowhere at all. As a practical man he realized that he had benefited from the job all that he could benefit if he remained money-honest. Certainly Calvin Coolidge did not desire either a well paid job or a political client who would bribe him politely with a legal fee. So he turned his back upon Boston for a time and ran for mayor of Northampton.

CHAPTER IX

Wherein Again Our Hero Creaks Through His Predestined Role of Cinderella

APPARENTLY Calvin Coolidge in 1910 ran for mayor of Northampton chiefly because he was unanimously nominated by the Republican city committee, known as "the organization," and was urged to run. It is likely however that his decade and a half of Republican service made it possible for him to be sure that the committee's urge was strong enough to justify his acceptance. No one can doubt that he desired to be mayor. He was a young man in his late thirties. Ordinarily men become mayors in their forties and fifties. The escalator of Massachusetts politics whereon he rose was moving him steadily. Yet he could not have been wise enough in that day to know that promotion from one trivial office to another would make him President of the United States. Somewhat he wanted the office of mayor because he knew it would please his father.[1] His father had been state senator in the Vermont legislature and an office-holder all his life. The young man knew that the father understood the value of a political coin which was stamped with the mayoralty. In fact, in his autobiography, Coolidge remarks that he celebrated his election by going to see his father who was serving a term as state senator in Montpelier, and adds:

"Of all the honors that have come to me, I still cherish in a very high place the confidence of my friends and neighbors in making me their mayor."

It sounds unreal. It is. He made himself mayor and then idealized his memories and forgot the realities. We may look at them without discrediting him. Ecce homo! a young man with the twinkle of a hangover adolescence in his eyes when he opened them wide, the father of two boys with

[1] See "Autobiography."

81

a pretty young wife often doing her own housework in a duplex residence, a young man who having lived fifteen years in a Massachusetts country town had mastered its politics. His mastery of politics reveals the man. Coolidge was essentially a countryman, and neighborly. Not being talkative, he liked to listen. As a young man, foot-loose and without family ties in the late nineties, he loafed at the blacksmith shop, Phil Gleason's for instance, or hung around the stone-cutter's shed, Bill Godfrey's, or chatted with Mike Lucey, the plumber, or visited with Jim Maloney who ran the baker's cart, or palavered at Jim Lucey's shoeshop. Jim was like many of the Irish in the eastern states, a Blaine Republican because Blaine twisted the British lion's tail. And when he called the spirits from the mighty deep, these Irish came running.

Calvin Coolidge, the young Yankee, was by way of being a local boss. Now all a local boss is in American politics, whether in the country town ward or in Tammany, is the man to whom the under-privileged go in trouble—and the over-privileged come in fear! Easily and quickly both learn to go for help only to those who are effective. The poor give votes. Coolidge helped and he got them. And he knew the tricks in his game. He knew that if you take responsibility in politics, you can rule. They tell of him in his younger days that at one Republican committee meeting in Northampton where there was trouble about electing a committee chairman, a secretary and a treasurer, Calvin Coolidge, who for the moment was not a member of that committee, slipped into a meeting where no one knew exactly what should be done, rose with an authoritative air, nominated the chairman, put the vote, saw that his man was elected, took care of the nomination of secretary, and then, in a lull, himself was nominated the treasurer and slipped out before anyone realized that he was not a member of the committee and had no authority. Probably he needed that committee organization. He got it. Then Coolidge would help them in trouble, any kind of trouble short of first degree murder. He was effective. As for instance, Ed Harris' wife's aunt, getting off a trolley car fell and broke her hip. The Company offered to settle for fifteen hundred dollars, she to pay all expenses. Ed Harris stepped in to see Cal [2] about it. In three days Cal called Ed on the phone and said the Company would pay all expenses and two thousand dollars, expenses including hospital and doctor bills and lawyer's fee. When Harris came around to get the check, his aunt signed on the dotted line. The lawyer's fee was included in the sum. Cal handed back to Ed his personal check for fifty dollars, saying briefly: "You got most of the evidence. Here's something for you!" That was the sort of thing that bound Northampton to him with hoops of steel.

He refused offices of profit that he might occupy powerful offices of

[2] "Cal" was what they called him, but rarely "to his face"!

service without profit. He managed local political campaigns for others. He had been five times before the people, four times successfully. He was laconic of speech, his wit being the dry, caustic juice of brevity rather than of conscious clowning, though to clown a bit sometimes amused him. On Sunday he wore a black, braid-bordered, four-buttoned coat with a blue tie and a low collar. And for week days, he wore a gray everyday suit. Punctiliously he went on Sunday to the Congregational church where Arthur Green, the janitor, was "Arthur" and the young lawyer was "Cal"—but only when his back was turned—and to Mrs. Coolidge, the janitor was "Brother Green" and she was "Sister Coolidge" after the fashion of the time and the place. The tough hickory bark of his exterior covered some pretty warm sap.

The saloons ran wide open. At stated sessions, that he might not be thought a prig, he sat down in a saloon and solemnly drank his glass of beer. He was the attorney for a brewery at Springfield. He gave the brewery no political service; merely looked after its barkeepers who often were haled into the court for violating the regulatory liquor laws. Occasionally he appeared for the brewery's drunks in police court, and advised and defended the brewery's representatives in their local controversies with the rulers of Northampton. But because he was an exemplary young man who was at home every night at nine as though a curfew had sent him there, who never wassailed, who avoided the town's frugal poker game, who gabbled, though not often, with his Yankee twang, who manifested his habit of repression and who minced along his modest way of life, no one held the fact against him that he was the brewer's attorney. He continued to live in Ward Two, Northampton's silk-stocking ward. There in his day lived a self-sufficient race, white, Nordic, sophisticated, on an average family income of something under three thousand a year. Massasoit Street, where Coolidge rented his residence, was a typical residential street in a typical Republican ward in a typical American country town. Here lived the merchant, the doctor, the lawyer, the teacher, the preacher, the professor, the head clerk, the superintendent, the banker, the retired farmer. Here in that day the average wife in the average house often lived without a maid. In the evening in a decent gown she did her duty at the head of a well-provided table.

Now in the interest of realism in politics, let us see how the people of Northampton made this brewer's attorney their mayor. The Republican city committee, which made the nomination, was Calvin Coolidge's committee, but early in the proceedings someone suggested the name of Alexander McCallum, President of the McCallum Hosiery Company, who paid high wages and was the ideal prominent citizen. But quickly up jumped George Spear, Cal's neighbor, and said he knew for a fact McCallum would not accept the nomination, being too busy. Spear nominated Calvin Coolidge.[3] A vote was taken. It was unanimous. Now here enters L.U.E. Lady Luck, who always was to be the chum and friend of Coolidge's mistress—politics. This is the story:

A man's club organized in the Jonathan Edwards Congregational Church was debating topics of current interest. The pastor, in 1909, desired a debate upon the question of license or no license for the saloons. Mr. Harry E. Bicknell, a promising young Northampton statesman, being a member of the church, told the pastor that it would be impossible to get any one to take the anti-saloon side of the question—what later was known as the dry side. On the day that Bicknell first talked to the pastor about the debate, Bicknell promised, by way of being a good Congregationalist and a good sport, to take either side of the debate. After a week's hard trial, the pastor could find no one else to talk against saloons and turned to Bicknell to help the club. Bicknell took the dry side. Being a man of earnestness and ardor, he debated with eloquence and emphasis. To him it was a purely academic debate, but it aroused the enmity of the wet element in the town. Within a few months, having forgotten all about the debate, Bicknell declared himself as a Democratic candidate for mayor. But the wets hadn't forgotten his eloquence and emphasis. The more Bicknell tried to explain the circumstances of the debate, the more he lost the few dry votes in town and the more the wets concluded that he would not do to tie up to—a rather preposterous conclusion under the circumstances, but one which many voters make in trying to simplify issues in terms of men.

Coolidge entered the race. At the election, one of his favorite quotations was from Franklin's declaration that public office should neither be sought nor refused, but he did not make the wet issue. He avoided that issue—which on the surface was the chief issue in the campaign. Coolidge talked about principles of government, about city administration in the abstract; when he talked at all, which was precious little. He relied upon the Republican silk-stocking respectable vote in Ward Two to back him because he was of it; and his reliance was wise. Then he depended upon his friends, the Irish, who had more or less been with him in the other wards, to stay with him; and his reliance there was wise. And he knew that his friends, the barkeepers employed by the brewery company and the drunks and the down-and-outs whom he had defended and advised, would probably be with him; but he took no chances. He was thorough. So he began a house-to-house canvass in the Democratic industrial wards. In that heated campaign when he was getting wet votes by the hundreds, he did not change his technique of vote-seeking. In simple, direct language which the men

[3] Letter from Edgar Harris to author, January 27, 1934.

whom he had befriended could not mistake, he asked them for their votes. He made no plea about principles of administration, indulged in no tall talk about reform, did not refer to his adversary in any remote way, but spoke in his dry, harsh, nasal voice, saying:

"I want your vote. I need it. I shall appreciate it."

Then, if debate or controversy arose while he was canvassing a voter upon the disagreeable issue of the campaign, he answered questions in monosyllables, being careful to inject the issue no further into the discussion. He kept the heart of his host and friend upon the Coolidge need for a vote. It was an undramatic pilgrimage for a knight-errant of democracy, but Coolidge won by a rather larger majority than usually is given to mayoralty candidates.

In the midst of these quacking alarms and nasal tocsins that he was sounding, he had time to detour into a mischievous prank. To his death he was mischievous—and often with a bit of craft in his mischief. The Democrats were betting two to one that they would carry Ward Seven, the Democratic ward where the Irish lived. Coolidge went to a janitor whom he knew and took a hundred dollars, told him to put fifty of it for one of those Democratic hundred dollar bets and take the other fifty down into Ward Seven and use it in the manner made and provided by democratic usage, a custom for convincing the voters that Coolidge was the people's friend. The janitor spent fifty on drinks and cigars; and bet the other fifty on Coolidge. Ward Seven went for Coolidge.

Coolidge carefully folded away the fifty dollar investment, split the hundred with his janitor friend and invited him and some other friends to a turkey dinner at the Rathskeller which made a hole in the third fifty dollars. So everyone was happy. Lady Luck smiled. For here was politics in one of her ironies: the brewer's attorney, who was dry from principle, elected by the wet votes against a liberal Democrat who only accidentally fell into the dry position. Northampton at the time had eighteen saloons and eighteen churches, her full quota of saloons and all the churches the traffic would bear. The liquor question was always a close one. But with the support of what in those days was known as the better element, meaning the upper strata in the economic structure and what with the help of his Democratic friends among the Irish, this quacking Vermont Puritan who made no speech, wrote no letter, made no promises, took another upward step on the round of the ladder. But before he wrote his serious little inauguration speech he sat him down and sent this letter to his unsuccessful opponent, also a member of the Jonathan Edwards Church:

My dear Harry:

Good friends—the high esteem of your fellows without regard to party —these you have. They are more than any office.

Northampton tradition declares that little John Coolidge met the little daughter of Harry Bicknell, his opponent, on the sidewalk, each laden with a note to the other's father; Bicknell's daughter carrying a congratulating note, John Coolidge this note of friendly condolence. The story is one of those which should be true but probably is not. For John Coolidge was scarcely old enough to bear the message unless it was at the second election in which Coolidge overcame Bicknell again by an increased majority. Then another letter was sent. This one:

My dear Harry:

My most serious regret at the election is that you cannot share the entire pleasure of the result with me. I value your friendship and good opinion more than any office and I trust I have so conducted the campaign that our past close intimacy and good fellowship may be more secure than ever.

Respectfully,

Calvin Coolidge.

If this is good sense and good feeling, it is also good politics. And for good sense and good politics, Calvin Coolidge was indeed renowned. Generally he exhibited that good taste which is the reflection of sound morals.

2

His services as mayor left no great impression. He was interested in the tax list and had the bookkeeper set up the books to indicate a reduction in taxes which was probably real. He was methodical, exact, fair, but none too amiable as mayor. It was obvious that he had a vision for a bigger and better and more beautiful Northampton. For a city-planning project took form under his mayoralty. The only thing his bitterest enemies held against him is a charge that he stopped a rollcall during a vote on a contract for electric light, but the result saved the city two cents a kilowatt and the charge against him really meant little. It was the day's work and if it was politics so much the better.

Mrs. Coolidge, however, salvaged one pleasant experience as the first lady of Northampton. As mayor's wife she chaperoned a group of North West high school girls on a trip to Washington. The official guide in the White House took them through the public rooms of the mansion and there she saw the gilded piano which the Theodore Roosevelts had brought to the East room, a gorgeous piece of furniture. To see whether it was real metal or gilded wood, she stepped over to touch it and the guide rudely elbowed her away to her immense embarrassment.[4] The memory of that piano aroused a little rage in her, and recalling the episode after twenty-five years she told a reporter that often when she saw the gilded piano

---

[4] Story in the *Christian Science Monitor*, June, 1937.

and was its mistress, she would give it a little sly kick to relieve her feelings when she needed release, which indicates that Grace Coolidge was a most human person.

Here it may be wise to set down Calvin Coolidge's opinion of politics and his place in the political scheme of things. This was not written at the end of his career but well at the beginning of it. Some one asked him if he felt there was any special obligation of a college man to be a candidate for office and he answered: "I don't think so!" And then he continued with dry, caustic Yankee sarcasm:

"It is said that although college graduates constitute but one per cent of our population, they hold fifty per cent of the offices, so this question seems to take care of itself. But I do not feel that there is any more obligation for a college man to run for office than there is for him to become a banker, merchant, or teacher. Some men have a particular aptitude and some have not for politics. Experience counts here as in any other human activity. . . . If an individual finds he has liking and capacity for his work in politics, he will involuntarily find himself engaged in it. There is no catalogue of such capacity. One man gets results in his life in one way, another in another. But, in general, only the man of broad and deep understanding of his fellow men can meet with much success in politics." [5]

The significant thing about his view of the politician is that he stresses the need of "broad and deep understanding" of one's fellow men, as an essential for political success. He had that. He knew the roughest and the worst of American politics at first hand. He knew and did business with self-seekers big and little, scoundrels, top-hatted as well as frowsy, obvious weaklings seeking to sugar-coat their bribe taking, and hardened cynics who knew what their treachery meant. All these he met as fellow workmen from the day he went into his first caucus in Ward Two in Northampton until the day he walked out of the White House, nearly thirty-five years later. Probably no other American excepting possibly Abraham Lincoln or Theodore Roosevelt had survived so long unspotted and unscathed in the thick of American politics, who knew so intimately its lower, darker levels. The public may have been shocked to see him hobnobbing with an occasional prophet of Baal. But at the end, scandal had not touched his administration from Northampton through Boston to Washington.

He played politics well in the mayor's office by keeping his own organization intact and serving the people honestly and substantially the while. But he was always a Republican. It was not surprising then to find that when he was serving his second term as mayor he had the unanimous indorsement of his county Republican organization as a candidate for the nomination as state senator of his senatorial district. The district included

[5] Speech at Amherst about 1915 or '16.

the city and the adjoining county. It was Northampton's turn to have the state senator. Coolidge was nominated without opposition and elected in the year 1912 when Theodore Roosevelt was devastating the Republican party. Coolidge weathered that storm. He did not take sides. He was regular but not mean about it. His few speeches in that campaign contained small reference to national issues. He did not cry out against the way Theodore Roosevelt was treating Taft nor did he clamor against the way Taft had treated Roosevelt. If he had views about either outrage no one knew it. Probably he did not care. It "wa'n't his stove!" Always he had but small capacity for indignation. From the first to the last he was static. Looking back on those years he wrote:

"When you grow up you will be the same kind of a man that you are as a boy."

He was changeless.

His senatorial contest was difficult. The district straddled the Berkshires. His friends called his attention to the difficulty which faced him. After they had exhorted, he replied:

"It will be just as hard for the other fellow."

In the direct primary the Northampton vote decided the election. Coolidge, a Northampton man, got the Northampton vote, and with the Northampton vote, he was elected. He was a vote-getter who was always strong at home; the best testimonial for a politician.

With his election to the state senate, Calvin Coolidge may be said to have begun his real political career. The escalator of Massachusetts politics began to accelerate speed. In the early winter of 1912, a lonely man, he writes:

"My old friends in the House [6] were gone. The Western Massachusetts Club, that had its headquarters in the Adams House where most of us lived that came from beyond the Connecticut, was inactive. The committees I had, except the chairmanship of agriculture, did not interest me greatly, and to crown my discontent the Democratic governor sent in a veto which the senate sustained, to a bill authorizing the New Haven Railroad to construct a trolley system in western Massachusetts."

The year 1912 marked the high tide for the progressive faction in the Republican party, and the progressive movement as an organized political force in American politics. For ten years under the leadership of Theodore Roosevelt, and under La Follette's—quite as independent but never coordinating—the progressive movement had been changing American political life. State institutions had been made over. To an extent there had been a political revolution, a return of certain fundamental functions of government from an oligarchy to the people. The direct election of United States Senators had been achieved in most states through a trick of the

[6] The Massachusetts House of Representatives.

primary for which Senator Coolidge voted, and which soon was to come nationally as a constitutional amendment. Workmen's compensation laws were generally accepted. A law providing for a minimum wage for women was found in more than a score of states. State bank deposits were being guaranteed. Massachusetts was establishing a form of state life insurance. It was the day of the agitator. State Senator Coolidge ignored the agitator and went his serene way.

That was the year of the Lawrence, Massachusetts, strike. Coolidge who voted against the legislative investigation of the strike was made chairman of a legislative arbitration committee to go to the strikers, to hold hearings, to try to effect a settlement. He disbelieved in the legislative arbitration idea. But he went. He worked. With great pains he heard strikers and employers. His investigation was thorough. He made his award. It angered the strikers. But it was probably fair. For one of their advocates writes a dozen years after:

"I was that year a candidate for Governor, backed by the forces of the strikers. Mr. Coolidge was chairman of a legislative arbitration committee which brought about a resumption of industry. Labor did not get all it wanted; at that time we were resentful at the awards. But looking back after the controversy has abated, and reviewing the situation, I have felt that the conduct of Calvin Coolidge in his position was entirely fair and his conduct of the judicial hearing courteous to all. Certainly he was not a standpatter though he did not give us all we wanted." [7]

Here is a political nondescript with the protective coloring of what? Sublimating shyness? Timidity? Cautious foresight? Uncertainty of conviction? Something of all of these? Who can say.

He came home to Northampton at the end of his first senatorial term, dignified by his job. "Senator Coolidge" sounded good to his ears, but he was not puffed up. His neighbors say he discouraged the senatorial title. He encouraged "Cal." He knew then what he always realized, that a senatorial handle to his name might emphasize his inadequacy to handles. The handle might get bigger than the name and people would grin. For never was he able to strut up to his capacity and to his honors. So he "Cal"-ed it and denied himself by discouragement the democratic title of nobility implied with the title senator.

His law practice was not growing so fast as his political reputation. The Coolidges still lived simply; not meanly but probably no other member of the Massachusetts Senate, chairman of so important a committee as the railroad committee, lived in a duplex apartment. And the wife of no other senator with a standing comparable to Coolidge's ran her house with one occasional servant. His law office was a flight up from the main street, a

[7] Roland D. Sawyer, "Calvin Coolidge President," Boston, 1924.

cheerless but not dingy place, orderly, never crowded with clients. He kept few extra chairs for loafers, preferring when he wished to loaf to go where loafers were. There he was able to get rid of loafers by leaving them. He was full of Yankee tricks like that—canny knowledgeous ways—given to folk habits of Vermont, which added to his essential "Cal"-ness. He was not highly esteemed as a coming young man. Opinion in Northampton of that day sensed that he had gone as far as he would go, as far as his capacities could take him. He was entitled to a second term by the courtesy of his first. One good term deserved another. That was all. He made few speeches for his second nomination. He missed few political meetings, but how he shook hands on the outskirts of a crowd and stood rather dumbly, inarticulate beside a voter, or before him after making his direct personal appeal: "I need your vote. I want it!" That was all.

So he rounded his first term in the Massachusetts State Senate. He was just forty years old. His figure was as full as it ever grew. His carrot top was drabbing into a dark chestnut. His lean Vermont jaw was slightly larded. His senatorial dignity added no frills to his gait. He took short steps rather quickly paced and sometimes carried his head at a slight rightward angle and just an inch or two forward from the perpendicular when he was preoccupied, which was much of the time. For he was given to cerebration. He was not one to whistle in vacuity nor to drum on chair arms, nor to preen.

He was a curious mixture: outwardly modest, self-effacing, inwardly pretty complacent. Like a fox he kept to his old haunts in Northampton; still saw the plumber and blacksmith, and the shoemaker, the carpenter, and the mill-worker as often as ever: said little, stood around in the more dignified loafing places, not on the court house steps but inside where someone was working, replied when he was spoken to and freely did odd jobs for those who had no other well-placed friends. He always promised less than he performed, and performed generally for people who were grateful, knowing human nature well. Over and over in his "Autobiography" he repeats the statement that he was always prepared for the next promotion. But he might have added, he knew it was coming if he could manage it. For by this time he was dedicated to political service.

He knew what tricks he had in his bag. He capitalized taciturnity and was not above, then nor ever, a little subtle clowning, pretending to be dumber than he was, cackling dry quips or, when he dared, putting an edge on them that ripped the skin as it precipitated a giggle—but never his giggle. Always he gave furtive evidence that he carried some sort of cripple complex. Perhaps it was a recompense for his shyness; a haunting, subconscious inner ghost of a dead inferiority that persisted without reason. During those first few years when he left Northampton to climb the ladder from the state senate to the Presidency there are few anecdotes about him, though anecdotes blossomed like dandelions in his path from the time he became lieutenant governor until the end. The anecdotes all revealed what seemed dumbness to the dumb but shrewdness to the wise. In his heart he knew it was shrewd. Perhaps in these senatorial days he had not discovered fully how to use this defensive weapon of artful clowning to advertise his wares. Or perhaps he had so little distinction in those days that his sharp repartee was forgotten. This might well be the case. For he went into his forties far from the shadow of his destiny. With all his meekness he did not expect in 1914 to inherit the earth. A second term in the Massachusetts State Senate was as much as he could ask.

BOOK TWO

PREFACE

While State Senator Calvin Coolidge, of Northampton, was plodding his way up the road from law-maker to executive-assistant, to governor, to the nomination for *Vice President* by his party, the forces in America that smoothed his path and lightened his burden on the way to fame were working in the world beyond the boundaries of the United States. Sparks from the gun at Sarajevo set off the powder in the social dynamite that had been under compression for fifteen years.

We cannot say even today that the World War was inevitable. Diplomacy might have found a way out. For the remoter causes of the war, we must look to mistakes and selfish policies in Germany, France, England, Russia, Austria-Hungary, and the Balkans. For the immediate outbreak of the war, Germany must bear the heaviest blame. Bismarck would never have taken Germany into war under conditions that made world sentiment outside overwhelmingly hostile. Bismarck, the diplomat, respected world opinion; overruled the German military strategists of his day. And the German diplomats of 1914, likewise, respected world opinion, and feared the course that Germany was taking. But the military strategists of Germany were, by this time, in the saddle, and they overruled the diplomats. It is not fair to say that German diplomacy failed in 1914. It is fairer to say that German diplomacy never had a chance, and, therefore, that British and French diplomacy likewise did not have a chance, since in dealing with German diplomats, they were dealing with impotent men.[1] Bethmann Hollweg admits this.[2] Speaking of General von Moltke, he says: "I had to accommodate my view to his." Bethmann Hollweg also emphasizes the intrigues of the militarists in Russia [3] in preventing a peaceful settlement in 1914.

False economic theories contributed a great deal in bringing about the World War. That economic tendencies inevitably produced the war, of course, cannot be proved, but many historians hold to this thesis. But this much is certain: America did become part of the economic chaos of those

[1] Munroe Smith, "Military Strategy vs. Diplomacy," *Political Science Quarterly*, 1915.
[2] Theobald von Bethmann Hollweg, "Reflections on the World War," translated by George Young, London, 1920, pages 137-38, 147.
[3] *Ibid.*, especially page 130.

first years of the war. America was in the war a year before America declared war. Perhaps in the midst of the combat, a peace without victory
might have been salvaged—a peace in which mutual good will would have
permitted compromise and so would have quenched the fires of hate. The
doom which debt and disorder were holding over the middle classes of
Europe might have been averted. One spring day in the midst of carnage,
thousands of soldiers of France on the Champagne front threatened mu-
tiny. In America Wilson was counselling peace. The fury of France led
men out by the hundreds to mass slaughter to punish the revolt. The war
went on and the middle class in Europe from Siberia west to the English
Channel went unwittingly to its doom.

The war ceased. Millions of men tried to go back into industry. Industry
was shattered. Credit was exhausted. Commerce was sick. The currencies
of the European belligerents were sinking. Industrial strife burned in strikes
like a prairie fire across Christendom. A score of little nationalities were
created at Versailles, which sought to be economically self-sufficient and
could not be. Tariff barriers, trade restrictions, the result of cramped nationalism that rose from a punitive peace, preserved the hates of the war
in the malignant peace. The war went on along the industrial front. Men
cried: "Peace, Peace!" But there was no peace. Credit was inflating; currencies were shrinking; hunger was stalking over the earth.

But in America we went about our daily tasks apparently unscathed.
The warnings of reformers were unheeded. The wisdom of our academic
seers was ignored. We had a seven-devil urge to get back to the old days,
to forget the nightmare of the war, to ignore, as though nothing had happened, the realities of Europe. How natural it was that the man who above
all others in his country believed in the divinity of the horse trade as a
national symbol, should move slowly to a place of power!

CHAPTER X

Herein We Meet Our Fairy Prince

WHILE the first Roosevelt revolution was in progress, the forces of conservatism were not entirely upon the defensive, though their defense sector was as busy as the pie venders at a fair. Behind the lines of the progressive attack, capital was entrenching itself. While Coolidge was entering Massachusetts politics, the holding company was forming in American finance. Those pole climbers stringing wires across country, whom he saw in the landscape as he rode to and from Boston, meant something new in the world, a new capital structure for the electric industry. The Holding company bought the stocks of competing companies to throttle competition. The sale of stocks and bonds became a primary interest of the holding company. Service to consumers was an incidental concern of the manipulators of remote and frequently irresponsible corporations. The Northern Securities Company holding the stock of the Northern Pacific and the Great Northern, two competing parallel railroads running from the Great Lakes toward the Pacific, was the classical instance of the holding company. President Roosevelt, through his department of justice attacked and dissolved the Northern Securities trust. Yet over America, and particularly in Senator Coolidge's New England, capital still was concentrating. Railroads were getting together, public utilities merging under secret common ownership. In Massachusetts one of the vital issues in the days of Coolidge's legislative service, was the mergers and combinations of the New Haven Railroad. Rumors of the pernicious anti-social activities of Charles Mellen fogged the air. Mellen, a New England product, born in Lowell, Massachusetts, a high school graduate from Concord, New Hampshire, rose in his profession, went west, learned financial magic, came east. And finally the elder J. P. Morgan adopted Mellen, made him president of the

97

Northern Pacific, and brought him back to New England and installed him president of the New York, New Haven and Hartford. Mellen,[1] politically distinguished as a delegate to the Republican national convention which nominated Theodore Roosevelt in 1904, was consulted and quoted by Roosevelt in his annual message to Congress the next year. From 1907, when Calvin Coolidge, the "singed cat," went to Boston, until he began serving his second term as state senator, Mellen, in the eyes of economic liberals, was the head devil of the plutocracy in Massachusetts and New England. Mellen had elements of greatness. He gathered the fragments of the New England transportation system under the Morgan control. He improved the railroads he merged, beautified the railroad yards, gave good service, was effective; and he was probably temporarily a profitable investment for the Morgans. Yet Mellen had the unconscious habit of extravagance that men inevitably have who have access to vast sums of other people's money. In politics Mellen walked to his ends directly, justified by the conscience of a plutocrat, which held in contempt the scruples of democracy. But for this narrative, the point is that while Mellen lobbyists had their way with weak men and fought their way with strong men, he symbolized success when Senator Calvin Coolidge came directly into the Mellen picture in the session of 1912, as Mellen took his exit. While Coolidge was gestating in the womb of politics, Mellen "marked" him—gave the embryonic president the perfect plutocratic heart.

By 1912 the Morgans had decided that they must get out of New England politics. Their political organization remained bi-partisan, superficially chastened, but allied for the most part with the legislative watchers of other private interests which felt that Calvin Coolidge was dependable on the side of their particular angels. Here were representatives of the brewers, the public utilities, the insurance companies, the distillers, the banks, the official representatives of the Republican party, its chairman, its elder statesmen. It was not what was loosely called the plunder bund—a low group of blackmailers and bribe-givers. But the so-called "interests" had a common interest with that lower lobby. Senator Calvin Coolidge rarely cast a vote or used his influence against this common interest.

Coolidge, in those days and always, distrusted reformers. When they appeared with bills in legislative committees where he met them, he treated the sponsors of the reform measure with a sort of embalmed courtesy, heard them, promised nothing. If the reformers really controlled votes he surprised them by helping their measures. He was educating himself politically in those senatorial years. "Education," he wrote in one of his pat phrases, "after all is the process by which each individual creates

[1] Not to be confused with Andrew Mellon of Pennsylvania, another breed.

his own universe and determines its dimensions." [2] We may therefore examine profitably some odd bit of the Coolidgean universe as it was revealed in Boston in those legislative days, to wit: He was on good terms with the lobbyist of the Mellen group. J. Otis Wardwell was Coolidge's friend and supporter. In the legislative counsel return of 1914 we find Wardwell's law firm had one thousand dollars from the Prudential Insurance Company, ten thousand dollars from the Massachusetts Electric Light Association. He was representing the Boston and Maine, the Boston Elevated, the New York, New Haven and Hartford. Arthur *P*. Russell, who was vice president of the New Haven Railroad, and Charles Hiller Innes, commonly accredited as the Republican boss of Boston, and of course called "Charlie," were fairly close to the Northampton senator, and, according to the tradition of the day,[3] in a pinch Innes could deliver Coolidge's vote. Innes testified in 1919 that he received forty thousand dollars in three years from the New Haven Railroad.[4] Also near Coolidge was Tom White,[5] a legislative member who was for years Coolidge's political mentor, a devoted friend and a smooth operator on the political ironing-board, but honest. Coolidge was a member with these men and others of the Knockers' Club—a luncheon and dinner club where he met and mingled with such Massachusetts stalwarts as Senator Lodge, John W. Weeks, William M. Butler, Frank G. Allen, Louis A. Coolidge, James A. Bailey, Walter Perley Hall, Justice of the Superior Court—all fine, upstanding Republican guardians of the party ark of the Covenant. These names are used here because they mark fairly well the dimensions of his political universe as Calvin Coolidge was creating it. The periphery of that universe was horse high, hog tight and bull strong around the citadel of prosperity. In the political education of Coolidge, he was creating a universe. Prosperity was the justification of privilege. Too often privilege in Massachusetts was bolstered by class-conscious arrogance tempered only when necessary by corruption. Coolidge was not corrupt. His personal ideals were high. But he was serene in the presence of this corruptible body in Boston even though he put on the incorruptible—a quickening spirit. He played a clean game with the run of the dirty cards!

About this time in his senatorial day, Coolidge made the acquaintance and attracted the attention of Guy Currier. Guy Currier, in Coolidge's Boston days, was the fairy prince. As such he deserves more than passing

---

[2] "Freedom," p. 38.

[3] Conversation with Robert Washburn, Coolidge's first biographer, and Robert O'Brien, former Chairman, Federal Tariff Commission.

[4] He claimed on the witness stand that he had been hired by Tim Byrnes, the vice president, who was in California and not available for questioning and that his relations with Byrnes were privileged.

[5] And in 1938 is secretary to young Henry Cabot Lodge.

mention. Currier was an older man than Senator Coolidge by five years, born in Massachusetts, a graduate of the Massachusetts Institute of Technology and the Boston Law School. Before Coolidge came to Boston—indeed a decade before—Currier had been a Democratic member of the Massachusetts House of Representatives and of the State Senate. When Coolidge met the fairy prince, the prince was decently and comfortably rich. He had enough money and wanted little more. He kept accumulating it slowly until he died a millionaire.

He represented a passing species of American politician. When the primary system came in, the generic state boss—using the term not in a venal and corrupt sense but in its connotation of purely political leadership—began to modify his survival qualities as America changed its type of political life from the convention system to the primary. Currier was one of the old leaders. He was a nonpartisan leader of leaders, controlling men who controlled masses. Generally when party lines are taut among the voters, they slacken among those who really rule. In Coolidge's time, Massachusetts' racial and religious lines were, and are today, more real than party lines, though party lines are much more visible to the carnal eye. Currier, being a member of an old well-placed Protestant family, in the legislature as a Democrat was an anomaly. He was suave, brilliant, audacious, sophisticated and rarely went indoors with the curfew. For he knew his way around on Beacon Street and was not a stranger in Back Bay. He was literary enough to sit on the board of the public library trustees. He had a villa in Florence, was a mild addict of Franciscan literature. But when Jim Curley needed something for the final uphill pull in his mayoralty race, Guy Currier knew where to find it, and when Malcolm Nichols ran on the Republican ticket, Currier was his friend; Nichols won. Such a man could not escape the admiration of Calvin Coolidge. He had social distinction, wealth, personal charm to the verge of glamour, a brilliant circle of friends, indeed everything that Coolidge lacked. And Currier's task, as one of the Massachusetts rulers of the ruling class, was to know the Coolidge kind. Guy Currier dominated New England public affairs during the whole of Calvin Coolidge's career. That each respected the other to the end and that the two were friends would indicate that Currier appealed to the best in Coolidge, and Coolidge was patient with the faults of his friend.

But something more than an agreeable friendship bound these two men in a common cause. Mrs. Currier,[6] writing to Frank Buxton, Editor of the

[6] Mrs. Currier was Marie Burroughs, a famous Shakespearean actress of the eighties and nineties who retired from the stage upon her marriage with the Massachusetts statesman. She was a young woman of unusual talents and intelligence. Their home was a center of artistic culture three or four decades in New England.

Boston Herald, in 1935, recalls that in his senatorial days, Mr. Coolidge came to her husband and told him that "he wanted to go into politics seriously." James Lucey symbolized Coolidge's education in democracy where he was as wise as a serpent on a rock. Guy Currier taught the Northampton acolyte "the ways of an eagle in the air!" For Guy Currier was a member of the American samurai of the second decade of the twentieth century—those who ruled the governors, guided the legislators, and stood conveniently at the elbow of the courts with subtle frankincense to soothe and then suggest most obliquely, most suavely indeed, the purpose and philosophy of the law. It was natural that in Coolidge's state senatorial days, President Taft should be considering Guy Currier as a candidate for the Supreme Court.[7]

These state bosses of the first quarter of the twentieth century were amiable men, honorable men, many of them men of grace and of noble parts, members of our American ruling classes. Typically they were attorneys for whatever corporate organization stood in need of their services, somewhat extra legal. They ruled with charming insouciance—sometimes in states, sometimes regionally, rarely with national suzerainty. They entertained with taste where taste was required, with prodigality where a lavish hand was exigent. No dirty money crossed their palms. They gave appropriate expense accounts to hirelings, asking no questions but loathing the men who took their bribes. Those who handled dirty money never flourished—long! The samurai survived—many were patriots according to their ideals. Their sympathetic tears when their country was extolled were not hypocritical. They loved their America with all the ardor of their expansive natures. Cold, calculating, mean and greedy men do not grip and hold their fellows as these men bound their henchmen to their train. Their capacity for affection, their loyalty to those who served them and their devotion to those they served revealed in these matters of public affairs such traits of zeal and passion as have made the heroic courtesans of history. The country they loved, the ideals they served were not the beloved country and the common ideals of democratic Americans. The samurai was a cult, serving a secret flame. Their altar was the Hamiltonian plutocratic aristocracy. When Calvin Coolidge made his pilgrimage to Guy Currier to say that he desired to go into politics seriously, the slim, shy, cautious yet essentially resolute little Yankee became by that act the child of destiny walking in the light of the secret flame. Then when Murray Crane pointed his finger at the trim, grim young state senator from Northampton, Coolidge was electrified with power.

After Currier had watched the Northampton politician for a few years, Currier realized that Coolidge was money honest. Currier also knew that

[7] Editorial in the *Boston Herald* after Currier's death, June 23, 1930.

one who was so cautious, so circumspect, so thrifty and diligent about his business was predestined to a life of eminent respectability whatever might happen to him. In the midst of Coolidge's Boston career, Currier took his protégé upon a mountain and offered the young Northampton statesman a place in a thriving established life insurance company.[8] Its president was stricken with an incurable infirmity. His tempter explained to Coolidge that he was invited into the organization that he might be next in line for the presidency when the incumbent would pass. The salary suggested was twenty thousand dollars a year to begin, with an increase as Coolidge might suggest it, and with an ultimate expectation of fifty thousand dollars with extras! Coolidge's annual income, as he sat across the table from Currier, was less than a tenth of the initial salary offered. Slouched in his chair not batting but opening his wide handsome eyes, he quacked: "No, that doesn't lie along my line of influence." Just that.

It takes wisdom, courage and some indomitable faith in his own talents for a man to know his way and to go his way past a temptation like that. He was marked and dedicated to politics—a self-consecrated public servant.

2

By that time Coolidge had pointed his political compass. Four years in the lower house of the legislature, two terms in the mayor's office and two terms in his state senate had given a definite slant to Coolidge's political career. Primarily he was for economy. This meant incidentally that he instinctively opposed widening the activities of government where widening those activities would make government the militant champion of the common man who was trying to regulate, to control and perhaps to direct the railroads, the public utilities, the banks, the insurance companies, and the brewers. His bent and direction turned him conservative. Now and then he cast a vote for labor. Occasionally he favored some legislative measure which would benefit his own constituents as in the case of a bill called the "fair sales bill" directed at ice and oil dealers. He was for woman suffrage because his district felt that way. And he had before him Mrs. Coolidge, an intelligent woman, even though he did not respect her education [9] and kept her out of his counsels. The memory of his mother, who also must have been an intelligent woman, may have influenced his vote for suffrage. Sentiment always moved him and if he felt that way he would vote that way for he was no hypocrite, never a coward in the ordinary course of politics. But in other matters—matters social and economic, his

[8] Conversation with R. M. Washburn, first biographer of Coolidge and Republican nominee for United States Senator, 1934, Massachusetts; also obituaries of Currier in Boston newspapers, 1931.

[9] Her own statement in *Good Housekeeping*, March–June, 1935.

belief in the moral government of the universe bolstered by the Garmanian philosophy—the righteousness of working and saving—anchored him on the rock. This belief gave him a distinct bias in favor of the hypothesis that the sheer right of the force he knew as God would certainly guarantee forever the title to private property and maintain the distribution of income as it was, in order to maintain the Massachusetts that he knew, the lovely landscape through which he went week by week to and from his legislative labors. This Massachusetts was to his eye the utopia of the philosopher's dream. Labor was decently employed at a wage sufficient to guarantee plain food, plain clothes, plain homes, and common schooling for childhood—enough for labor. Capital was rewarded in Massachusetts with the cream of the common income. Capital could live in the big white house with pillars and green blinds. Capitalists could travel to far places and get wisdom, live with some degree of luxury, educate their children at New England colleges, and pile up for each succeeding generation just a little larger gain of accumulated wealth than the fathers had possessed. For this the Massachusetts of his weekly panorama stood. It was this Massachusetts upon which he pinned his faith. It was this same Massachusetts that the lobby employed by the railroads, the public utilities, the banks, the insurance companies, the distillers and the brewers was pledged to maintain in statu quo "one and inseparable now and forever."

Murray Crane guarded the West. Lodge was the sentinel on the East. Charles S. Mellen was its prophet. Albeit to the visionaries, to the doctrinaires of a new social science who were yearning vaguely and talking loosely about social and industrial justice in those days, Mellen, the prophet, was a prophet of evil. If Coolidge was affronted at the candor of Mellen's greed he spoke no word against the prophet. The followers of the prophet knew Coolidge for their own.

If Mellen was the head devil of plutocracy in New England and particularly in the Boston legislature in Coolidge's legislative years from 1907 to 1912, the apotheosis of democracy was Louis D. Brandeis. Brandeis was a corporation lawyer who became the people's advocate. He was the descendant of a Central European Jew. The elder Brandeis was a "forty-eighter" who fled Germany in the turmoil of the revolution in the middle of the nineteenth century. Louis, his son, was born in Louisville, Ky., studied for a few years in Dresden, was graduated, before he could vote, from the Harvard Law School in 1877, went to St. Louis to practice in 1878, returned to Boston to become a partner in law with his friend and fellow student Samuel D. Warren, who later left law and became a paper manufacturer. But Brandeis stuck to the bar. Brandeis was everything that Mellen was not.

"I have only one life to live," Brandeis told an interviewer [10] in those Boston days when he was fighting the domination of the New Haven Railroad, and he added: "Life is short. Why waste it on things that I don't want most? I don't want money or property. I want to be free."

He had a mathematician's mind and worked out a formula which was the basis for the Massachusetts Savings Bank Life Insurance Act adopted in 1907 before Coolidge came into the legislature. In the meantime, Brandeis, fighting bribery and corruption in Boston, helped to regain valuable franchise rights that had been given away by the legislature. Street railways, gas and electric companies all felt the power of Brandeis as a champion of the people. When Coolidge first came to Boston, Brandeis was exposing the designs of a Wall Street group to monopolize the railway system of New England by merging the New Haven Railroad with the Boston and Maine. By digesting and assimilating statistics he won his fight by the sheer logic of mathematics. A year before the Interstate Commerce Commission gave Brandeis the victory, and nearly two years before Mellen resigned, politically discredited, from the New Haven Railroad, Mr. Brandeis wrote a private letter to the editor of *Harper's Weekly* in which he said: "When the New Haven reduces its dividends and Mellen resigns, the decline of the New Haven and the fall of Mellen will make a dramatic story of human interest with a moral—or two—including the evils of private monopoly. Events cannot be long deferred and possibly you may want to prepare for their coming.

"Anticipating the fact a little, I suggest the following as an epitaph:

" 'Mellen was a masterful man, resourceful, courageous, broad of view. He fired the imagination of New England; and being oblique of vision merely distorted his judgments and silenced his conscience. For a while he trampled with impunity on laws, human and Divine. But as he was obsessed with the delusion that two and two make five, he fell a victim of the relentless rules of humble arithmetic.

" 'Remember, oh stranger, arithmetic as the first of sciences is the mother of safety.' "

Isaiah at his best could not have done better as a prophet than Louis D. Brandeis. Brandeis was a lone leader in Massachusetts while Calvin Coolidge, galvanized by Murray Crane, shepherded by Guy Currier, was trudging up the moving incline of his legislative career. Coolidge seems to have avoided Brandeis as he publicly avoided Mellen. They two, the militant reactionary, the consecrated crusader of liberalism, were the Scylla and Charybdis past which Coolidge had to go. To have turned to either wholeheartedly would have been fatal. For that way lay political death—for a man of Coolidge's type and talents. Mellen fell when Coolidge was be-

[10] Norman Hapgood.

ginning his senatorial career. Brandeis' name at the close of Coolidge's legislative career was before the United States Senate for confirmation as a member of the United States Supreme Court. But the only record Coolidge made to indicate that he and Brandeis lived on the same planet, in the same land, in the same town at the same time, is a characteristic remark made to one who mentioned the crusading lawyer: "Brandeis is unsafe!" [11]

In the session of 1913, Senator Coolidge was chairman of the commission on railroads and his most important fight was over a bill called the Western Trolley Act. It provided that the New York, New Haven and Hartford Railroad might build trolleys between the smaller towns in the vicinity of Northampton west of the Connecticut River. Northampton would be connected by trolley if the Western Trolley Act became law, with a dozen similar industrial communities. Coolidge steered the Western Trolley Act through the senate, a difficult job when the shadow of Mellen, though receding, still darkened the political circumference of New England politics. A gubernatorial veto met the Western Trolley Act. But the bill was repassed over the governor's veto by a large majority. Coolidge's committee also reported a bill transforming the railroad commission into a public service commission with a provision intended to define and limit the borrowing power of railroads.

The bill became a law, but in its becoming a law probably the railroads and utilities got more than they lost. The original author of the bill voted against it and denounced it in final passage. The forces which seemed to be organized to conserve the public interests accepted the bill grudgingly, realizing that it gave the Mellen interests much that they did not deserve; hoping that the good would counteract the bad. Coolidge accepted it, supported it and later, looking back across twenty years, was proud of it. But in 1913, his cast of mind was set. He was safe—the phrase ran "safe and sane." The struggle over this Western Trolley Act was long and vicious. Coolidge, as much as he ever became in his legislative career, was the hero of this contest. He wrote:

"It was the most enjoyable session I ever spent with any legislative body." [12]

In spite of Garman's moral mottoes—he surely loved victory! Behind the veil of his immobile face how warm he was; how human; how hungry for success!

Senator Coolidge certainly was not discredited by his activity for the trolley bill. It was popular among the farmers of his district. The trolley

[11] Letter to author from a former chairman of the ways and means committee of the Massachusetts House of Representatives.
[12] "Autobiography."

was supposed to be effective in collecting and distributing milk. Western Massachusetts farmers were going into the dairying business professionally. The new policy of the New Haven road and perhaps new methods by its lobbyists took the curse off Coolidge in his trolley enterprise. In writing of this episode nearly two decades later, he declared that he desired to be on the railroad committee because he wished "better to understand business affairs." Then he added with a characteristic naivety:

"I made progress because I studied subjects sufficiently to know a little more about them than anyone else on the floor. I did not speak often but talked much with senators personally and came into contact with many of the business men of the state. The Boston Democrats came to be my friends and were a great help to me in later times." [13]

No one can draw so perfect a picture of the mousey, competent little countryman edging his way onward and upward on the path of glory as that he drew of himself. He won his way by diligence in Amherst. He won his way to the top in Massachusetts politics by tempering his diligence with kindly patience and a lively sense of gratitude. And he knew it.

Incidentally he was making a colorless record. The Reverend Roland D. Sawyer, his colleague and early biographer, wrote ten years later that Coolidge was "uncomfortably progressive for some of his constituents in Northampton." But in Boston the predatory lobby always allowed a friend leeway to make a record that would maintain him in power! He voted for a state income tax, for a law to legalize picketing, for the direct nomination in the party primary of the candidate for United States Senator, for woman suffrage, and for most of the progressive measures before the senate in his first two terms. More than that, he was forever helping his friends—individuals in trouble or politically a-hunger. In his weekly day-coach ride from Northampton to Boston, sitting silently smoking his cigar, looking out at the landscape which entered into his cosmos and became a part of his mental processes and his spiritual attitude, short shift did a stranger get who tried to talk to him. But if a fellow senator going to Boston on the train asked a favor, though Coolidge was crusty he did not forget the request. He did the favor if he could. So in his way he was beloved. It was in his first senatorial session when he was sponsoring the trolley bill that he sent a laconic note across to his colleague, the house chairman of the railroad committee, "Sand your tracks, you're slipping." It was quoted over the State House and became part of the Coolidge myth of the day as he was emerging from obscurity when he had to be reckoned with.[14]

[13] "Autobiography."
[14] Also he had to be written up. More books had been published about him when he arrived at the end of his rainbow as Vice President in Washington than about any other Vice President: "The Boyhood Days of Calvin Coolidge" by Ernest Carpenter;

3

In his second senatorial term he was made chairman of the committee on resolutions in the Republican state convention. At one of the hearings of the resolutions committee someone asked him a question, probably an embarrassing question, certainly puzzling. He whirled his swivel chair around, looked out a window, and a friend waiting for Coolidge to answer said to Governor McCall standing nearby, "I could take dictation from Coolidge in longhand." In that convention tradition says that he did not mix and mill with the delegation in the hotel lobby after bedtime but slipped away, hung his trousers to press on the door, left it ajar and went to sleep where they found him after midnight. He was a source of delight as well as a very present help in trouble. But even in the rough and tumble of his legislative career in Massachusetts, he never recovered from his shyness. A fellow senator wrote of that period:

"One day he came in here and after sitting for the longest time in silence he said out of a clear sky: 'Do you know I have really never grown up. It is a hard thing for me to play this game. In politics one must meet people. That is not easy for me.' He sat silent for a long time after that; just looking out of the window. Then he went away without another word. He has never mentioned the subject since." [15]

But despite his inner warmth, despite the well-shielded flame of his personality, behind the curtain, it is interesting to note how little he seemed to realize the significance of the pageant in which he moved. That Western Trolley Act which he sponsored was evidence of a deep change in America's economic and industrial life—the change wrought by the newer power begot by the scientist and the mechanic who could distribute electricity at long range. At the end of the old century electric current could be sent profitably less than two hundred miles. In the first decade of the new century the laboratory experimenters had found how to transmit electricity for commercial purposes, profitably, over a radius of five hundred miles. They had also learned how to cheapen the cost of electricity. Hence interurban trolleys and interstate trolleys were becoming possible and profitable. Financial reorganization of public utilities was going on with unprecedented swiftness and magnitude under Coolidge's nose, as he

"The Preparation of Calvin Coolidge" by Robert A. Woods; "Cal Coolidge, President" by Rev. Roland D. Sawyer; "Calvin Coolidge, His First Biography" by R. M. Washburn; "Calvin Coolidge—From a Green Mountain Farm to the White House" by M. E. Hennessy. These books were my sources of information about most of the material incidents in this biography. For it must be obvious to any one who reads them all that Calvin Coolidge inspired all of them.

[15] "Calvin Coolidge, His First Biography," R. M. Washburn. The friend was Frank Stearns.

served in the legislature from 1907 to 1917. The financial structure of most of the distributors of electricity, including telephones, was undergoing profound change. In Calvin Coolidge's childhood, few Americans dreamed that a corporation could own stock in another corporation. But before he was a boy grown, several American states had handed over to the corporations rights and special privileges which gave corporations this new and deadly power. So the holding company was born, creating a ruthless collectivism —the collectivism of purely acquisitive capital. Ostensibly it was building the material framework of an expanding civilization, but really it was creating evidence of obligation—corporate debt, destined to be the plaything of gamblers menacing the stability of the world. Problems were assembling which Garman's mottoes did not help Calvin Coolidge to understand. Yet riding from Northampton to Boston in the day coach in those early days of the second decade of the new century, watching workmen stringing power lines across the hills, even realizing that electricity as power was cooling hundreds of furnaces in small Massachusetts factories, that once generated their own power, he saw no political significance in the appearance of all these wires and poles. He was pleased. To him this was progress. Yet it was common knowledge in his Boston, that huge aggregates of capital-controlling industrial plants were swirling centripetally into unified control, which was to institute a financial and industrial order in which ownership and operation were divorced.[16] A new, hard, anti-social group —the utility operators, the operators of the plants owned by the widely separated stockholders of the great trusts—was about to become something like a caste in our industrial system: the managers who hired themselves by controlling a small minority of a corporation's stock, and then went mad with power. The workmen stringing wires along the lines whom Coolidge saw out of the day coach windows, he hailed as the soldiers of the army of progress. Alas, they were instead the pioneers of an army carrying vessels of wrath fitted unto destruction.

Yet Calvin Coolidge was not vegetating in those day coach hours. After he had read his pile of Boston and Springfield papers, he never idled. But no doubt he was thinking of the day's obvious job. He thought clearly. He acted when necessary, promptly—if cautiously—and generally upon minor matters. For he had to get little jobs for constituents, small favors in the State House, to angle other favors out of small officials. He had the bill of this colleague, the measure of that member of the lower house of the legislature to help; this man to see, that man to interest in another man's scheme. He took few favors. If he helped a utility lobbyist, and he did help where he could decently many times, he asked no immediate return.

[16] See "Main Street and Wall Street" by W. Z. Ripley for a clear view of the economic reorganization going forward in that day.

He put his service upon time deposit. He was hoarding his political savings, a political miser. And so he sat there clinking his political coin, adding up his assets, his stock in trade, his political savings, as he gazed out of the window of the day coach, blind to the vast political change that was growing—a miracle in steel and iron and cement, along his way. Little did Calvin Coolidge dream that this miraculous machine was but the skeleton of a monster—an intangible corporate debt-making monster that would bring trouble to his country in another day.

CHAPTER XI

Our Hero Dances at the Grand Ball and Meets One Grand Old Duke

IN 1913 Frank W. Stearns, a trustee of Amherst College, went to interview Senator Calvin Coolidge in his office during the last few days of a legislative session. Tradition declares that Senator Coolidge kept the extra chair in his room locked in a closet. Certainly he wanted no casual loafers bothering him. But he graciously unlocked the closet and brought out the chair for Stearns. Stearns called to interest Senator Coolidge in a bill providing for some sort of sewer construction in and around Amherst College. Senator Coolidge never had seen Stearns before. To Coolidge, Frank Stearns was only a merchant prince, partner and head of the advertising department in one of the large dry goods stores of Boston. He had heard of Stearns as a trustee of Amherst. The Senator was busy. He knew that the legislative day for introducing new bills was past. He let Stearns sit in the precious chair and make his speech. Then the Senator snapped the merchant prince off with five words, "I'm sorry! It's too late!" [1] and turning around resumed his work.

Stearns found his own way to the door as Coolidge put away the chair and sat down. Stearns looked at Coolidge's back as the narrow-chested little man bent over his task in the senatorial ashes. Here was a jolting insult to a man who was to become Coolidge's benefactor.

Senator Coolidge's second term was closing. His legislative performance had been unimportant. But out of it two assets had been garnered for his career: first, the reputation based upon a solid character for being a safe partisan Republican, an acolyte at Guy Currier's altar, a follower of Uncle

[1] All of the early Coolidge biographies agree on this story.

110

Murray Crane. His second asset was a thorough knowledge of his own
strength in the state senate. He had been running errands not entirely for
Crane and Currier but for Coolidge. Men were under obligations to him
for legislative help in promoting their pet measures. He promised little and
performed much as an errand boy. And he knew exactly the size of the
political surplus he had deposited in the bank of his future career.

So definitely he decided to be a candidate for a third term. In American
politics third terms are unpopular, almost forbidden. Anyone running for
a third term takes a risk with his political future. Coolidge took the risk.
He believed that with his Republican status, with his Crane backing, supplemented by Currier's benediction, and with his own senatorial assets he
could be made president of the senate; in Massachusetts a most important
office. As president of the senate he would name the committees of the
senate. He would send bills coming from the House or introduced by his
fellow senators into committees of his own naming. He would make up
the senate calendar committee, the most powerful committee of all. Also
he would name committees on conference to consider disputed measures
between the two Houses. He would be a power in Massachusetts politics,
and he had a zest for power. Power repaid him for much that life was
denying to him.

He took this chance for a third term because he understood that the
president of the senate, Senator Levi Greenwood, would be a candidate for
lieutenant governor. But after Coolidge got into the senatorial race for a
third term, President Greenwood changed his mind. He ran for the senate
again also expecting to be reelected president of the senate. Senator Greenwood was from western Massachusetts. He also was a liege man of Uncle
Murray Crane and not unfriendly to Currier. But Coolidge was in the race.
He could not withdraw. Then Lady Luck began working for Coolidge.

Two young women, advocates of woman suffrage, came into Senator
Greenwood's district, opposing his reelection. Senator Greenwood was the
typical conservative Republican state senator, one of the squires of the
Western Manor. He had made a bitter attack upon woman suffrage and
had put stumbling blocks and secret pitfalls along the path of state suffrage in Massachusetts. Also unlike Senator Coolidge, Greenwood had
definitely displeased labor. He held views about the divine right of capital
to rule because forsooth capital has and is "brains." His record reflected his
views. The young woman suffragists attacked Senator Greenwood in his
district vigorously, taking his labor record to the mill hands and his suffrage record to the women of all social strata interested in suffrage. They
painted him black. Coolidge knew what was going on.

The election came. The Republican State ticket went down for the first
time in many years. Greenwood was defeated by a Democrat, but Coolidge

was elected senator when the polls closed Tuesday night. Coolidge's story of what happened was this:

"Again I was ready. By three o'clock that Wednesday afternoon I was in Boston. By Monday I had enough written pledges from Republican senators to insure my nomination for president of the senate at the party caucus." [2]

The other story is this: that during the Tuesday night following the election, Murray Crane and his friends, and the Boston handymen of Currier hurriedly met. Among others present were the political lawyers for the railroads, banks, insurance companies and public utilities. They realized that Senator Greenwood who was their mainstay in the race for president of the senate was going to defeat. They turned to Coolidge. He was their logical candidate. They telephoned to him. Coolidge took a night train, not an early morning train, upon summons of J. Otis Wardwell, herein before mentioned leader of the cohorts of organized conservatism who was generally friendly to Uncle Murray Crane and William Butler, Crane's alter ego. According to the later story Coolidge actually arrived in time to eat breakfast with Arthur P. Russell who represented the New York, New Haven and Hartford Railroad. Later Russell and Wardwell picked up Charles Innes, the Republican boss of Boston. They and Coolidge went to work. They had the support of the Crane organization. Coolidge, sitting in a room with Russell, Wardwell and Innes, took charge of a swift, successful campaign on the long distance telephone to secure for Coolidge the pledges of Republican senators who guaranteed to him the Republican nomination for president of the senate which was safely Republican. The nomination was equivalent to election.

But this also must be considered. Calvin Coolidge himself was in the senate a first rate power. His power came from his intimate, shrewd knowledge of every Republican senator and the accumulated gratitude these men owed him for entirely proper services rendered during two terms past. He had helped these senators to secure the passage of their personal bills, helped them get their friends on the public payroll. He scratched their eager itching backs in many satisfying ways, and when he needed a back-scratching they returned the favor! Back-scratching is the coin that oils much commerce in American politics. He hypothecated all his political assets. So partly under his own steam, partly by guidance and aid from the emissaries of organized capital and largely under the piloting of Murray Crane's organization, Calvin Coolidge, in 1914, took an important upward step and became one of the titular leaders of his party in the state.

That he served his supporters well no one can question. Never was he accused of ingratitude or never was he convicted of duplicity. Timidity?

[2] "Autobiography," page 105.

Possibly. Caution? He was proud of the accusation. But direct conscious duplicity, no. So at the age of forty-one he was occupying a position second in power only to the governor. More than any other man in the legislature he could direct and in a measure control legislation.

It was a day of pause in political progress in Massachusetts, a slowing down from the Roosevelt tempo. Republicanism was reasserting itself in its ancient New England stronghold where Republicanism meant conservatism. Thus when Coolidge began paying his political debts, he paid them to conservatives in conservatism. So in paying them (for such is the way of a wise man in politics) he accumulated rather more assets than he dissipated. To those in command of the Massachusetts Republican party, Coolidge was the good and faithful servant, the instrument exactly suited to their ends. His dumb manners, which were a mask for exceptional political perspicacity, his sly, shy, subtle clowning which gave him distinction as a "character," his sterling qualities of heart and mind, his nerve under fire, his honesty of purpose, his competent New England education, his unfaltering gratitude and his profound belief in the moral government of the universe under the direction of the Republican party in the latitudes of Massachusetts, all made him the living, breathing spirit of the hour in the politics of that day and place.

To top it off, Calvin Coolidge was impeccably respectable. In a day when prohibition was gaining ground he abstained from all liquor beyond the bounds of rigid temperance. He revealed unconsciously an utter ignorance of the wiles of women. And to make a waxworks figure of ideally pure Republicanism which was the handmaiden of capitalism, this man had no desire for money as a counter in his life's rewards. He was the precious gift of a Garmanian god to his party in the second decade of the century when in New England, at least, reaction from Theodore Roosevelt's liberalism was running rampant. That liberalism in that day seemed to Coolidge radicalism. He was out to check it lest it prove destructive. He said so later, often.[3] Here is his definition of this liberalism: "It consisted of the claim in general that in some way government was to be blamed because everybody was not prosperous, because it was necessary to work for a living, and because our constitution, the legislatures and the courts protected the rights of private property owners, especially in regard to large aggregations of property." [4]

When he took his office as president of the senate he made a short address, a conservative appeal contending that government could not relieve us from toil, that progress was expedited by the formation of large concerns in which labor and capital both had a common interest. He defended representative government and the integrity of the courts. It was

[3] Conversation with author in 1924.
[4] "Autobiography," page 107.

the crowning plea of scholarly conservatism in a day of reaction. The address was printed far and wide and was known as "Have Faith in Massachusetts." Coolidge had great faith in that address. Looking back at it across the decades, recalling that an expectant commonwealth was eager for this Messianic utterance, he remembered that: "The effect was beyond my expectation. Confusion of thought began to disappear and unsound legislative proposals diminish." [5] Moses smote the water and the sea gave way.

It was on the day that young Senator Calvin Coolidge returned from Boston as an acknowledged victor in the contest for president of the senate, that Northampton began to ask itself "What manner of man is this?" Until that fateful Monday when he stepped off the afternoon train from Boston with the papers acclaiming him victor in a first-rate, state-wide political fight, Northampton had regarded him rather casually. A state senator, taken by and large in American politics, and particularly in the politics of a small town, carries little weight by reason of his office. Most state senators are honored by their office. They generally seek it for the honor or for a chance to get a good political state job. Few of them are heard of after their term ends. Northampton having elected Coolidge three times to the senate, and having honored him before nearly half a score of times by electing him to minor offices of trust without much pay, gave him small attention. At the bar he was one of a dozen or twenty minor attorneys struggling into second place. The financial fathers of the village knew that Coolidge had exemplary commercial habits, and had something less than ten thousand dollars saved up as a justification for his vice presidency of a small savings bank. In those days, he was not a leader who was consulted seriously about town matters outside of politics. He avoided many civic responsibilities. His caution led him away from the local chamber of commerce leadership that was steaming ahead trying to make Northampton a bigger, busier industrial center than it was. He still represented the Springfield brewery legally, certainly not, at least at home, politically. And he was a brother-in-law of the Congregational church in which Mrs. Coolidge took active, happy, useful leadership. He was a platform sitter at Republican gatherings, glad to introduce the speaker of the occasion, and wasting few words on his task. Men realized that he was modest and saw no reason why he should be otherwise. But when he returned that Monday after election in November, 1914, the responsible leaders of Northampton began to scratch their heads and wonder how he did it. Then and there they recast their estimates of the man. The common run of folks in the town did not forecast what this upward step in Massachusetts politics might mean. But the men in the banks, in the court house, in the city hall, in

5 "Autobiography," page 108.

Smith College, and over in Amherst College had a sense of puzzlement. Realizing that their judgment of Coolidge had erred, many of them began to endow him with capacities which he possessed only scantily. At least they quit laughing at him on the olympian heights of Northampton, where the real rulers of the town held their high councils.

He presided over the Massachusetts senate in the session of 1915 with dignity and most efficiently from the viewpoint of his party. No radical measure passed the senate under his suzerainty. It was the year before this time that Coolidge quacked Frank Stearns out of his office when Stearns came to him about the sewer bill for Amherst College. And it was in his third term as state senator that the sewer bill passed as Stearns desired it. Carpers declared that it was not Coolidge's bill, that another man sponsored it. But Coolidge, as president of the senate, doubtless gave it his necessary blessing and that without a word to Stearns. So it was in those days that the Boston nobleman—known [6] in later years as Lord Lingerie—adopted this Yankee bound boy as his political godson.

Amherst graduates in Massachusetts politics felt keenly the fact that Harvard was taking more than her share of the political plums in the state pie. In the winter of 1915, at an alumni dinner in Boston, Amherst men were bewailing their political inconsequence. Someone suggested that as an organization they should get behind some alumnus and push him to the fore. None seemed obviously available. Then up spoke Judge Henry Field, of Northampton, in whose office Coolidge had studied law. Field had no illusions about the young man, but Coolidge's election as president of the senate was a conspicuous proof that Coolidge had moved into the King row. Being politically wise, Judge Field knew that Coolidge had been jumped in there by grace of Crane's and Currier's organizations using Wardwell, Innes and Russell of the New Haven. Also he knew that Coolidge was a local vote-getter, that some way his character attracted citizens at the polls. So Judge Field threw Coolidge's name into the Amherst alumni dinner council. Only the political alumni knew him. Frank Stearns, a trustee, and a loyal Amherst alumnus of the seventies, the leader and the most important figure at the dinner, spoke briefly thus:

"Well, if you say Coolidge, it's Coolidge. But the only time I ever met him he insulted me."

And then he told the story of the miraculous passage of the sewer bill without his assistance at the next session of the senate.

So it was Coolidge. Lady Luck touched him with her wand and off he danced. In the Coolidge calendar it was a red letter day when Frank Stearns said: "Well, if you say Coolidge, it's Coolidge."

---

[6] Festively, but only in a limited circle of local wits.

2

So now let us consider for a moment this noble lord. Being a twentieth century noble, he was a merchant prince and his lordly lance was advertising. He was head of the advertising department, and partner in one of the important Boston department stores. He knew advertising men and editors all over Boston and throughout the suburban centers around Boston where Stearns Company advertised, and in the near by industrial cities where sometimes the Company placed its advertising. Being a methodical man and a New Englander, though without political experience, he began organizing the Coolidge campaign; for just what, no one had exactly decided at the moment. It began as a propaganda campaign for Coolidge. Stearns saw advertising men and they saw editors and talked about Coolidge, interested the press and so later the people of eastern Massachusetts in Coolidge. Frank Stearns did a thoroughly fine piece of publicity. It was all honest, all ethical as these things go in the newspaper offices, but also carefully worked out, effectively organized, a first rate Yankee job with no nonsense about it.

Frank Stearns, when he became to Coolidge the fidus Achates, was in his late fifties; Coolidge was in his early forties. Between the two men sprang up a deep affection which lasted until death parted them; not fraternal but a father-son relation. They were complete apposites. Stearns was a short, barrel-shaped man of about five feet six. He moved deliberately, walked with measured step and slow, spoke briefly and without emotion. He had keen eyes, rather heavily lidded, a heavy voice, and a heavy hearty hand. He had inherited his job. Little luck and much work and worry had been the price of his success. Naturally he was smitten with the agile, lean, sharp-spoken, keen-eyed, short-stepping, quick-moving Coolidge who seemed from a casual view to have come to his success, a darling of the gods, blessed with luck, and with a kindly leading light which beckoned him to his destiny.

The two men met at a dinner given by an Amherst Society for Coolidge. Stearns whose heart was for some reason lonely, whose capacity for affection was immoderate and who loved the pageantry of politics for the power it veiled, adopted the younger man and gave him unstinted devotion for twenty years. Coolidge became Stearns' major passion. His faith in Coolidge was saintly, adoring. As early as 1915, Stearns told the Boston politicians that Coolidge, who was then president of the senate, some day would be President of the United States. Their ribald laughter is remembered still by old men who sit around the lobbies of the political hotels when Massachusetts statesmen meet and tell old wives' tales of the giants' days. But Stearns was not diverted by the ribald cackling of the politicians.

He went on with his task, built up his machine and in the winter and early spring of 1915 began insisting that Coolidge should run for lieutenant governor. Coolidge was busy with his senate session. He would not discuss with Stearns his prospects for the lieutenant gubernatorial nomination, yet he knew, of course, that he was what in politics is called, the logical candidate.

We must remember that during Coolidge's three terms in the senate, a Democratic governor and Democratic lieutenant governor were sitting in the State House. Coolidge was the ranking Republican, in effect the titular leader of the state. He wrote Republican state platforms. In a state which was rather meticulous in matters of party rank and title, Calvin Coolidge as president of the state senate, with broad powers, held a real leadership in his party that could not be challenged save by United States Senators. And one of those senators, at least, Murray Crane, was Coolidge's liege lord, a faithful friend. The Crane organization was for Coolidge for lieutenant governor. The invisible government of the state was for Coolidge for lieutenant governor. Its lobby liked him. The railroads, the public utilities, the insurance companies, the banks were generally for Coolidge for lieutenant governor. That solid respectable block of public opinion led by Guy Currier, by Lee, Higginson, by Kidder, Peabody—all the solid forces that centripetally adhere to those privileges that go with heavily accumulated wealth in its corporate forms wherever it is invested—nodded mandarin heads in aloof approval of the "ouden" of Amherst! These staid Boston financial patricians, whose class consciousness was reflected in the unconscious arrogance of their economic security, piously realized that they had come through the troubled waters of the Theodore Rooseveltian rebellion by a narrow squeak. Somewhat they were chastened. They were willing to go to western Massachusetts for a winner. After the Roosevelt wave had subsided, Coolidge had their full permission to fill the Republican platform with kindly words if not actual pledges for measures that if achieved would have promoted a considerable degree of social justice in Massachusetts. Read this: [7]

"The continued support of every means of compulsory and public education, cultural, vocational and technical, merited retiring pensions, aid to dependent mothers, healthful housing and fair protection, reasonable hours and conditions of labor, and the amplest protection for public health, workingmen's compensation and its extension to intra-state railroads; official investigation of the price of necessities, pure food with honest weights and measures; homestead commissions, city planning, the highest care and efficiency in the administration of all hospital and penal institutions, probation and parole, care and protection of children and the

[7] Republican state platform drafted under Coolidge's leadership.

mentally defective; rural development, urban sanitation, state and national conservation and reclamation; and every other public means for social welfare consistent with the sturdy character and resolute spirit of an upstanding, self-supporting, self-governing free people."

That was written in the early days of the reaction against Theodore Roosevelt's liberalism. These social beatitudes would not have crept into a Republican platform much later than the middle of the second decade of the century. And they are set down here to indicate how so thoroughgoing a conservative as Coolidge was could give service, probably something more than lip-service, to a creed that soon was out-moded and forgotten by his party.

In his last term after his customary reelection as president of the senate, and his fourth term as state senator, Coolidge still lived at the Adams House, the second-rate political hotel where members of the legislature who came from west of the Connecticut were wont to forgather. He still paid a dollar and a half a day for his room, a little inside room looking out on the court. Probably he lived within his per diem. His ways were frugal. Possibly he saved a little. For with all his place and power, he never forgot that there are only a hundred cents in a dollar and that honesty is the best policy. He accepted no favors that could be coined into cash. He lived upon terms of intimacy and worked harmoniously with the political moguls who dominated his party, rich and powerful men for whom he did substantial favors that were worth much to them in financial returns. No scandal touched him. In Northampton he was still living in the duplex apartment that cost him twenty-seven dollars a month. And Monday was wash day at the Coolidge house. The little boys were in the public school.[8] They needed many things which they could have only through the practice of a close economy which Coolidge's Boston cronies might have called the meanest parsimony. Yet those were happy days for Calvin and Grace Coolidge. He had his self-respect. The father and husband was violating no conviction when he gave his blessing as president of the senate to measures which bolstered the rights and principles of the rich and well-placed. They and their proud circumstances were a part of the orderly universe in which he believed. Twice a week on the train he considered the picture of that universe mapped on the country-side and it was good. He would have brightened it up here and there with workingmen's compensation, with widows' pensions, perhaps with the minimum wage for women. Yet that was Brandeis' pet measure during those years. And Brandeis was "unsafe." Certainly Senator Coolidge would have done nothing to the footing stones upon which his benevolent plutocracy was founded. He would have changed no walls, lightened no really dark places. For in his soul he saw

8 Robert Woods' Biography.

the things that were as a part of a static cosmic order. And he saw clearly, eye to eye with Guy Currier, and with Frank Stearns.

It was in those days that the young men around Coolidge in the senate and the Amherst group in the state began to refer to Stearns as "Papa Stearns." Always by his own admission Stearns was an outsider in practical politics, but nevertheless he was a power for Coolidge. The young senator refused to announce his candidacy for lieutenant governor until two things had eventuated: First, until the legislative session was over and he could not be blackmailed or bludgeoned into trading his official votes for political support; and second, until he knew definitely in his shrewd heart that the higher powers in Massachusetts were with him. Then on the last day of the last session of his term as president of the senate, Calvin Coolidge walked into Frank Stearns' room in the Stearns drygoods store, stepped daintily across the floor rug to the massive desk, and when Stearns looked up he saw the sharply chiselled flinty Vermont face of Coolidge with his lips properly pursed and then watched him pull a folded note from his vest pocket and lay it on the big oak desk—without a word. The senator turned around and prissed out of the room before Stearns could open the note. It read:

> I am a candidate for Lieutenant Governor.
> Calvin Coolidge.

Also there was some backstage work. Shortly before he put this note on Mr. Stearns' desk, a delegation of his friends which included Thomas Green, State Civil Service Commissioner, called on Governor Walsh to urge Coolidge's appointment as a commissioner of public utilities. Walsh told them that they were fifteen minutes too late as he had just named another man for the position. Whether this was a move to keep Coolidge out of the race for lieutenant governor, or whether it represented a genuine desire of Senator Coolidge for a five or six thousand dollar job which in Massachusetts he might have held practically for life, no one knows. But the episode indicates how the wheels of fortune rolled a lucky number for the Northampton statesman.[9]

[9] Letter from Frank W. Buxton, Editor of the *Boston Herald*, to author after conversation with Thomas Green. Also confirmed by Governor, now Senator Walsh.

CHAPTER XII

And Tries On the Crystal Slipper

IN THE year 1915, Calvin Coolidge's name first appeared on a state-wide ballot in Massachusetts. In twenty years, he had come up the political stairs slowly but steadily. He was elected member of his precinct Republican committee in Northampton in 1896. In '98 he was elected from his ward to the city council. The next year and the following year he was twice elected city solicitor, widening his sphere a little; then was appointed clerk of the court, and in 1904 named as chairman of the Republican city committee. In 1905 he was defeated for the school board, in 1907 and 1908 he was elected member of the legislature, in 1910 and '11 elected and re-elected mayor of Northampton. We see him broadening his reputation and how carefully he has kept it clean! In 1912, '13, '14, and '15, four times he was elected state senator, the last two times by courtesy that he might serve his first and second terms as president of the senate. But his electoral campaigns were confined during all of these years to Northampton and its environs. Generally he led his ticket in election. Frequently his majority increased at his second election. He was popular with the voters. When the women came to the ballot box he still retained his percentage lead over opponents. He became "'the undefeatable." As president of the senate he had constitutionally much power. Also as luck would have it, during his years as president of the senate, he was the highest ranking Republican state officer in Massachusetts, where the State House was the habitat of Democrats largely as the result of the Theodore Roosevelt national bolt of 1912. Thus he came into state-wide fame. He was consistently conservative at heart. Politically, he was not narrow and rarely if ever vindictive. He helped Democrats and Progressives where helping them did not jeopardize the Republican party nor the vital interests of his conservative backers. He had few enemies and few but powerful friends and many casual supporters who were grateful for benefits actually received. Politically he was a substantial person who asked few favors and granted many. He had money in the political bank, little or none elsewhere. Certainly he was not "on the make" in politics. Apparently raw cash did not excite his desires. But when raw cash became property with a vested interest, it commanded his abiding loyalty and respect. Yet he was a poor man and one of the considerations [1] which urged him to become a candidate for lieutenant governor other than the fact that it was a stepping stone to the governorship, was the salary, two thousand dollars a year with an expense account which he never padded.

The accession of Frank Stearns, a publicity man par excellence, to his candidacy was valuable to Calvin Coolidge. It is notably curious to read in that particular portion of his "Autobiography," which recalls those days of his first candidacy for lieutenant governor, grateful tributes to Frank Stearns and Senator Crane.[2] He devotes a page of praise to Mr. Stearns and a succeeding page to an encomium of Senator Crane. In those days apparently Senator Crane and Coolidge were intimate. He writes that he often had breakfast in Crane's room at the hotel and doubtless they discussed pending matters of local politics.

"Although he had large interests about which there was constant legislation, he never mentioned the subject to me or made any suggestions about my official action." [3] Further Coolidge wrote that even if he had asked Crane how to vote on those interests "about which there was constant legislation," Crane would have told him to consult his own judgment and vote for the public interest. The fact that the two men never discussed Crane's personal interest in legislation is significant. They enjoyed that deep intimacy which made them understand each other. Following this paragraph is another statement about Crane's good taste in discussing legislation with the president of the senate. Coolidge wrote:

"He confirmed my opinion as to the value of a silence which avoids creating a situation where one would not otherwise exist." [4]

A perfect illumination of political thought transference! Indeed, across this radio-active silence most American politics is transmitted. Continuing his tribute to Crane, he wrote:

"In all political affairs, he had a wonderful wisdom and in everything he

---

[1] Statements to author by many Massachusetts friends of Coolidge.

[2] "Autobiography," page 114.

[3] But never a word to Guy Currier. Currier was a familiar spirit to Coolidge but never an intimate of Coolidge. This is true of all the forces of the invisible government of Massachusetts. Instinctively he knew their place in the political order. Instinctively he shunned personal contact with them, so far as was prudent.

[4] "Autobiography," page 114.

was preeminently a man of judgment who was the most disinterested public servant I ever saw and the greatest influence for good government with which I ever came in contact." And then indicating how deeply his affectionate faith abided in this elderly fatherly patron and prince, he adds: "What would I not have given to have had him by my side when I was President!"

No cold-blooded man could have written that. Only a sentimental friend could have recalled with such manifest emotion the days when Crane gave him a lift.

In the spring of 1915 after the legislature was out of the way, the Stearns publicity organization began to grind, the Crane political machine began to move, and all Guy Currier's Daughters of Joy, the same being the lobbyists for allied corporate interests in Massachusetts, began to sing for Coolidge. Stearns had one strong propaganda document, the address of Senator Coolidge to the senate entitled "Have Faith in Massachusetts," on the occasion of his election as head of that body. It was a notable utterance, his first public address to attract the attention of the people. It was in effect his platform in that campaign and in many others. It became an important chapter in his collected speeches published six years later under the title "Have Faith in Massachusetts." It will bear reading and rereading.

This senatorial address was circulated by the tens of thousands by the Coolidge committee, and because Murray Crane's machine had many cogs and all well oiled, the Coolidge platform fell into the right hands, good seed upon good soil.

Senator Coolidge made few speeches during that campaign and they were not important. He was opposed by an eloquent man, an accomplished orator—Guy A. Ham, an orator "as Brutus is"; an old-fashioned spellbinder. Ham had been in the legislature. He had talked all over the Commonwealth. It was believed that he had the backing of the Church which he probably did not have. But he was of the oratorical, emotional type, in that hour of emotional congealing from Theodore Roosevelt's ardor, exactly the type which was needed as a foil for Coolidge. Coolidge was slow of speech, addicted to statistics, given to a hard polished style without adjectives and without climax in his rhetoric. Also in that year the primary was fairly new. Men's names appeared alphabetically and Coolidge was near the top. He received a certain amount of froth votes but he won triumphantly. By all the rules of the game a rabble rouser should have beaten this repressed, cackling, flint-faced Vermonter with a Yankee twang. But Crane said: "That Yankee twang will be worth a hundred thousand votes" and it was. Because he talked poorly and did not try to spread the eagle's wings, he convinced those who heard him of his sincerity. The primary vote stood: Ham, 50,000; Coolidge, 75,000.

He was careful to explain in his autobiography that his campaign expenses were kept within the legal limit, $1,500, a sum "which was contributed by numerous people." Thus he declared that he was under no special obligation to anyone for raising this money. But he added in his life's story this information; that the news of his nomination "reached my father on the hundredth anniversary of the birth of his father." The sentimental old coot! As he dwelt in marble halls "with vassals and serfs at his side" Calvin Coolidge remained the Vermonter; his heart turned back to Plymouth and to the old grandfather giving him a finger as he led the child about the little village.

In his election campaign after the primary, he toured the state with Congressman McCall, the Republican gubernatorial candidate, making open-air speeches from automobiles during the day and every night finishing with an indoor rally. This was something new to him! And his kind was new to Massachusetts. His teammate, Governor McCall, told a story about their campaigning. Stopping over night with a Republican family they listened to the most amazing unabashed heresy! Admiration of Wilson, praise for the Wilsonian policies, of peace without victory in the World War, of tariff control by commission and, of federal regulation of trade under the trade commission; for all sorts of Theodore Rooseveltian doctrines promoted by the Wilsonian democracy. The host, his wife, his sister, his daughter, his mother, his maiden aunts, all joined in the heretical clamor. McCall, uneasy, was astounded. Riding the next morning to their appointment, McCall poured forth his astonishment and at the end of an excited sentence cried to Coolidge:

"How do you account for it?"

On they rode, and on for half an hour, an hour in complete silence. Then as they were approaching the environs of the village where they were to speak from the car, Coolidge opened his thin-pursed lips and ejaculated:

"Weemin!"

And further he saith not.

It was said of McCall and Coolidge in that campaign that McCall could fill any hall in Massachusetts and Coolidge could empty it. Yet he learned public speaking as far as he ever learned it, of a dry, cackling, statistical, factual sort, embellished here and there with close-knit, well-chosen language in short, snappy sentences. He was, for all McCall's polish, eloquence and erudition, probably the more convincing speaker of the two. McCall was the target of the Democrats. He had a lively opponent running for reelection, Governor David I. Walsh. McCall barely squeaked through to victory by a plurality of 6,313, while Coolidge being a Republican in a

Republican state, having no one in particular against him, not being on the target, got the party vote and was proud to relate sixteen years later that his plurality was over 52,000. This also was vanity; vanity of vanities; but put down in cold figures, it leaves the impression that Coolidge was a vote-getter extraordinary in Massachusetts. Later, perhaps he was; but not then. He merely polled the party vote. He knew the psychology of the Republican voter in Massachusetts. It wasn't luck. He was the type of man for the day. He personified his times—then and always.

Let us have a glimpse of what time and public life have done to the man in twenty years. A fellow citizen of Massachusetts who saw him in those days writes:

"There is no sense of quick and eager response. It is a pinched drawn face, not hard but anxious, the face of a man perpetually faced with problems too big for him. The face has New England written all over it." [5]

So he was always thinking—sitting in his office for long minutes motionless, gazing for hours out of the car window at the Massachusetts landscape.

"Not mind wandering," declares French Strother,[6] nor "casual consciousness, but hard, disciplined, purposeful thinking upon his problems—thinking ahead. . . . He is never hurried, never off his guard, never excited."

And Gamaliel Bradford remarks: [7] "This constant mental activity is in no way incompatible with difficulty and slowness of thought. . . . It may well be a result of such condition. But it is an undeniable characteristic of Coolidge the politician."

And Coolidge, the politician, from that hour when he became lieutenant governor, began to submerge and inundate Coolidge, the man. His law books grew dusty. Main Street, in Northampton was weaning him. Mrs. Coolidge and the boys were seeing him only at weekends. They knew few of his new friends. He acquired a bodyguard—Ned Horrigan,[8] a Boston police officer—who remained with him five years, and who insulated him from mortal contacts so that he could ruminate in his vast silences. He was becoming a habitual walker—he and Horrigan. He told a friend [9] that walking alone gave him "a sort of naked feeling." "The real lover of nature," observes Coolidge's friend, "like any other lover, wants to be alone with the object of his affection." So he and Horrigan hoofed it together over Boston; through the parks, along the Charles, in the lovely elm clad suburbs. Sometimes they chatted—if Coolidge wanted gabble, but mostly

---

[5] "The Quick and the Dead," by Gamaliel Bradford, p. 225.
[6] A contemporary magazine writer and later White House literary secretary.
[7] "The Quick and the Dead," p. 227.
[8] *Boston Evening Transcript,* January 6, 1933.
[9] Kenneth L. Roberts, "Concentrated New England," p. 51.

they trudged on in silence.[10] But the joy and the beauty of it, the stimulation to the senses in the imaginative chambers of the heart, never came to this man in those days of his youth. He walked that he might think—might work—without joy in it—without even a love of it. He walked and thought and worked because a stern and terrible New England god prodded him on, would not let him dally. In his walks he visited no galleries, heard no music, sought no noble statues. And though Ned Horrigan was Irish and knew where the fairies lurked in the dells, they followed no hidden paths to gay persiflage or shimmering sylvan delights. The plowman plodded his weary way—to work.

2

The office of lieutenant governor of Massachusetts was a worker's paradise or a lazy man's delight—as its incumbent looked at it. In most American states, the lieutenant governor is president of the senate, and while the senate is adjourned the incumbent is a sort of plug-hatted nuisance in state politics, a compromise between the Prince of Wales with his eye on a throne and an unwanted stepchild. But in Massachusetts, the lieutenant governor is a deputy governor functioning as administrative inspector. He is a member of the governor's council. He is chairman of the finance and pardons committee. It was part of the duties of his job to visit the state institutions. Moreover apparently Governor McCall, who was an orator, had many invitations to speak upon various formal occasions and generously accepted them all, delegating to his private secretary and to various members of his official family, and especially to the lieutenant governor as his representative, those places into which time and space prohibited him from projecting himself. So Coolidge went about the state making formal addresses. Naturally he could not announce policies, but he could defend the McCall administration and did. Robert A. Woods [11] declares that these speeches "were all written out by hand and delivered without manuscript. The audience heard nearly word for word what was given to the press." It was hard work. Whatever grace and oratorical power he achieved was honestly earned.

In the governor's council, Lieutenant Governor Coolidge lined up with the governor. Samuel W. McCall liked to be known as "the scholar in politics." That was Senator Lodge's particular moniker. Lodge's friends feared that McCall, having defeated Walsh, a rising young Democratic statesman, would ride a popular wave from the governor's office to the United States Senate. A cabal arose among the Republicans in the council.

[10] See Horrigan's interview, *Boston Evening Transcript*, January 6, 1933.
[11] "The Preparation of Calvin Coolidge, an Interpretation," 1924.

In his memoirs, McCall sets down the fact that Coolidge was with him in the council "even when the vote stood seven to two."

It is likely that Senator Murray Crane, who represented organization and the practical end of politics, would have been pleased to have McCall as a colleague, perfuming the political atmosphere of Massachusetts Republicanism with Bostonian erudition, aloofness and the salty Bostonian pickle of academic aristocracy which New England loves; just as pleased with McCall as with Lodge and perhaps a tickle or two more so. And Coolidge, being psychic sensed it. So McCall could always count on Crane's friend, Coolidge. Loyalty was one of Coolidge's sentimentalities, a source of pride.

The governor's council met once a week, which meant an official trip to Boston every week the whole year round for the lieutenant governor. While he was a member of the legislature, he could be at home nine or ten months in the year, save when he was campaigning; but what with serving one day on the executive council, making speeches, inspecting state institutions, considering pardons, wrestling with tax problems, Coolidge found that Northampton was fading from his cosmos. He was living —and certainly saving—on his salary of two thousand dollars a year and expenses. He took a law partner for his dwindling law practice, and though he did not foresee it then, in 1916 Calvin Coolidge was done forever with the active practice of the law.

He usually spent his weekends in Northampton, but Monday morning he was on the wing.

"It was a familiar sight, the Lieutenant Governor, sometimes with Mrs. Coolidge, and on occasions with the boys in a day coach on the railroad that runs all the way through rural regions, beautiful now with apple blossoms, now with autumn leaves, now with snow, the train seeming to stop at everybody's garden gate. Northampton was at the end of the line. The conductor was the next door neighbor. Mr. Coolidge read the papers; Mrs. Coolidge knitted. For years she had knitted the boys' stockings. As a member of the Red Cross, and this was the only organization to which she belonged aside from the church, she knitted fifty pairs of socks in addition to sweaters and helmets during the winter when they were most needed. She also secured subscriptions and kept the chapter journal." [12]

Here is another picture of him in those days in the midst of the century's second decade, a picture drawn by the Rev. Roland D. Sawyer, a colleague in the legislature and an associate in the state government of Massachusetts. He writes: [13]

"I had been introduced to him the week before and had not been

[12] "The Preparation of Calvin Coolidge" by Robert A. Woods.
[13] "Cal Coolidge, President."

favorably impressed by his cold exterior. The next morning I was walking down the car and merely nodding to Mr. Coolidge when he said: 'Representative, I see you have a road bill in. Sit in here a minute, perhaps I can help you a little on it!' Gladly I sat in and in a few words Mr. Coolidge explained to me, a green legislator, a member of the opposition party, the methods of legislation and the way of getting favorable consideration. The brief and well-chosen sentences finished, Mr. Coolidge turned his head to look from the car window and to smoke his stogy."

There he is—all flinty-faced, sugar-cured and hickory-smoked, the word-less Yankee joss sitting cross-legged in the cosmos, a stogy smoking seer, but always New England, always the country town man. Let us go into his house with Mr. Woods: [14]

"Welcomed at the door by Mrs. Coolidge. She has just been downtown getting something for the boys' lunch. Here are some of the properties found in the background of the family life: a book-strewn table, rug of standard pattern, framed photographic copy of Sir Galahad, photograph of Plymouth Notch, Vermont, a Bible picture, sewing bag, a lidless phono-graph with no operatic records, bric-a-brac, bay window with blinds tied back, the ice card, parchesi board, hooks with the boys' hats, baseball bat, gloves, dog, gas stove, fireplace with wood ashes, and over it this sentence—

> " 'A wise old owl lived in an oak,
> And the more he saw the less he spoke.
> The less he spoke the more he heard.
> Why can't we be like that wise old bird?'

"Mrs. Coolidge is a companion to her boys. John, the elder, is like his mother; Cal, Jr., like his father. John plays the violin, is mechanical, and has built for himself a quite recognizable automobile. Young Calvin essays only a mandolin and is interested in erecting a flagpole. Both handle the typewriter. They go to public school where Mr. and Mrs. Coolidge believe boys learn to work out their own salvation. They not only help their mother about the house but undertake various services on a business basis for the neighbors,[15] contributing a part of the income to the church. This is in accordance with their father's views. . . . At night their mother reads "The Swiss Family Robinson," "Robin Hood," and "Ivanhoe" to them for fifteen minutes after they are tucked in.

"Mr. Coolidge is the kind of a father who has the absolute respect of his sons but he does not easily become their playmate; occasionally reads to them, takes them on walks; has even gone skating and fishing through the

---

[14] "The Preparation of Calvin Coolidge."
[15] This means they had chores, mowed lawns, cleaned out furnaces, swept porches, ridded up yards and did all sorts of little odd jobs.

ice with them. He makes common cause with his wife in some exceptional ways. He enjoys going shopping with her. He will assist in the selection of gowns but not of hats. He is quite ready to join her in getting the dinner and can wash the dishes if occasion demands."

Here is a perfect middle class home north of the Ohio River anywhere from Philadelphia to Honolulu. This is not the log cabin of the pioneer, but it is close to the American soil and dear to the American heart of a generation which no longer knows the rigors of heat and cold, of long hours at labor, of forest terrors and prairie dangers. Here is a picture of secure middle class America in the first quarter of the twentieth century.

In the middle of the century's second decade, the Theodore Roosevelt rainbow was fading. The European war had stimulated industry, business began to boom and Calvin Coolidge in Boston felt the impulse of a new prosperity. It seemed no part of his concern as president of the senate, or as lieutenant governor that the allies were placing tremendous orders for steel and for farm products with America. These orders seemed a godsend. The Wilson administration was insisting upon neutrality. Short term loans to the allies were being made in Wall Street to pay for the goods and munitions they were buying. Even while he was in the governor's chair, far beyond his horizon, the banks of Boston and New York were beginning to find difficulty in financing their industrial friends. Submarines were cutting off imports. Foreign securities in gold to pay for American goods and keep the American steel hot were no longer available. It was announced in Washington by the Wilson administration that our neutrality would forbid long time American loans to the allies. But the French government persuaded the National City Bank of New York to find a way around Wilson's ban on foreign loans. Secretary of State Bryan, who was adamant against the French plan, resigned. Robert Lansing came in. Under the advice of Colonel House, Wilson's alter ego, the bankers invented an international commercial credit hocus-pocus under which American bankers bought bonds to pay themselves for the goods "sold to the allies." [16] Those bonds began to pile up. As the fortunes of the allies went down, America became the creditor nation of the world. Even while Calvin Coolidge was sitting in the lieutenant governor's office, in Boston, the allies became heavily our debtors. If the allies failed, America was due for a commercial and industrial panic.[17] This situation alone did not bring us into the World War. Our creditor position was merely one of many contributing causes. But Calvin Coolidge ten years later was to face our position as the world's creditor, and facing it, characteristically, with common sense when a super-sense was needed, lead his people into the second

[16] Charles A. Beard, "Peace for America," March, 1936.
[17] See published reports of the Nye Committee, Washington, 1935–36.

phase of the great war, the economic debacle of Christendom. Of course this is looking ahead. Lieutenant Governor Coolidge, in Boston, apparently was busy inspecting state institutions carefully and with intelligence. He was making speeches where the governor sent him, sensible, essentially Republican speeches proclaiming the conservatism of New England.[18]

Calvin Coolidge surely saw ahead as far as the governor's office. He had every right to cherish his gubernatorial ambition. First of all he was prepared for it in the best sense. He was closing two terms as lieutenant governor, which had given him administrative experience, familiarity with the state institutions, its servants and the ways of its politics. He had learned industry, tact, capacity for team work, a knowledge of the strength and weaknesses of men, a knack for honest compromise with dishonest or visionary adversaries, and a kindly understanding heart. These qualities Coolidge had shown. In 1917 he had come to the threshold of the governor's office. So Murray Crane's organization backed Coolidge. Guy Currier's forces of business conservatism, having followed Crane when he picked Coolidge for lieutenant governor, had no fault to find with Crane's gubernatorial choice. Again they lined up for Coolidge. It was at this time that Lodge and his group first became Coolidge conscious.

Frank Stearns precipitated the episode which brought Coolidge, as a state leader, definitely to the attention of Lodge's temple pharisees. It happened thus: At the Republican national convention in June, 1916, at Chicago, Senator Lodge at the head of the Massachusetts delegation was chairman of the committee on resolutions, not at the moment an important place. But Lodge was none the less powerful in the convention. Sitting half a mile from the Republican convention, was the Progressive national convention, threatening to nominate Theodore Roosevelt over Roosevelt's protest, again to inject a third presidential ticket into the election of that year. Among Republicans at Chicago, there was much casting about for a compromise candidate who would satisfy the Roosevelt progressives. Theodore Roosevelt himself had suggested Lodge, and then facing revolt in his Progressive following Roosevelt had even suggested United States Senator Weeks, of Massachusetts. The Progressives were aghast at Roosevelt's suggestion of Lodge, disgusted with the name of Weeks. Then into the holy of holies, where the leaders of the Massachusetts delegation sat, walked Frank Stearns and suggested the name of Lieutenant Governor Calvin Coolidge for the Presidential nomination of the Republican party. Stearns was a rich man and a liberal giver; as such

[18] In the midst of this Eden of Massachusetts conservatism appeared the flaming sword. Louis D. Brandeis, who had for ten years and more been the leader of constructive liberalism in Massachusetts, and who had taken his place as one of the leaders of liberalism in the nation, was appointed by President Wilson to the Supreme Court of the United States. Massachusetts, politically and socially, was shaken to the core.

he was entitled at least to silence, when he had spoken. But when he had left the conference, it broke up with ribaldry. Men went to Lodge in the holy arcanum of his bedroom and cried:

"Whom do you think Frank Stearns has suggested for the Republican Presidential nomination? Calvin Coolidge!"

And to the reiteration of the name in questioned bewilderment the reply reechoed from Lodge: "Calvin Coolidge! My God!" [19] Whereupon laughter roared through the room.

In the meantime, Calvin Coolidge ran for governor of Massachusetts in the Republican primary and easily won.

[19] Conversation with Henry Field of Northampton. But fate had her rod in pickle for Senator Lodge. Even in that proud day in Chicago his qualities were undermining his aspirations. In his "Across the Busy Years," Nicholas Murray Butler writes:

"The crowning revelation of Lodge's political character was made by his conduct at the Convention of 1916. In an impassioned speech he placed the name of John W. Weeks of Massachusetts in nomination for President and cast his vote for him on the first ballot. On the second ballot he deserted his candidate and voted for Theodore Roosevelt. When the third ballot was taken he remained outside of the Convention Hall because of excessive modesty, since he hoped that by reason of Theodore Roosevelt's intervention the nomination might come to him. A few minutes afterwards when Hughes was nominated, Lodge appeared upon the platform and made a speech in high praise of him. This rather swift boxing of the political compass was the subject of well-nigh universal comment and universal hilarity."

CHAPTER XIII

Fate Begins to Shift Scenery for a New Day

But Frank Stearns, y-clept Lord Lingerie, lost no ardor because of the rebuke from the leaders in Lodge's stronghold. He returned to Boston to take up the promotion of the Coolidge gubernatorial candidacy where he had left it. He and Murray Crane and Guy Currier had their plans well developed before the Chicago Republican national convention. It was Stearns' part of the work of the triumvirate to impress the newspapers with Coolidge's peculiar talents. Stearns spent his money giving dinners in Coolidge's honor, to which he invited the right sort of people. They were social affairs. Coolidge was never a hale fellow, but he did appeal to the sort of men who would gather around a rich man's board. Here Stearns enlisted the services of Dwight Morrow of the New York House of Morgan and at one of those dinners, more or less crystallized around the Amherst crowd in Massachusetts politics, Morrow presided. He had invited Colonel John C. Coolidge, Calvin's father. "To prove to you that Calvin is a chip off the old block," declared the toastmaster, "let me read to you Colonel Coolidge's reply:

"Gentlemen: Can't come. Thank you.
"John Coolidge"

At these dinners Frank Stearns was host, but never speaker. As an expert in the new profession of political publicity, Mr. Stearns planned to have at his table men of power with an interest in politics. They represented the ruling class. Of course in the Massachusetts legislature and in the Commonwealth at large many men realized the evils of the times. A few men sought to protest. Fewer still sought to make their protest effective by sponsoring corrective legislation, some of which was enacted. But the more or less predatory powers of Massachusetts knew their man and appreciated his work. Honesty compels the admission that their appreciation of his service was probably deeper than his own realization that he was serving them. Coolidge bent no limber knee to mammon, consciously sought few favors of its priesthood, joined few of its cabals, rarely moved in its intrigues. And that in a day when cabals and intrigues were common and somewhat necessary to promote the plans of the masters of Massachusetts finance. All Coolidge did to promote the friendship of the financiers was to follow his natural bent and belief that life was a Garmanian project in which wisdom, foresight, the acquisitive faculty and the constructive imagination of accumulators if given free play would produce in the alchemy of God's moral government of a complicated universe, a fairly perfect world. Garman's thesis still held good.

The drag in American business that preceded the war, seems now to have been due to financial preparations being made on the other side of the water for the war.[1]

[1] After the mild reaction in business in 1912, production of goods seemed to have caught up with demand for goods. But whatever seeming manifestations of underconsumption appeared were due to causes not apparent at the time.

The farm problem, as it relates to the equilibrium between agriculture, manufacturing and mining was at this time probably in its beginning. The farm was in as good a position as it had been in the history of the country. In the nineties there had been a relative excess of agriculture (so farmer John Coolidge had to scrimp to send Calvin to college) due to the overexpansion of farming in the West, following the building of the transcontinental railways. But between 1897 and 1914, manufacturing had expanded, and agriculture had not, seriously. The fertile, free western land had been taken up. Agricultural prices improved then compared to prices of manufactured goods. But in the years immediately following the World War, a land boom appeared, notably in Iowa and Illinois. Farm lands were bid up high, which incidentally piled up the mortgage debt. In many cases, after the farmer had strained himself to borrow for buying more land, he was short of working capital for operating the land.

Which reveals one of the worst phases of the farm problem—the farmer's financial technique. As farming population grows and further subdivision of the land is undesirable, inheritance involves a growth of mortgage debt as one heir takes the farm and gives mortgages to the other heirs. Hence farm debt multiplies.

I quote the following from B. M. Anderson, Jr.'s book, "The Effects of the War on Money, Credit and Banking in France and the United States," page 143:

"The outbreak of the war found the United States with a very safe credit position. Trade was dull, merchants and bankers were moving under shortened sail, no great new enterprise had recently been undertaken and the general situation was thoroughly solvent. There had been indeed no real boom since the panic of 1907. The 'business cycle' as we know it, is an alternation of prosperity, crisis, depression and prosperity again. But following the panic of 1907 there had been a steady drag. In 1912, there was a substantial rise in wholesale prices followed by a substantial setback in 1913—but there had been nothing following 1907, that could be called a real boom.

"In retrospect, it is possible to offer an explanation of this, though some shrewd observers, as Mr. A. D. Noyes, had seen the explanation before the outbreak of the war. For a good many years before 1914 Europe had seen the war coming. The Banque de France as early as 1899 began its policy of accumulating gold, primarily as a war chest. Between 1899 and 1910 the Banque de France increased its gold re-

So Calvin Coolidge running for governor of Massachusetts in 1918 went about within the periphery of his cosmos virtuously pursuing a decent ambition to serve his gods and to help mankind while the noble triumvirate Crane, Currier and Stearns carried him inexorably starward through a booming land where war babies were smiling cherubically their golden benediction upon him as he wended his upward way.[2]

In the summer of 1918, Coolidge was nominated by the Republicans for governor of Massachusetts without serious opposition. Here is the story of the political mechanics of his gubernatorial nomination. Until the early autumn, he had not announced his candidacy. But the lieutenant governor had been Governor McCall's willing errand boy. Time and again Governor McCall left Massachusetts with the lieutenant governor in charge, as acting governor, and Coolidge did many disagreeable tasks cheerfully. But, as he himself has remarked, he was careful not to encroach upon McCall's domain. In his Biography, discussing those times and days, referring to his relations with Governor McCall, he wrote one of his most illuminating sentences: "While I have differed with my subordinates, I have always supported loyally my superiors."

Stated cynically and therefore not entirely truthfully, this means that Calvin Coolidge always knew on which side his bread was buttered. He did not serve McCall for naught. Now incidentally, America in the autumn of 1918 had been in the war a year and a half. The influenza epidemic was raging; they could not hold a Republican state convention in Massachusetts. Ordinary political meetings were abandoned. Governor McCall desired to go to the United States Senate. There was some feeling that he should continue his gubernatorial task. But McCall understood Coolidge's ambition to be governor. McCall realized that Coolidge had not announced his gubernatorial candidacy out of deference to McCall. The governor apparently desired the excuse of opposition to retire gracefully as much as Coolidge desired to be governor. So McCall took his lieutenant governor aside and told him to announce. Coolidge declared later that no one knew that McCall had told him to run and "some supposed I would run against him." But Coolidge was not of that stripe. McCall started to run for the Senate but for some strange reason gave up the race in the midst of his primary campaign and John W. Weeks entered the senatorial race. It was a sad campaign in Massachusetts. President Wilson had made his unfortunate plea for Democratic support which antagonized the

serves by 75 per cent, but increased its discounts and advances during the same period by only 5 per cent. In general, for several years before the war, European investors were becoming more cautious in their purchase of American securities and the United States was increasingly obliged to provide its own capital for industrial developments."

[2] Incidentally the automobile industry was making a new frontier as real for that decade as the Missouri Valley was in the seventies.

West but was not a sufficient handicap to defeat David I. Walsh, the Democratic leader, who overcame Senator Weeks. And Coolidge, who had been unanimously nominated for governor, was elected by only 16,773. His personal political adviser in that campaign and for many years thereafter was Tom White, a fellow member of the legislature whose loyalty and acumen attracted Coolidge. But he was wise on his own account—all but psychic in his sense of ballot estimates. Tom White recalls [3] that in Coolidge's first gubernatorial election, White was in the *Boston Herald* office the night after the election when the returns were coming in. Naturally the political wiseacres of Boston gathered there and it was obvious by nine o'clock that Senator Weeks was swamped. Coolidge came in, asked White what his average showed. White stepped to a telephone booth, called the *Globe* and found that they were estimating Coolidge by 15,000. White's averages showed Coolidge ahead 10,000. He told Coolidge this. Coolidge turned and left White and paced circumspectly across the floor of the editorial room to the managing editor's desk, took a paper from his inside coat pocket and handed it to the editor:

"Here's my statement!"

"What statement?"

"My statement thanking the people of Massachusetts for electing me!" And then he trotted primly out of the office.

White says that in that campaign Weeks's managers were not telling Weeks the truth about his status. Women, labor and prohibition were weakening him. Coolidge knew this and avoided tying up with Weeks as much as possible. He told Tom White, his manager: "Don't go down there until I tell you"—meaning Weeks's headquarters. Finally two or three weeks before the election Coolidge said to White: "Guess you better go down now." White went and told Weeks's people the truth and promised as he was making a final roundup for Coolidge he would report on Weeks, which he did the Saturday before election. It was hopeless. It was out of that background that the episode in the *Herald* office grew. It is set down here to indicate how canny Coolidge was about people and how profoundly he knew politics. No lucky innocent was he. He knew the game from the super-boss back of all the banks, who ruled from the top, all down the line to the elector in the booth alone with his lead pencil and his God. A year before Coolidge had been elected lieutenant governor by more than a hundred thousand. The slump was not personal. Probably it came out of the turmoil of the times, a revolt against the war profiteers, and partly because the Republican organization in Massachusetts, potent in getting votes, was unable to function because of the war and the influenza epidemic, and because in an emotional moment when war passions

[3] Conversation with author, June 1935.

were deeply stirred, Coolidge, an utterly unemotional candidate for governor, unable to rouse the rabble, was running with a senatorial teammate who in the popular eye represented all the sins of American high finance; a bond and stock broker with a most conservative senatorial record.

Here is an odd and characteristic footnote to this story. Up at Plymouth, Father John Coolidge who had always taken the *Boston Globe*, a Democratic newspaper, continued in this campaign to take it as his only source of news of the Massachusetts campaign and he had no idea that Calvin would be elected governor.[4] It would not occur to any Coolidge to change his newspaper subscription to get a new slant on political events. The *Globe* was good enough in the nineties, it was good enough in 1918. Conservatism was bred in the Coolidge bones.

Three governors of Massachusetts had come and gone since Calvin Coolidge came to the state senate. Coolidge persisted. Democrats infested the State House. To restore Republican harmony after the split of 1912, a cautious, cold-blooded leader was needed. Coolidge was on the whole the most cautious and could be the most cold-blooded of Massachusetts Republican leaders in effecting and preserving harmony in his party. He had no enemies to punish, few friends to reward whose claims would embarrass him. He was ever a grateful man, yet it was one of the minor marvels of his life that his gratitude rarely led him into trouble. He was as parsimonious in asking favors as he was meticulous in repaying them. As a candidate for governor this was his stage setting. And as just indicated, he was ever loyal to his superiors. Here were his superiors: America for the first time in a generation was beginning to inhale the dangerous gases of war profits, from a major war. The whole eastern seaboard of the United States was suddenly transformed into a market place for munitions and for the bones and sinews of war. War profits made war politics. By 1916, the men at the head of the major financial houses in Massachusetts, Kidder, Peabody and Lee, Higginson, one vaguely supposed to represent the Morgan interests, the other popularly presumed to be loosely allied with the Rockefeller financial group of New York, keenly and definitely realized that Calvin Coolidge who was "Crane's man" was also a safe man. Crane was no political ogre. Crane was the political clearing house for his satellites. Also Crane represented them at court; court being the powers that be, the financial powers, these same Lee, Higginsons, Kidder, Peabodys, and other smaller bankers who were acquiring control of New England industry. The wires and cables which Calvin Coolidge saw as he rode to and fro between Northampton and Boston from 1907 to 1917 had become industrial veins and arteries feeding the capital structure in the financial heart of Massachusetts.

[4] Conversation in 1934 with Robert O'Brien, Chairman, Federal Tariff Commission.

By 1917 the great industrial families west of the Connecticut River were losing control. In the average mill office that Governor Coolidge visited beyond the Connecticut he saw on the wall the picture of the founder. Typically he was a sturdy, old, chin-whiskered robber-baron born in the eighteen thirties or forties who fought his way up until he owned or built the mill. His son was of his kind. The old baron and his sons defied the federal government which sought to regulate and control New England industry. But alas! The grandson of the house too often became addicted to polo and yachts. Being hauled before the Federal Trade Commission during the Wilson administration, the grandson chattered. Congressional investigation overawed him and he turned to Murray Crane or to the bankers for refuge. And so Crane thrived. He tied his overlords and lieges into trusts and mergers, and made them safe against the ravages of the industrial upheavals of the hour and time!

The commercial revolution changing Massachusetts, New England, indeed America and the world, was wresting control from individual captains of industry and passing it to the financial oligarchy, interstate and international. The old industrialists looked longingly back at the days of tariff subsidies and could not understand the new internationalism which Murray Crane and his banking friends were proclaiming. The bankers allowing for certain leeway in the control of industry—shortening the hours of labor, safety devices, some sort of social insurance, protection of women and children—felt it was better to standardize these things through the various mills which they controlled, under the leadership of young efficient engineers who went about managing half a dozen works, than to have wrangling in half a dozen mills about half a dozen standards. Politically, Calvin Coolidge did for the coördination of the middle class and its rulers in Massachusetts a service not unlike that of the bright young engineers who were taking over amalgamated industries. The middle class was producing survival qualities against the power of the acquisitive social forces which, lacking a better word, we call plutocracy. This plutocracy was not the master. Politics which the plutocracy assumed to control, more or less controlled the plutocrats. Back of the acquisitive forces is the electorate, the workers, the farmers, the professional classes, the small business men, the creative forces of society. The top-lofty rulers rule only as they are ruled —only as they submit to the vague, muffled, instinctive, inarticulate voice of the people who have in the end actually the power of veto in the government of the land. Calvin Coolidge was middle class, certainly never a rebel against possessive forces of society; but also quite certainly he was never the conscious protagonist of the predatory powers. One searches in vain through his speeches and writings, one looks futilely into any deed or record he made to find evidence that he was conscious of any other class than the middle class. He associated in a friendly way but never upon terms of anything like social intimacy with the very rich. In Northampton the mill hands and casual workers represented the only proletariat America had. As chairman of the committee which investigated the strike of the Lawrence Textile Mills workers, Senator Coolidge had first-hand knowledge of an incipient proletarian revolt led by Big Bill Haywood, I.W.W. Communist, an experiment in class consciousness. In Boston he was just another governor who had no social standing with the Brahmins nor with the class-conscious aristocracy. Casually he knew who the aristocrats were. He even met them officially, and had no quarrel with them. For his mind definitely seemed to reject the presence of classes in his country. The mobility of the citizen, the fact that man moved easily upward and sometimes slipped downward, that heredity was no mark of social distinction, seemed proof to him that America was without class distinction. As for those who enjoyed special privilege by reason of birth or by reason of their endowment with exceptional qualities, Coolidge regarded them as mere middle class, plus—God's problem and not his. Of course, he was not the ideal democrat. But playing the game with Crane and Currier, with Stearns and Jim Lucey, with Main Street in Northampton and Beacon Street in Boston, with Tom White and with Charlie Innes, a satrap of Boston politics, and with Smith College and Jonathan Edwards Congregational Church, how he did put into vital flesh and blood the common impulses and realizable ideals of his fellow Americans in his generation. Through him and his kind the middle class waxed fat and governed the world.

CHAPTER XIV

Our Hero's Stage Is Set

THE Friday after the election, Governor-elect Coolidge went to Maine for a few days' rest. He was awakened Sunday night to hear that the Armistice had been signed. Sadly he trekked back to Boston to join in the celebration; scarcely a heroic figure, rather dolefully and dutifully following the spotlight officially accorded to him amid the general rejoicing. Certainly his election as governor did not go to his head. For a decade and a half he had been occupying that dollar and a half room at the Adams House, the forgathering place of western Massachusetts statesmen, a dingy, political hotel. When he was elected governor, he ordered the next room, and paid two dollars and a half for the two rooms. There he lived in solemn state without an executive residence. On high days and holidays Mrs. Coolidge came down to Boston. But for the most part she remained in Northampton with her boys. She was little known in the Boston of those days. There in those two connecting rooms with a bath, Governor Coolidge lived in Jeffersonian simplicity. In the right-hand lower compartment of the washstand he kept the harmless necessary bottle of Bourbon whiskey with which to regale his thirsty visitors after the custom of the day. It was in that throne room and from that washstand that he took the bottle to serve drinks to the visitors 'the night his election was assured. And there arose the famous story probably apocryphal of the incoming visitor who took his drink and noticed sitting on the bed an old friend to whom the Governor offered no drink. And when the later visitor expostulated, Coolidge replied:

"Bill's already had hisn!" The retort courteous of the representative of a parsimonious Yankee people!

Massachusetts makes a pageant of the inauguration of her governor.

Then uniforms appear. Much punctilio is handed down from colonial days. So on January 1, 1919, Mrs. Coolidge and the two boys and Colonel John Coolidge came to Boston. They were seated in the governors' section of the gallery. Near-by was Frank Stearns, surrounded by Amherst friends.

Northampton tradition declares that Coolidge's first high silk hat appeared when he became president of the senate. Styles in silk hats changed but little. He wore that hat as governor when he appeared at the State House to present himself to the outgoing governor. He walked nimbly across the room to the desk where Governor McCall, surrounded by his military staff emblazoned beautifully—for it still was a war day—stood while the new governor appeared. A Committee from the joint session of the legislature came into the room to inform Governor-elect Coolidge that the legislature was in session and awaited his pleasure. And repeating the formal words that had been used for more than a hundred years, the new governor told the committee that he would attend forthwith to take the oath of office and communicate his views to the legislature on public questions. The line formed with the governor and the governor-elect walking together and appeared at the door of the convening legislature. The presence of the governor-elect was announced by a silk-hatted, mace-bearing sergeant-at-arms in colonial knee breeches, an imposing figure. Back of the governors were the members of the executive council, the justices of the supreme court, accompanied by the sheriff of Suffolk County in full uniform—and what a uniform—and wearing his dress sword; a personage worthy of the panoplied pageant with which Cleopatra, sailing down the Nile, met Mark Antony. Following the sheriff were the judges of the superior and municipal courts, officers of the regular army and navy detailed in Boston, foreign consuls, United States senators, congressmen, the presidents of Harvard and the other Massachusetts colleges, meticulously clad—silk-hatted, frock-coated, dressed like oiled and curled Assyrian bulls for social slaughter. The presiding chief justice of the Supreme Court raised his hand, Coolidge's hand went up. He stood there stiffly aloof, of medium height for Massachusetts men, slim, trim, lean but not gaunt, for he was in his early forties and his face was slightly full. Twenty years in politics, watching other pageants, had steadied his nerve. His uplifted hand did not tremble. He spoke the words that swore him to his duties with that clear cut nasal voice which had made him ten thousands of votes. He was like any of the long procession of governors, nearly a hundred who had stood there for a moment under the white light that beat upon the gubernatorial substitute for a throne, and only one thing gave him distinction. He opened his eyes wide and looked around. There in that open-eyed glance he revealed the unusual man. It was in that day that Barrett Wendell, a high caste Back Bay Brahmin, described Calvin Coolidge in these words:

"A small, hatchet-faced, colorless man with a tight-shut, thin-lipped mouth; very chary of words but with a gleam of understanding in his pretty keen eye."

All that Coolidge was then or would be flashed in those eyes. They were ever the beacon of his highest intelligence, rising out of the kindliest recesses of his heart. Until one saw and understood those eyes, Coolidge remained a puzzle.

He delivered an unimportant but entirely adequate common sense inaugural message competently and the scene dissolved. He had reached the peak of his ambition.[1] His inaugural message stressed the need for public health, education, and the "right of the people to be well born, well reared, well educated, well employed and well paid." He emphasized the need of keeping government expenses as low as possible, to which latter need he gave his best endeavor. To have paid serious attention to the public welfare would have required the use of taxation as a weapon of distributive justice, a political device which always filled him with distress and alarm.

After the inauguration, when the crowd had gone and Mrs. Coolidge and the boys had brought his father into the office, the new governor sat down in the governor's state chair and remarked, with his wise eyes open, looking across the years: "Well, I guess they won't turn us out this time," [2] and so began his upward journey at that turn of the road.

But now consider an odd and characteristic thing which the new governor did: He discovered in going over matters that his friend, Frank Stearns, was a member of two honorary unpaid state commissions. Stearns had been appointed by Governor McCall as a member of the Pilgrim Tercentennial Commission, and as a member of the commission to welcome home the returning Massachusetts soldiers from the World War. When he met Frank Stearns after acquiring this information, he said:

"I think it your duty to resign from both places!" No other explanation, no word of regret, no trimmings on the order. It would have been easy to explain to the voters Frank Stearns' place on these commissions, serving under the man whom he had so heavily supported for the governorship, but resignation was simpler than an explanation, so he required a resignation. He was just that primitive—playing safe. Frank Stearns used to say that he never could get a direct reply from Coolidge in conversation. If he required one, he wrote him a letter. The relation of these two men, both more or less silent, remained upon the simplest terms through the years. Sometimes Coolidge would peremptorily send for his friend to come to the governor's office. He had nothing to say. The two men smoked their

---

[1] See "Autobiography," where this fact is stressed.
[2] See page 63.

cigars in silence. Possibly Coolidge worked, probably Stearns looked out the window. When the cigars were finished, they made some innocuous remark about the weather and the visiting was over. Probably in addition to shyness Coolidge had "a sort of naked feeling" [3] when he was alone. Time and again such spiritual congress occurred between them. Two farm horses in a fence corner,—haunch to neck—flapping flies with their tails, must have the same spiritual satisfaction that gave these two men their sparse delight. Yet undoubtedly the bond of affection between them was strong and certainly it never was strained from the beginning of their contacts to the end. Before Coolidge went into the governor's office, Frank Stearns had envisioned his friend as President of the United States. Soon after, the book "Have Faith in Massachusetts," which had done such splendid service in state campaigns, was reprinted by the hundreds of thousands and started out across the land to politically minded men who might be in a position some day to help Calvin Coolidge rise.

In those gubernatorial years, Coolidge moved about Boston rather an odd, lonely, natty little figure. Sometimes it was his wont to climb to the editorial rooms of the Herald, the leading Republican organ of Massachusetts, paw over the Massachusetts newspapers, talk with the editorial writers, read the Associated Press dispatches as they were coming in. Robert O'Brien, who was then the editor, remembers him in the office one evening after a mayoralty election. The Independents of Boston had made heroic efforts to overthrow the Democratic machine of Boston, a little New England Tammany, and had succeeded that day in electing a rather unusual mayor, Andrew J. Peters. The Herald had contributed what Republican influence it could command and was proud of itself. As the precinct returns came piling into the office, and Peters' lead increased, Coolidge stood silent, mulling over the figures. The men in the Herald office were exalted and of course excited. When it was certain that Peters had won, Coolidge turned and went over to the editor's desk with this precious comment:

"Be careful what you say tomorrow about Peters. Remember you are just buildin' up another Democrat!" And so paced grimly out of the office into the street where all decent Boston was rejoicing in a victory for good government.

He was to meet Mayor Peters in a few months at the turning point of the Coolidge career, when the tide that leads on to fortune was in flood.

The first official act of Coolidge, as governor of Massachusetts, was to approve a legislative appropriation of ten thousand dollars to defray the expenses of the official welcome home when the Yankee division returned from France. On a bitter gray drab day in late April, this Yankee division

[3] Kenneth L. Roberts, "Concentrated New England," page 51.

came home. A cold wave was coming; a biting north wind was blowing. The sky was overcast but occasionally the sun broke through the fast flying clouds to glint on the bayonets of the marching men. Five New England governors stood beside Governor Coolidge: Governor Carl E. Milliken, of Maine, Governor John H. Bartlett, of New Hampshire, Governor Percival W. Clement, of Vermont, Governor R. Livingston Beeckman, of Rhode Island, Governor Marcus H. Holcomb, of Connecticut. Coolidge stood fourth in the line, immovable as the Bunker Hill Monument, and as speechless. People stamped and threw their arms about to keep warm. Coolidge was rigid. Others sat huddled along Commonwealth Avenue too cold, too numb to move. Men were selling newspapers to be read and wrapped around the feet and legs of the buyers. It wasn't quite cold enough to congeal the cheering. Enthusiasm mounted and at the end of the afternoon everyone was hoarse, and there were big gaps in the stand along the Avenue.

For five hours the Yankee troops passed the reviewing stand where the five New England governors stood beside Governor Coolidge, the Massachusetts host. On and on marched company after company, regiment following regiment, upon the heels of brigades, one division and another and still another. The Governor of New Hampshire remembered [4] that he stood "by Coolidge's side five hours and he spoke to me just once between the time he greeted me and the time he courteously said goodbye. When we had stood half that time, Governor Coolidge turned to me and said in his drawl:

" 'Governor, I think you will find that if you put one foot on the rail and lean in my position a while and then change to the other foot, you will find it will rest you.'

"I tried it, and sure enough it was a relief. But I could not, and cannot now, comprehend a man who could stand five hours and have nothing else to say."

It was three months later, in the early spring when Governor Coolidge was sloshing across the Common in his rubbers, that he met General Edwards, who had been having trouble to avert a reprimand for his incidental remarks about some army matter.

"Hello chatterbox!" said the General approaching the Governor.

"Well, General, I notice what I don't say gets me in less trouble than what you do say," retorted the Governor as he passed on to work. His habit of silent cerebral cogitation made him conspicuous sometimes, but never notorious.

[4] "Cal Coolidge, President" by Roland D. Sawyer, Boston, 1924.

2

He felt deeply the thrill of patriotism that all America felt when the victorious heroes came home from across the ocean. He comported himself with the dignity which moves men who feel genuinely. In February, 1919, Governor Coolidge welcomed President Wilson home in Boston when he returned in recess from France to present to America the first draft of the Covenant of the League of Nations. The Mechanics Building was crowded. It was a memorable time. Coolidge rose above partisanship in that hour. Senator Lodge was preparing for the assault upon the Covenant of the League of Nations and the Versailles Treaty which afterwards defeated those instruments in the United States Senate. Distant rumbles of that Senate battle were becoming audible and it was no discredit to Governor Coolidge, feeling as he did toward Senator Lodge, that he took his place rather with Taft, Root and Hughes, Republicans who had been fostering the League to Enforce Peace during the years of the War and for several years before 1914. At that meeting in Mechanics Hall, Governor Coolidge unquestionably lined up with those Republicans who were willing to accept the Covenant of the League with the Hughes-Root-Taft amendments. He chose his words carefully. The core of his speech is in these lines:

We have welcomed the President with a reception more marked even than that which was accorded to General George Washington [said he]; more united than could have been given at any time during his life to President Abraham Lincoln. We welcome him as the representative of a great people, as a great statesman, as one to whom we have entrusted our destinies, and one whom we are sure we will support in the future in the working out of that destiny, as Massachusetts has supported him in the past.

Those words probably represented his matured convictions. He was not then and never became an isolationist. When he came to say the final word on the episode in his "Autobiography," he wrote of his address as "supporting President Wilson, helping him settle the remaining war problems." He recalled that this meeting began "a friendly personal relation between the President and Mrs. Wilson which has always continued." [5]

[5] It is odd to recall that President Wilson there in Mechanics Hall looked into the eyes of two of his successors that day: Governor Calvin Coolidge and Franklin D. Roosevelt, Assistant Secretary of the Navy. Nineteen years later in the White House a visitor saw President Franklin D. Roosevelt at his desk idly drawing a map of the Massachusetts coast. He explained to the visitor how President Wilson's boat, on which the young Assistant Secretary of the Navy was a part of the Presidential party, nearly ran aground near Bass Rocks close to Eastern Point. The young naval secretary knew the coast so well, having cruised there as a youth, that he located the land for the ship's officer. But let him tell the rest of the story as he told it to a friend:

"Wilson got a grand reception, didn't he? And that meeting in the Mechanics Building

No word while Wilson lived was uttered by Coolidge which could be taken as a retraction of the sentiments he expressed that February night at Mechanics Hall.

Always lurking in his heart seems to have been a desire to justify his secret sins in following Theodore Roosevelt even the little way that he went. One of his earlier biographers, an Amherst man who knew him well, Robert A. Woods,[6] records that "every item in the social justice plank of 1914 (which Coolidge drafted) which had not been substantially covered between that time and 1919, was taken up by Governor Coolidge and in nearly every instance, whether through legislation or executive action, he did his effective part usually with complete success" to fulfill the promises of his early flirtations with the Bull Moose leaders. He handled the legislative session of 1919 with unusual skill. He had been a dozen years in Boston as member of the legislature, as a state senator, as lieutenant governor. Under the tutelage of Murray Crane, he had come to know the legislative game. He was not a genius politically. But he was above the average in political experience, for a governor of an average state. Someway, because he was so canny, he was able to sense political trends and follow them. He incarnated democracy as it was staggering along from the old order to the new. He was just that, contemporary democracy incarnate in January, 1919.

Fifty new governors of our states and dependencies, colonies and provinces, appear every biennium in American life. During a decade probably two hundred American governors flash into the fickle light of fame, and fade into inevitable obscurity. On the whole also they do their work fairly

was very impressive. I shall never forget Governor Coolidge's welcome to him. A diffident little man whom nobody outside of Massachusetts had then heard of, the Governor had carefully prepared a very prosaic speech of welcome on three small, typewritten cards. It ran something like this [F. D. R.'s imitation of C. C. was good, but not up to Will Rogers']:

"'It is a great privilege today for me, as Governor of the Commonwealth of Massachusetts, to welcome the President of the United States. We honor him not only for what he has done but for what he is. The Commonwealth of Massachusetts, ever mindful of the contribution of her sons to the great war in which the nation has just taken part, greets its commander-in-chief, etc.'

"Then Coolidge seemed to realize the inadequacy of his formal set speech. And giving one of those rare, friendly little smiles of his to the President, he said something along this line extemporaneously:

"'We hail, moreover, a great leader of the world who is earnestly striving to effect an arrangement which will prevent another horrible war. He has gone across the seas to further his purpose. He has given of his strength and energy. I can assure him that, in all his efforts to promote and preserve the peace of the world, he has the hearty support of the people of Massachusetts.'

"Most of the press interpreted this as an endorsement by Governor Coolidge of the League of Nations—all of it, except probably the *Boston Herald!*"

F. D. R.'s memory was faulty there. The *Herald* was for the League just then.

[6] "The Preparation of Calvin Coolidge," page 140.

well. Scandal rarely touches them. On the whole they go out of office financially and also politically, poorer than they were when they came in. Calvin Coolidge was little better and little worse than the incoming crop of 1919. If all that year's fifty had met in convention on April Fool's Day of that year, a dozen would have seemed more important than he and a score, more famous nationally. He was in truth the average American, the normal man, a little more taciturn on one hand, somewhat less attractive, a little more intelligent on the other.

Yet this man in a score of ways defies simplification. If Coolidge was just one of the mill run of governors in the year's grist, why did Clarence Barron, publisher of the *Wall Street Journal*, pick out Calvin Coolidge as "a man of destiny," long before fame touched him in the Boston police strike? Barron was a sophisticated financial writer. He had interviewed men of national fame and influence before he met Coolidge. What was it in Coolidge's homely background and repressed personality that impressed Barron? He saw in Coolidge in 1917 what Frank Stearns saw in 1916—a coming hero.

Could Barron, by some psychic prescience, realize that the war would center munitions manufacture in New England? How could he foresee that the activity in the shipyards of the rock-ribbed coast was certainly pointing the finger of destiny at Coolidge? At any rate, the high wages received by comparatively unskilled labor in the factories and shipyards were probably the most potent force in stirring up insurrection in the heretofore staid Boston police force. Many of these officers living on less than twenty-three dollars a week were arresting shipyard and factory workers on their Saturday nights in Boston. Their spending was profligate. Wages of seventy-five dollars to one hundred dollars a week were common through the pyramiding of high overtime rates.

On the other side of the ledger, as an irritant to these same Boston police, was the steadily mounting cost of living superimposed on a fixed income. It is sheer madness to presume that Barron could rationalize all this before the event! Calvin Coolidge is supposed to have been in sympathy with the discontent in the ranks of the police. He talked to members of the Massachusetts legislature about it. But the legislature was slow to act, then refused to act under coercion. Barron, visiting Coolidge in Boston in 1917, must have been a clairvoyant indeed to see that chain of events projecting into the Presidency this man who was reputed to have cast wistful eyes, not a decade before, at a superior court clerkship in western Massachusetts. Yet there is Barron's record that Coolidge was a man of destiny, when he was shrouded in obscurity. The Barron soothsaying is interesting but inexplicable. For after all, the cold truth is that as a man and a local statesman that year, 1919, Governor Calvin Coolidge struck the average. Tradition says [7] that he had three suits of clothes: his Sunday suit which he wore with a silk hat—black frock coat with braided edges and suitable gray-striped trousers, a blue double-breasted business suit, and a gray travelling suit. After half a score of years of usage he wore his silk hat easily without being self-conscious. He had a little money in the savings bank at home, probably about twenty-five thousand dollars, for he had saved something every year. That sum would have been the average American gubernatorial wealth. He was paying thirty-two dollars a month rent for part of a duplex residence. He had a wife, two children and a father living, also a stepmother to whom he had been devoted for many years. Again in his domesticity he was the normal American.

As governor he was interested in exactly the same problems that were perturbing forty-seven other American governors in forty-seven other states. Somewhat these problems were war problems. In those Boston days when he was basking under the benediction of Guy Currier, Calvin Coolidge was curiously blind to the machinations of the leaders and directors of the great commodity industries; oil for instance. Being an inveterate newspaper reader, Governor Coolidge knew that the oil barons were trying under the Wilson administration to get Secretary of the Interior Lane to sell or swap the federal oil reserves which Theodore Roosevelt had taken over. He was to meet that plot later in life. But that year the incipient oil scandal did not interest him. It "wa'n't his stove."

Naturally Governor Coolidge felt that his first duty and major interest was in Massachusetts. There the high cost of living was becoming a familiar phrase. So governors, legislators, congressmen and statesmen of various ranks were trying to reduce the high cost of living. In Massachusetts in the spring of 1919, a check was made on landlordism. An attempt was made to regulate the sale prices of the necessities of life. The long fought labor measures were a forty-eight hour week for women and children which came to successful achievement. It was one of the Coolidge things held over from the Theodore Roosevelt days. Like other governors in that day, Governor Coolidge listened to social workers who had plans for the betterment of state education for the defective, the feeble-minded, the insane, the inmates of the penitentiaries, and various reformatories. He was likely to show up at any time in the early part of his gubernatorial career at a state institution and go through it from cellar to garret and back. His experience as inspector when he was lieutenant governor, gave the governor a shrewd appraising eye. The growing cities and towns made the transportation problem acute all over America after the war. And in Massachusetts, cities and towns had grants of power to establish publicly owned street railway lines in certain

[7] "The Preparation of Calvin Coolidge" by Robert A. Woods.

areas. The Cambridge extension of the Boston subway was purchased by the city.

Problems like these were in the minds of all Americans. Democracy was thinking in terms of palliative measures to relieve unpleasant symptoms of the deep-seated disease that was sapping the blood of the economic order. And in Massachusetts, a review of the achievements of Governor Coolidge in his first term reveals the processes of political democracy clicking on cam and cog in rhythmic unison with the mind of democracy all over the nation—to an extent all over the world. But economic democracy, the world's heart blood, was showing disturbing symptoms. Only a few in the academic cloister sensed the nature of a trouble that was producing occasional economic chills and fever, the bad skin blotches, the pains of economic indigestion manifest in civilization. So Governor Coolidge of Massachusetts, in those first busy months of his administration, politically, mentally, spiritually was a typical American democrat functioning in the democracy as were thousands of his kind, doing his best under his light and leading.

During the late spring, strikes began to appear over the world and particularly in the United States. "The hitherto conservative railway workers came out for the 'Plumb plan' by which the government would continue to direct the railroads and labor would have a voice in the management." [8] Distinct evidence appeared that socialism was affecting the American political mind and a strange reversion to violence along the eastern seaboard startled the country. Bombs were discovered in the mail. A poor devil of an anarchist was blown up by his own bomb when he tried to destroy the residence of Attorney General Palmer. Reaction was quick. Mobs of young veterans returning from the war began to attack suspected communists. The soldiers became sensitive about the flag, how it was carried, where it was floated. Evidently American nerves were screwed up a little too tight. The country could not relax after the war.

Governor Coolidge helped to settle wage disputes, tried to reduce taxes, vetoed a beer bill which he felt was a violation "of the Constitution which I had sworn to defend." In his autobiography written a dozen years later he looked back and found that he had little opportunity to help the people with constructive action. Remembering that day, he felt "they were clearing away the refuse from the great conflagration preparing to rebuild on a grander and more portentous scale." Even in 1930, looking back on that day with its sinister warnings, that day's symptomatic indications of disease, he declared that they were beginning "a great work of reconstruction." Indeed he said [9] that he had seen the "people of America create a new Heaven and

---

[8] "Only Yesterday" by Frederick L. Allen, Harper & Bros., 1931.
[9] "Autobiography," page 137.

a new earth. The old things have passed away giving place to a glory never before experienced by any people of our world."

This was his view of democracy, proud of his party and ignorant of the virus in his country's veins—the virus of debt that was overcapitalizing American agriculture and industry.

If Calvin Coolidge had died that summer when he went up to Plymouth for his vacation after putting on his smock to help his father in the hay field, and to help the photographers earn an honest dollar, he would have been one of thousands of honest, competent, conscientious, courageous American governors who have functioned in the last hundred and fifty years, served their people well because they were from the people, and have gone back to the people. The people gave and the people took away, blessed be the name of the people. But destiny was going down the road bringing opportunity to knock upon Calvin Coolidge's door. He rose and opened and went out to meet fame. Fame had rarely flirted with a creature so strange and shy—and yet so eager and worldly-wise in his heart's recesses!

CHAPTER XV

Enter the Hero R.U.E.

WHEN he had reached the Governor's office, and had established his mother's picture on the smooth glass-covered mahogany top of his desk, Calvin Coolidge supposed he had climbed to the summit of his possible political preferment. He declared in his Autobiography [1] that at that time he would have been quite content to close his political career as Governor of Massachusetts. He realized what a long way from Plymouth it was to Beacon Hill. The calf that started to Boston when Calvin went to Ludlow came quicker, but Calvin stayed longer. He had changed little since he rode down the mountain that raw September morning with his father and the calf. Essentially he had the same spirit; attacked life with the same grip as a man that he used as a boy, while he struggled with that arm of destiny known as environment.

Luck sometimes gave him a lift. But despite his luck his character was his destiny. It may be well here and now to take a reporter's [2] estimate who saw Coolidge in that gubernatorial day and wrote for that day. Most happily the reporter is not describing Governor Coolidge in a heroic mood, but as he was in his daily walks in Boston. Read this:

"It was as good as a show to watch him cross Tremont Street. The traffic was thick, of course, and sometimes Coolidge came to the street before the traffic cop was out in the morning. He always stopped, glanced, birdlike, up and down the street, measured the distance to the nearest car, and if he thought he could make it, he started across. If that car brushed his coat tails, he would not run. He had calculated the distance and the time. He had faith in his calculation. And evidently he considered it the driver's fault if

[1] "Autobiography."
[2] From the *Boston Globe.*

149

he went faster than the Coolidge calculation provided. Having escaped, he did not exult. He never emotionalized. It was one of his few self-praising aphorisms that 'the Coolidges never slop over'! Presiding over the state senate, he sat and saw his pet measures triumph by one scary doubtful vote, or go down suddenly by a miserable mischance, and never did he flush or pale. He cared—he cared a lot. But wild Indians could not have tortured a groan or a grin from him. It seemed to be a part of some proud family tradition, to stoicize. He had just one vanity, as we reporters saw him in Boston. It was his writing. Men said Marvin, of the Home Market Club, later of the tariff board, wrote Coolidge's tariff speeches. He did not. Coolidge was too sure of himself ever to ask help in that direction; but anywhere else—yes, if he needed help. He had no other vanities."

There is a snapshot picture of the man, in his square-cut business suit and square cut business government, who never had been in business save in the business of governing, but nonetheless developed executive talent as a member of the governing if not the ruling class.

Calvin Coolidge walked into the turmoil of the times in the year 1919 in much the same way that he crossed traffic in Boston. He would not be hurried. He sighted his course. He knew his speed. He was not afraid of the cars. Caution, courage and intelligent honesty were his rather simple virtues. While of course his traits and talents wove the pattern of his life— made his destiny—the economic currents of the times took him to the bourne of his success. The war and the post war phenomenon, the gradual slowing down of American industrial production caused by the sudden cessation of the European demands and the glut in the labor market by the demobilizing of the American troops, created the unrest of unbalance in American labor. Capital became arrogant in its attitude toward labor on a falling labor market. Strikes became infectious. They broke out all over the land. Early in the year at Seattle the I.W.W., infiltrating into the A.F. of L. unions, started trouble. It looked revolutionary. Mayor Ole Hanson met the strike with a famous statement:

"Any man who attempts to take over the control of municipal government functions here will be shot on sight."

The strikers had attempted to put exemption tickets on municipal trucks. The Mayor added: "We refuse to take exemptions from anyone. The seat of government is the City Hall." Hanson's flare into fame flickered and faded, and Coolidge being canny, knew why. Hanson talked first and acted afterward. The strike broke in three days and Hanson was forgotten in three weeks.

During the spring of 1919, in Boston, a telephone strike was called. It was serious. It may have been a part of the technique of revolution in the brain of some mad communist. For revolution strikes first at sources of communication, then of light and power and finally of fuel and food. Coolidge kept out of the telephone strike publicly. But privately he wrote to Washington asking for authority to take over the telephone and telegraph wires if the strike became serious. He had probably charted his course and stood looking at the troubled traffic in the social highway before him. Then a strike in the Boston Fire Department was threatened and averted. In late spring, Boston policemen began complaining, and complaining justly, at their low wages and at certain incidentally threatened wage cuts. Their pay was based on an annual minimum of eleven hundred dollars out of which they had to buy their uniforms. The policemen complained also justly about living conditions at the station house. The Commissioner of Police, Edwin U. Curtis, agreed in the main with the complaints of the men. He even induced the city government to raise the pay two hundred dollars a year, but he contended that the city could not make the station house improvements which the men demanded. The Governor was more liberal. In a letter to the Mayor, Governor Coolidge suggested improvements in station house conditions. Stations were over-crowded. Men had to sleep two in a bed and use the same lockers. Prohibition had come. Boston at first attempted rigorously to enforce the law and did so. Policemen's perquisites from the saloon keepers were cut down. The "policeman's lot was not a happy one." The A.F. of L. organized the policemen.

Governor Coolidge left Boston in mid-summer on his vacation. Reading the Boston papers he knew that trouble was winking like heat lightning on the horizon in the vicinity of the police. He did nothing. He continued his vacation. He was sighting his way across the traffic. It was not time to start. The police undeniably had a just complaint, but real trouble arose when their right to join the American Federation of Labor was tacitly challenged by the community. The police strike, as a weapon in the armor of organized labor, brought forth protests from the press, from conservative politicians, from many pulpits, from various organized agencies of public opinion. But the policemen and their friends answered and defended the right to strike. Clamor arose. Emotion gradually overcame reason. A "situation" was created. The situation must be briefly mapped. Three men, the Commissioner, the Mayor, the Governor, dramatized curiously the struggle for power that was engaging Massachusetts politics.

Commissioner Edwin Upton Curtis has been described by one of Coolidge's biographers as a "stiff-necked martinet." He was not quite that. He was a large, serious man addicted to long double-breasted coats, with a reserve which passed easily for dignity in the pre-Civil War era; a reserve which men in his own day were inclined to call pomp; one of those solemn, self-sufficient Bostonian heroes who apparently are waiting in the flesh to walk up the steps to a pedestal and be cast into immortal bronze. Boston's parks are peopled with them. A full-faced, full-chested, deep-voiced man was Curtis whose weak heart was indicated by pink pouches under his eyes and a pallid skin. Of the old stock and of the old school. A "gentleman unafraid" was he; with a Jovian knowledge of a large number of things which once had been dependable but no longer were true; yet a real man.

Commissioner Curtis had served as mayor of Boston at an earlier period. He was the youngest mayor that had been elected and was remembered somewhat because of his youthful dash and charm nearly a generation before he served as commissioner. He, in the dramatis personae of the piece embodied the spirit of traditional inherited wealth, traditional inherited Republicanism, traditional inherited skepticism about the capacity of democracy for self-government, and a profound faith in the divine right of the propertied classes ultimately to rule.

Mayor Peters represented politics, urban politics, Boston politics, the struggle between the thrifty Yankees and the Heaven-endowed Irish. So let us next consider Mayor Peters, a youngish, middle-aged man, son of a rich manufacturer, eager for social distinction through political fame, eager enough for power but shy when he had to use it. Peters was a Harvardian realtor in those days, beginning to grow a paunch and cultivate what is known in politics as the glad hand. A nervous man with a high-pitched voice and shaggy eyebrows was Mayor Peters; and under his eyebrows, keen friendly eyes, never removed far in that day from the main chance of politics, thinking in terms of threatening minorities. Remember that he was that good-government Democrat against whom Lieutenant Governor Coolidge had warned the men in the *Boston Herald* office a year or so before. He had been elected more or less in opposition to the Boston Irish hierarchy, therefore he was anxious to please them, being that kind of politician. The Mayor was mentally and politically agile, with no exceptional acumen, willing to serve without much realization of what to do. He had been born full-panoplied out of the ballot box. Personifying the politics of his day he functioned with an honest intention to please the majority, with a keen sense of public sentiment and with a lively desire to win at the next election. He looked upon his constituent supporters as givers of all good and perfect gifts. Naturally Mayor Peters was skittish about public sentiment. He had little real power in the situation. When the police strike began to impend, he called out of the electorate of Boston an advisory committee, thirty-four citizens, representatives of various accredited Boston organizations. Its membership was nicely divided among the estates of the community: Protestants, Catholics, labor leaders, bankers, merchants, professional men, college men, men in the streets and about town—surely a political group even though it was recruited outside of the political arcanum.

It thought in political terms, emphasizing peace and law and order. It was known as the Storrow Committee.

The Storrow Committee met occasionally at the Union Club, a most respectable rendezvous, a place of meeting befitting a committee appointed by a man elected mayor as a good-government candidate and not averse to political advancement. The Mayor was money honest, and brave without audacity, smart rather than wise. Not a drop of dictator's blood coursed in his veins. When the police strike first threatened Boston, Mayor Peters did not realize that by reason of a forgotten statute he had power to call out the members of the National Guard who resided in Boston. His ignorance of that law accounts for much of his preliminary sparring for position. When he ascertained his powers, his attitude changed.

The third figure was Governor Calvin Coolidge, waiting in the wings, entirely unconscious of the fact that fate was about to give him a cue that would call him to the stage, shower spotlights upon him and begin to petrify him for his monument. Even if to recall it is repetitious, we must not forget his background as he listens for his cue. Remember he is New England, cautious, shrewd, more or less repressed, thrifty of many other things than money—of time, of energy, of emotion, of morals, hoarding them like a string-saver against a necessitous day and time. He is in his late forties as he shivers for the first time under the national spotlight; still spare but not gaunt. Tiny pads of fat upholster his jowls. Curves here and there adorn a body once angular. A faint, evanescent crescent seems to flicker in front of his torso. Slowly he has risen. Because he was what he was, his character typifying New England, perfectly mirroring his environment, has pushed him forward into the governor's office. Always the respectable elements of society have been his anchor and his strength. He has learned definitely that money symbolizes respectability. He has come to realize that those who control money are powerful forces in the destiny of democracy. And money has stamped him with its approval. The industrial aristocracy of western Massachusetts, beyond the Connecticut River and the financial Sanhedrin of Boston, the important bankers of the town, the men who control the public utilities, who can give orders in Massachusetts Republican politics— these have said sesame at the gates of his good fortune. Once, twice, three times have the gates swung open to let him pass—upward. There blindfolded he stands in the wings beside the mayor who is embodying politics and the commissioner who is incarnate conservatism, a nineteenth century Republican. The three representing distinct phases in the struggle for political power in the America of that day are waiting for fate to give the beckoning signal to call them to their little wink of fame. Each is shrewd enough to know that a crisis portends; each is solving the problem before him with the inevitable equation of his own inner qualities that mold his fate. And now, lights, music, curtain!

2

Police Commissioner Curtis had been appointed by Governor McCall. Mayor Peters had no control over the Commissioner of Police and little control over the police. But he knew Boston better than the Commissioner. The Commissioner had forbidden the police to join the A.F. of L. When they joined in spite of his orders, Commissioner Curtis put nineteen leaders of the policemen's union, mostly officers, on trial for violation of his orders. A strike threat came. The Mayor's Storrow Committee of thirty-four citizens recommended that the police be permitted to continue their own long established independent organization and that the men be permitted to withdraw from the A.F. of L. without prejudice. The Committee also recommended an arbitration of differences between the police and the Commissioner. Commissioner Curtis refused to consider a solution that might be construed as a pardon to men on trial. His type naturally would contend "for regular military channels." The trial of the nineteen police officers went forward. They were found guilty. Again the Storrow Committee pleaded for delay—for anything to postpone a crisis. The police were in no mood to go caps in hand, pulling forelocks, asking the Commissioner's pardon. He and they represented different estates in the government. Incidentally they were Irish, many of them fighting Irish. He belonged to the ruling aristocracy—Back Bay and Beacon Street, more or less—invested wealth.

After a long session with the Storrow Committee, when Commissioner Curtis had stood and battled for his convictions with the pouches sagging under his tired blue eyes, he clutched his heart, staggered and fell to the floor. The doctor ordered him out of combat for a few days. He disappeared.

That was the situation Calvin Coolidge found when he came home from his vacation early in September, and took a look at the strike menace. Naturally his sympathies were with Commissioner Curtis. Also naturally he suspected Mayor Peters. For Peters was a Boston politician and in the Coolidge catalogue, worse: a Democrat, one who was obviously trying to avoid trouble by placating the policemen. But the Mayor and probably Governor Coolidge knew the police temperament. Commissioner Curtis did not. He lived on another planet, in another era. As an industrialist he had an early Victorian habit of hiring and firing when and where and how he chose. But Curtis forgot that policemen and certainly higher police officers are embryonic statesmen, not operatives. They are not to be fired with impunity. Mayor Peters informed the Governor that when the nineteen officers were discharged the force would follow them out of office. The Commissioner felt sure the force would stand by him; probably reasoning from a bread and butter angle, not knowing politics, certainly unmindful of the Irish. The Commissioner reported one thing to the Governor, the Mayor another. Events proved the Mayor's report was correct. However, the Governor, feeling profoundly his duty to maintain law and order, stood by the Commissioner in all of the preliminary maneuvers leading to the strike. Governor Coolidge had no direct responsibility for the conduct of police affairs in Boston. His predecessor appointed Commissioner Curtis, but officially the Governor and the Commissioner were more nearly in agreement than the Governor and the Mayor. Thursday, Friday, and Saturday, September 4th, 5th, and 6th, daily the situation grew more and more tense. Saturday the Governor disappeared. He did not tell the newspaper reporters where he was going. He did tell his secretary, the adjutant general and the attorney general.

A story, possibly apocryphal, but at least credible, declares that the Governor left word with two or three people whose business it was to know his whereabouts, to call him on the telephone at Northampton and ask for "Grace" as the password.[3] Monday he was billed for a speech at Greenfield, a town near Northampton. He went to Northampton by motor, staying there Sunday overnight, spoke before the state convention of the A.F. of L. at Greenfield, did not mention the threatened strike in Boston, and left town for Boston after the speech. He called his office on the long distance phone from Fitchburg, to know if there were any new developments and suddenly, out of a pouring rain, he bobbed into his office the afternoon of Monday, the 8th, as crisp as a dill pickle and about as sour. A conference of the powers was assembled—the Mayor, the Police Commissioner, men from the city government, the Storrow Committee, the elder statesmen of the day and time. The conference met at the Union Club on Park Street. According to its own report, it threshed over old straw. Commissioner Curtis, whose patriotism and courage cannot be questioned, but whose judgment was poor where human nature outside of his class was involved, saw no danger of a strike. The Mayor was close to the Irish. They dominated the police. But because he had official authority they talked to the Mayor frankly though he was elected by a good government fusion movement over the regular Irish Democracy. All Monday morning the Mayor had been running in and out of his office, up and down the street, in the rain, fussily trying to compromise the issue, to avert the strike, to placate the policemen and to please the people. He was sure that the strike was coming and that it would bring trouble. The Storrow Committee appeared to favor the compromise with the police. Everyone knew that the absent Governor desired the Mayor to support the Commissioner.

[3] His secretary in a letter to the author says "Grace" was not his password; just Mrs. Coolidge.

The Commissioner rejected in toto the Storrow Committee's compromise. The Mayor had his own plan which did not include absolute subservience to the Commissioner, who under the law was the source of police powers in Boston.

There can be no doubt that the Governor was satisfied to be out of town that Saturday, Sunday and Monday. His absence was deliberate—a part of his strategy. The air was thick with rumors of compromise. Of necessity the Governor would be involved in these rumors—possibly he might be trapped into an improper position. The Storrow Committee did stand by a compromise which provided that the policemen should withdraw from the A. F. of L., form their own Union and present to the proper authorities their case for wage increases and better working conditions. When the compromise was drafted, Commissioner Curtis and the Governor were out of town. Thus from Saturday until Monday, the two men, Curtis and Coolidge, representing the other side of the controversy who contended that a police strike is not a strike but a desertion of duty, were not to be found. Their signatures were needed with that of the Mayor to make the Storrow Committee compromise a real treaty of peace. The strike thus might have been averted. But the principle at stake, as the Governor and Commissioner saw it, would not have been defined and settled.

Looking back on that day and hour, those who were nearest to him,[5] reporters, fellow workers, politicians, never doubted that Coolidge's mind was made up. At no time is there any evidence that he wavered from the conviction that the men as policemen had no right to strike. Around that he built his policy. His habit, however, was temperamental, constitutional in his makeup, to let matters ride, to delay action and let time settle the preliminary issues and possibly develop a moral issue—simple and understandable by the people.

Mayor Peters begged the Governor at the conference in a final plea to let the suspended police officers return to their jobs. The Mayor had behind him the business men of Boston—merchants who had something to lose in riots. Public sentiment so far as it was distinctly vocal favored compromise. Coming out of the conference in which the Mayor, who represented the compromisers, did most of the talking, the Governor gave out this statement:

"Governor Coolidge has taken under advisement what action he can take."

The Governor, however, did say to the newspaper reporters: "Understand that I do not approve of any strike. But can you blame the police for feeling as they do when they get less than a street car conductor?"

[4] Letters from, and conversation with, Robert Brady, *Boston Post* reporter who saw Coolidge hourly in those days; his private secretary; James Bailey, political associate; Tom White, political ally.

Smart as chain lightning was this gaunt little Vermonter, realizing the justice of the police demand for better economic conditions, but balancing it against the violation of a principle which he held dear, that public servants have no right to strike. He said just enough to hold the confidence and respect of labor and he held it, and just enough to attract public acclaim as he stood for an ideal.

The Storrow Committee, whose chairman, James Storrow, Harvard, '85, a well known banker, feared the general strike, proposed arbitration. James H. Vahey, attorney for the city and counsel for the police, felt that the strike was improper as well as futile. He sent one of Coolidge's best friends to talk with the Governor, Monday, outlining the arbitration proposal. Coolidge listened, then said:

"Have you said all you wish to say?"

"Yes," was the reply.

"What do you think I should do?"

"I think you might send for the Commissioner and persuade him to agree that the men could submit their troubles to arbitration."

"Well, I won't do it," said the laconic Coolidge.

The emissary took his hat and with a deferential bow started to leave the executive chamber. As he was about to pass out through the door, he was halted by a call from the Governor.

"And say, Eddie, you knew damned well I wouldn't when you came up here." [5]

The fear of the general strike chilled the heart of Boston. Governor Coolidge was serene and not without cause. Monday evening Diamond Jim Timilty, a Roxbury Democratic ward leader, slipped out of the back door of the Governor's office. He had been a member of the state senate when Coolidge was president of the senate. Though Coolidge had steamrolled Diamond Jim many times, he had done it on what the men of the ward called "the up and up." Also Coolidge had befriended Diamond Jim times without number. Years later, Timilty told a friend [6] what he was doing in the back room of the executive suite that day. Said the husky Hibernian:

"I just went in to see my little pal and tell him not to worry over all this mush about a general strike. You know, I'm president of the largest labor organization in the state, the city and town laborers' organization, with the largest membership of any union in Massachusetts. I just told Cal that 'we won't go out,' and we have more votes in the central labor organi-

---

[5] Letter from Robert Brady, reporter for the *Boston Post*, who had the story from "Eddie" whose name has appeared elsewhere in this narrative.

[6] Robert Brady, *Boston Post*.

zation than any of the others.[7] You see, Cal's my kind of a guy, and he's right about these damned cops."

Monday afternoon, the Mayor, knowing that the small force at the Commissioner's command could not hold back a mob, pleaded with the Governor to call out the National Guard. His friends declared that Peters said to the Governor:

"I know you are running for reelection and that you do not want trouble with labor. All right, let me take the responsibility; call out the troops at my request."

The Governor slumped down in his chair, dour and graceless, shook his head, said "No!", offered little explanation. Apparently he was determined to wait until actual trouble started before calling out the troops. And then, of course, we must remember that Commissioner Curtis felt there would be no need of troops, that his small force of metropolitan police and the motor corps of the state guard at the armory could keep the peace. A general trades union strike menaced Boston. So Monday night passed.

Tuesday dawned gray and cloudy and promised a muggy day. For three days rain had fallen intermittently. When the strike finally broke it broke in a September mist that occasionally thickened to a light rain.

To complicate matters, Tuesday morning the Bay State car men threatened to strike unless the trustees of the road gave the men wages equal to the wages paid the men on the El road. But even amid complications, the Governor gave out no interview. All Boston knew that with him the crucial point was not the addition of a car men's strike to the trouble, but the basic issue; whether or not the policemen would be allowed to affiliate with the American Federation of Labor. The question at issue was simple in the Governor's mind: Has a policeman a right to strike?

All day Tuesday the police were voting. Even in the face of the slowly accumulating returns from the election calling the strike, Commissioner Curtis was determined to punish the men. It was obvious that the central labor union, an A. F. of L. affiliation, was supporting the patrolmen. Finally the returns were announced. The ballot at Union Headquarters was overwhelmingly in favor of a walkout—1134 for the strike, 2 against it. The patrolmen decided definitely to quit at the 5:45 rollcall if their officers were suspended. But Commissioner Curtis still was sure an hour before the walkout that the strike would not occur. He could not accept the evidence of his senses when it affronted his deep patriotic wish—feeling that no American could strike against his government. But all day volunteer policemen were coming to Commissioner Curtis and receiving instructions

[7] About this time Governor Coolidge secured a confidential survey by the Stone and Webster Engineering Company to guide him if a general strike tied up public utilities —railroads, telephones, telegraph, gas and water.

for duty. Harvard offered seven hundred volunteer policemen that day. Probably the Commissioner could have recruited twice that number from his volunteers.

Tuesday afternoon, the *Boston Transcript* declared:

Plainly the die is cast in the Boston police situation. Police Commissioner Curtis has found nineteen officers of the recently formed labor union guilty of disobeying his latest rule, and the peace, comfort and convenience of the city will depend upon the action that the police union will take today or tomorrow.

The *Transcript* closed its editorial Tuesday evening with these words, representing conservative opinion in conservative Boston:

In rejecting the compromise proposed by the Mayor and his Committee, Commissioner Curtis has only added to the debt the people owe him in this emergency.

Late Tuesday afternoon after the evening papers were printed the Governor issued a letter to the Mayor throwing the weight of his office with the Commissioner, calling attention to the fact that under the law the Mayor had authority to call out the Tenth Regiment, the motor corps and the cavalries, totalling a thousand men stationed in Boston. The Governor also pointed out the fact that the Mayor might call on the police department of other cities and towns for assistance. "It is natural," concluded the Governor, "that the Mayor would exhaust all agencies at his hand before appealing to me for assistance."

The police, according to a headline in the *Herald*, "were jubilant." [8] Then Tuesday afternoon came the strike order. The strike was called. The metropolitan police force of more than a hundred and the city police of thirty or forty men were put on duty, and the motor corps of the state guard was held at the armory. These were the Governor's policemen. But he held that the Mayor must ask for them. Boston was in chaos. Only the tradition of law and order, the ghost of habitual authority patrolled the Boston streets.

[8] Possibly some solid foundation for police jubilation arose from the fact that they knew Governor Coolidge respected the justice of their grievances. Monday he had discussed these grievances sympathetically with Martin T. Joyce, head of the Massachusetts division of the A. F. of L. Answering Joyce's telegram demanding the removal of Commissioner Curtis whom labor saw as the man forcing the strike, Governor Coolidge wrote:

"The Governor has no authority over the appointment of the police force of Boston. I earnestly hope that circumstances will arrive which will cause the police officers to be reinstated. In my judgment it would be unwise to remove Commissioner Curtis. I thank you for the reception you tendered to me yesterday." Labor felt that in the end labor had a friend in the Governor's office. However the Governor might distrust the police strike, he was a friend of the policemen. His position was anomalous; but it was good politics.

3

Tuesday night in his room at the Adams House, the Governor, the Attorney General, the Adjutant General, and his secretary awaited developments until nearly midnight. The streets were calm. The ghost of orderly authority still held back the mob. Police headquarters reported everything quiet. Shortly after eleven, the Governor was left alone. He went to bed but during the night he tried to call the Mayor three times. The phone did not answer. After midnight that night, the mob rose. The forces of civilization weakened. The devil was loose in Boston.

While the Governor and his supporters were waiting early in the night in the Adams House, little knots of boys and young men began wandering through the streets. The old policemen were gone. Groups joined groups, at first hilarious, but acutely realizing that no one would bother them. Under the street lamps scores of games of craps began to operate on the Common. No one molested these games. It was evident that the new police were not interested in crap games. This emboldened the gamblers. The mob grew noisy, also offensive. Its voice changed from a mumble to a high-keyed, nervous falsetto. Sporadically, little mobs broke apart and gravitated to the larger mob instinctively. The treble yelp of all the mobs indicated rising hysteria. By midnight the coagulating crowds had formed one raging mob, a drunken, noisy, irresponsible mob, without grievance, without objective; an aimless idiot mad with its own sense of unrestrained power. Riots broke out in various parts of the city. Someone threw a loose paving stone crashing through a store window about one o'clock. The tension snapped. The mob was crystallized. It found its courage. Its desire took hazy form—loot! Sticks and bricks went whizzing into offices on the second and third floors. By two o'clock looting had begun. The situation was out of hand. The volunteer police were powerless. The Mayor's guardsmen, motor corps, and cavalry had not yet been called. Sixty thousand dollars worth of goods from stores and shops were scattered from the idiot's fingers along the streets in South Boston and the north and the west end, and in the downtown section. Still the Mayor hesitated. The Governor again tried to call the Mayor at 2:30. The telephone did not answer. The Wednesday morning *Herald* declared that it was not certain whether or not the State Guard would be called upon to serve during the police crisis. It depended "on the decision of the Mayor." Obviously it was a case of avoiding responsibility. The Mayor wanted the Governor to do it and take the blame. This was not the Coolidge way. He was moving automatically—after the pattern of his life. He was cutting across the street in the traffic. The Governor wanted the Mayor to call the troops and take full responsibility for it. For two days that question had been threshed out in many conferences.

About noon, Wednesday, after the night of rioting and looting, after the Superintendent of Police had advised storekeepers "to board up their windows," the Mayor rather suddenly decided to assume control of the Boston police strike, to call out the State Guards within the boundaries of the city and to request Governor Coolidge for at least three thousand additional guardsmen. The Mayor, apparently advised Wednesday of the provisions of an antiquated and forgotten law, invoked it. But he did act. In a statement issued that afternoon, the Mayor expressed his astonishment that Governor Coolidge should attempt to make the Mayor responsible for the rioting that occurred. It was obvious to all Boston then that the Mayor, on one side, and the Governor and Commissioner Curtis, on the other, had seriously disagreed as to the plan which should be adopted to avoid the strike in the first place, and by what measures to suppress the strike when it came. But even until the last, even until the hour the rioting and looting began, Commissioner Curtis was serene in his own patriotism, and sure that similar high motives would inspire the friends of the striking policemen. He declared Tuesday after the men had dropped their batons, that his police provisions were ample and that he did not desire soldier help.

And here is a revealing thing. Early in the week Calvin Coolidge addressed a family letter to his stepmother at Plymouth, Vermont; at least the political apocrypha of the day chronicles this letter. In it he is supposed to have told the family there of the impending strike and that he would stand by the Police Commissioner.[9] It was typical. His principles were rigid—copybook mottoes—probably Garmanian. Of course he did not outline exactly what he would do in the family letter; he was no prophet. He could not foresee circumstances. But he did know how he would act under any circumstances. He would wait. He would take no chances, and when he decided to move he would go in a straight line oblivious of danger. He mapped his course. He marked his track. He stood waiting in perfect type and character; not heroic, just methodical, unemotional, coolly logical—the Yankee on a monument smiling glumly at grief!

He appeared at his office, Wednesday morning, after the riots, cool and unruffled.

The State House, of course, was hectic. Reporters were hovering around the outer office in the executive chambers. Local statesmen were edging in with advice. Labor leaders were sitting hopefully on benches. Telegraph messengers were coming and going. Coolidge avoided the crowd, disappeared, was invisible for the morning, while all Boston was distraught. At noon, Wednesday, after Mayor Peters decided to act, under the ancient statute giving the Mayor power to call out the State Guards within the boundary

---

9 Robert L. O'Brien and Frank W. Buxton doubt if such a letter were ever written.

of the city and after he had asked the Governor for three thousand additional guardsmen, the Governor still issued no statement, made no sign.

In the meantime, a firemen's strike in sympathy with the police was strongly urged. Sentiment for the firemen's strike seemed to be gaining headway. The streetcar men threatened to join the firemen and the police. A revival of the telephone strike was attempted. A general strike of all trades unions evidently became more than a menace. It seemed imminent Wednesday. The technique of revolution was beginning to move; the paralysis of the police, the invitation to arson, the threat against transportation and communication. And so the night approached. In the September twilight idle mobs drifted along the streets. Uniformed guardsmen patrolled public places. Volunteer officers, green and ignorant of police tactics, appeared in the residential districts. The Governor had no other idea than to wait, to stand by his Commissioner and let eventualities develop. He was no hero in his own eyes. Probably he was frightened, but fright did not move him. He was more stubborn than scared.

When he arose Thursday morning these headlines greeted him, streaming across the top of his favorite morning paper, the conservative *Boston Herald*:

"RIOTS AND BLOODSHED IN CITY AS STATE GUARD QUELLS MOB

"Mayor Assumes Command, Calls Out State Guard—Brushes Curtis Aside— Asserts Authority Conferred on Him by Old Statute

"Volley in South Boston kills two and wounds nine. Cavalry sweeps Scollay Square."

The Mayor was in command. Curtis was out of the situation. At last the time for action clicked in the Coolidge mind. The worst had happened. Under the heading which greeted him was this sub-head:

Riotous mobs bent on plundering and destruction of property attempted last night again to plunge Boston into a turmoil of crime and general disorder; and only sobered down and went home after the State Guard had fired into a crowd at South Boston, killing two men, wounding nine persons and cavalry had charged into a crowd at Scollay Square with drawn sabers, scattering it in all directions. . . .

The story ran:

Six regiments of infantry, a troop of cavalry, the motor corps, two machine gun companies and an ambulance company of the Massachusetts State Guard are now on duty to preserve order in Boston. The state troops are reinforcing the police. The city is not under martial law. Mayor Peters is in control of the entire machinery of law enforcement.

The general strike brooding in Boston threatened New England. Street car men, firemen, telephone and telegraph operators all had grievances. The head of the state division of the American Federation of Labor obviously was holding back demands of radical union workmen for a general strike in all trades; Diamond Jim Timilty was as good as his word. He held off a strike that would begin with transportation, spread to communication, then to electric power. Such a strike would paralyze New England. The days and times all over America, all over the world were full of unrest. Labor troubles were in the air. Reactionaries were hysterical. Conservatives were puzzled. Liberals were confused and divided. Commissioner Curtis came to the Governor in distress. He and the Governor both felt that Mayor Peters who had taken charge of the police force of the state under the old statute might, if he chose, take back the discharged policemen, and so destroy a principle for which the Governor and the Commissioner had been standing.

In the meantime, the Governor's friends and political associates had been considering the situation. Senator Murray Crane for instance was alarmed. During the morning he had talked over the telephone with William M. Butler.[10] Herbert Parker, who was an old friend and special counsel and legal adviser of the Commissioner, by his appointment under the provision of a statute authorizing such an appointment and employment, remembered in 1935 [11] that Senator Murray Crane was directly interested in the attitude and conduct of the Governor and was anxiously watching the issues as they revealed themselves through the confusion somewhat attributable to the apparent indecision of the Governor. Mr. Parker felt sure on that day when he and Mr. Butler were in critical conference with the Governor that Butler was in these conferences by Crane's request and somewhat following Crane's instructions. Mr. Parker seems certain that Commissioner Curtis also was aware in the early stages of the impending disturbance, of Senator Crane's interest in the premises. The recollection of these witnesses makes it entirely warrantable to say that Senator Crane's interest, backed by his political association with Governor Coolidge, made a strong, probably determining impression upon the Governor's final decision to act after three days of hesitancy. Senator Crane appreciated his weakness as well as he knew his strength. Evidently [12] Senator Crane had complained to his

---

[10] Letter from Senator Butler to the author, July 24, 1935.
[11] Letter to the author, August 26.
[12] Letter of Chief Justice Wm. H. Taft, to Judge Learned Hand, May 3, 1927, filed in Congressional Library.

intimate friends in politics that the Governor was lacking in ability in crises to size up situations and to direct men. Crane seemed to sense that weakness. He understood the Coolidge caution, his habit of hesitancy, his slow, silent periods of what seemed like sheer timidity. So long as Coolidge was loyal to his chief, the chief could well afford to be patient with his protégé. But in this crisis Crane saw the need for caution. So Crane came into the picture. It was obvious that in the end Crane, not Coolidge, would be blamed by powers of Boston if the forces of law and order should finally surrender or even compromise seriously with the rioters. So the Crane machine began to move.

After talking with Senator Crane in the morning, Mr. Butler called up the Governor and "informed him what was going on in the downtown section of Boston and somewhat of the strained conditions and the general atmosphere which was exceedingly unpleasant and threatening." [13] Characteristically the Governor revealed no plan to his friend and sponsor and again later in the morning Butler called up the Governor to add some further information. The situation was growing dangerous.

The outcome of this telephone conversation was an invitation by Coolidge to Senator Crane's vicegerent to meet him at the Union Club at one o'clock. At the Union Club, Mr. Butler found Mr. Herbert Parker, counsel for the Police Commissioner, with the Governor "and we proceeded to have lunch and during the lunch talked matters over. When the lunch was over the Governor asked me what should be done. I said that the Governor should take over the situation, call out the militia and also take charge of the police affairs of Boston. Mr. Parker expressed enthusiastic approval of this statement and as a result the three of us left the Club and went to the State House and assembled in the Governor's office. The Governor then prepared executive order number one calling out the militia, and executive order number two for the purpose of taking charge of the police in Boston, and issued these orders." Apparently the Governor's mind was made up. No longer did he sit looking blankly, busily cerebrating as he did when he was thinking things out. He was ready to move!

He dictated his orders as though he were recalling constructed phrases, a lifetime habit when he had come to a decision. Mr. Butler remembered:

"There was no hesitancy on Coolidge's part. He had made up his mind what to do and had it clearly expressed in a short time."

Then he whirled around to Butler and asked: "What further damage can I do?" [14]

[13] Letter to the author, June 24, 1935.

[14] A letter from James A. Bailey, who was Water Commissioner of the City of Boston at the time, an influential organization Republican politician and an intimate of Senator Crane, declares expressly that he knew that Senator Crane made his views known to

It was then that Calvin Coolidge, looking at traffic each way, sizing up the speed and the tensity of events before him, ventured to cross the street. Immediately after Mr. Butler had left the Governor's office, Governor Coolidge issued a general order restoring Commissioner Curtis to his place in control of the police, called on all citizens to assist the Governor in preserving order, and directed all police officers to obey Commissioner Curtis. Then he took over the peace forces of Boston, deposing the Mayor. In that last demand of his proclamation reestablishing the authority of Commissioner Curtis was the meat of the whole controversy. Viewing the matter in retrospect, Coolidge wrote that he always felt that he should have called out the State Guard as soon as the police left their posts. If he had called the National Guard earlier without the Mayor's consent the violence of the strike might have been averted. But its violence was the dramatization required to make the people understand the principle which the Governor felt was involved, that a policeman does not strike; he deserts. The policeman's status is not that of a laborer but a defender of peace and order. It is a nice point; not easily demonstrable except through some such dramatization as that which came with the violence of Tuesday night and the fatal charge of the soldiers in Scollay Square. The Governor's order made him the hero of a national event. Modestly enough he writes in his "Autobiography":

"This was the important contribution I made to the tactics of the situation which has never been fully realized. To Mr. Curtis should go the credit of raising the issue and enforcing the principle that police should not affiliate with any outside body whether of wage earners or of wage payers but should remain unattached, impartial officers of the law with sole allegiance to the public. In this I supported him." [15]

And so out of this drama the hero for the first time under a national spotlight started on his journey to the White House. Across the American continent for three days headlines had been flashing the story of the Boston police strike. It was a national event. With the rumors of a general strike, a naval vessel ran up to the electric station with electricians ready to go

Governor Coolidge about this time rather definitely. In view of the certainty of Commissioner Curtis, in view of the statements of Mr. Butler hereinabove quoted, it seems fair to presume that the Governor's period of internal debate and doubt ended when he knew how Senator Crane and the Republican organization felt about the matter. The halfgod went, the gods arrived. For after all Senator Crane and the Republican organization would have to suffer if Coolidge had fumbled, if he had taken the wrong turning. It was natural that he should wait to hear from his friends, natural and characteristic. "The Coolidges never went West." Nor had Calvin Coolidge one drop of martyr's blood.

[15] Even after the police strike, Governor Coolidge's soft heart impelled him to help striking policemen to get jobs in other departments of the city and state when they could not be rehired by Commissioner Curtis.

over the side and keep the light and power plant going. President Wilson issued a statement condemning the police, congratulating the Governor. A federal wagon train of supplies, arms and ammunition, was rushed to the support of the Governor that day. Thousands of volunteer police appeared. A half-million dollars was raised by popular subscription to care for the dependents of the Guards and even to help the families of the police who had left their post. It was a typical American scene there that Thursday when the Governor took over government and made the Commissioner his lieutenant. He had leaped the borders of Massachusetts, even of New England, and became a national figure.

Telegrams rolled in. Newspapers applauded from Miami to Seattle, from San Diego to Augusta, Maine. Coolidge was national news. And the nation saw his picture flashing that night in a thousand daily papers, a wizened, sharp-eyed, sharp-featured, lean-faced man with thin down-curved lips, a high brow, a stubborn chin and a peaked, not very generous nose. He was copy. Reporters made his voice quack across the land in word pictures that all but talked—the new savior of his country. It was a great day for the simple mind of democracy when a fairy tale—featuring a real Jack the Giant Killer—came true. The sun of that day went down with a new American hero. It rose the next day to hear the hero defy Samuel Gompers. In the drama Gompers, head of the American Federation of Labor in Washington, was cast as the man-eating Giant and heavy villain of impending disorder. To Gompers, Governor Coolidge sent these potent words:

"There is no right to strike against the public safety by anybody, any time, anywhere!"

The last happy, necessary touch. A hero and a slogan rung the curtain down amid cheering millions. Mayor Peters the day before had expressed the same sentiment with some circumlocution and caution in addressing Gompers. No one heeded it. But the Garmanian touch, the Garmanian philosophy, and the Coolidge instinct for succinct public appeal, an instinct whetted and sharpened by sixteen distinct appearances on the ballot, in sixteen different kinds of electoral campaigns—all these guided the Coolidge pen, whittled off every superfluous participle, preposition, objective clause or conjunction and sent the arrow straight into the heart of the truth.

More than seventy thousand letters and telegrams flooded the Governor's office. These were "the voices" that whisper to men in modern times. These voices were saying over and over: "You can be President!" Most of these he answered perfunctorily. But he found one letter from an Amherst classmate,[16] and in less than a dozen words revealed with curious subter-

[16] Elmer Slayton Newton, Principal Western High School, Washington, D. C.

ranean warmth that romantic encrusted glow of affection, the real happiness in his new found fame. He wrote:

Dear Newt:

      I am glad you liked what I did. I knew you would.

                                  Cal

In those four words, "I knew you would," were distilled much of the best that lay deep in Calvin Coolidge's life; the essence of a proper pride, a decent modesty and the fragrance of a lifelong affection whose evidence he had repressed.

CHAPTER XVI

Our Hero's Unheroic Chores

The executive, any executive, township constable, city mayor, Governor of a state, or the President, under the American political system, is constantly pulled and hauled by two ancient devils that have persisted since the beginning of our American experiment three hundred years ago. These ancient jinns are on the one hand vested interests, property rights, expressed in modern terms the functional needs—perhaps privilege is the more exact word—of organized capital in a complex civilization, and on the other hand the protection of the common man against the inevitable and probably unconscious encroachment that follows a too rigid defense of the vested rights of property, or too lax an extension of the privileges of capital. In colonial days in Massachusetts, Governor John Winthrop stood on one side of this moat, the Rev. Thomas Hooker upon the other. Their contest was not economic but political, turning upon the question whether the governed should govern themselves or be governed by their elders and—betters! In Washington's first term, Hamilton was the guardian of property. Jefferson fought against Hamiltonian encroachments. Calvin Coolidge, in the governor's office in Boston felt the shock of these two forces constantly. He was not entrenched on either side of the ancient battle line. Sometimes, as in the case of his legislative program, he deeply offended, by his espousal of various housing proposals, the vested real estate and banking interests. But on the other hand he recognized his obligation to those forces which Guy Currier sheltered under his brooding wing; the public utilities, the insurance companies, the investment companies, the host of interests large and small which ten years before, Theodore Roosevelt was denouncing as aggrandized capital. Probably Coolidge, who in many relations with men was a careful accountant, did not keep books in his heart and balance nicely his gubernatorial favors for each side of the ancient feud. Evidently he helped labor and the common man when he had to and stood by capital when he could expediently. But at those odd times when he helped labor, the left hand told the world what it was doing. When he helped capital, the right hand was sometimes not so eager for publicity. In November, 1919, the Massachusetts branch of the National Industrial Conference Board was trying to offset the clamor of labor for higher wages, the demand of the consumer of those commodities sold by the public utilities for lower rates. In the *Boston Sunday Globe* appeared an interview with Governor Coolidge. It was cleverly written, described in detail the Governor's appearance, his force of character, his strength of purpose. It told how he smiled and answered the questions "as he looked thoughtfully down on the historical Boston Common." In its defense of the interests of the manufacturers in Massachusetts, it contained such obviously Coolidgean balanced sentences as this:

"We have curbed the cupidity of capital; we cannot yield to the cupidity of labor." And this: "Full stomachs and full hearts will be found to have their own answers for wild-eyed inciters to terrorism," and more of the same kind, setting forth by rather obvious implication the Divine right of corporate capital in Massachusetts to rule in its own realm. An interesting thing about the interview will be found in a circular of November 28, 1919, sent out by Magnus W. Alexander, the president of the National Industrial Conference Board, with a local office at 15 Beacon Street, and a national office at 724 Southern Building, Washington, D. C. The interview published in the *Boston Globe* was enclosed with the statement that it "will appear in many papers throughout the United States, Sunday, November 30." This was before the interview had been planted in the *Boston Globe*. The circular continues:

"This interview was prepared in our office and then submitted to Governor Coolidge who readily agreed to stand for it with a few small changes which he suggested and which we made."

Then furtively covering up his tracks, Mr. Alexander continues: "The National Industrial Conference Board does not appear in this case but editors of newspapers know that the information was prepared here and comes from us. What we are interested in, of course [and here the line is set in italics] is to put proper information across to the public [end of italics] rather than have the National Industrial Conference Board advertised. Quite a number of editorials on the industrial situation which have appeared from time to time in the public press, have been similarly influenced and often initiated in our office."

Which closes Mr. Alexander's letter. The letter was published. Some editor with a weak stomach gagged at it, and printed the letter in his paper. It is reproduced in this narrative as evidence that Coolidge did not always walk upon that high serene plateau where the innocent meander in fields of lilies, but that sometimes his feet were not above a bit of dalliance down paths where "the rich, the wise, and the good" [1] sowed, planted, plowed and garnered their profitable primrose crops. He knew where he was going. When he put his OK with a "few small changes which he suggested" upon that interview prepared in the offices of the National Industrial Conference Board, he knew what the Industrial Conference Board was. He knew exactly who Magnus W. Alexander was and what use would be made of his propaganda. He believed honestly in the sentiments attributed to him. He saw no reason why the propaganda should not be sent out under the auspices of an association of capitalists, made to appear as though it was the spontaneous outgiving of a patriotic governor. On the whole, he was a Hamiltonian. His spiritual line ran straight back to Winthrop, and he believed with the writer in the *Federalist*, in the beneficent despotism of the "rich, wise and good." He was a conservative Republican, only as liberal as the challenging times required him to be.

But it must not be inferred that in Boston from the first day he came there until the last, Calvin Coolidge was ever socially intimate with the high-placed Massachusetts financiers and industrialists. He did not frequent the Union Club, was never a familiar spirit at St. Botolph. His ingrained respect for the representatives of incorporated capital did not lead him, hat in hand, knocking at the doors of Back Bay and Beacon Street. The idle rich did not engage Calvin Coolidge at bridge, at golf or motoring around New England in his idle hours. First, because he had no idle hours, second, because with all his respect for their economic and political function as divinely anointed rulers, he had no liking for their kind, no place in his cosmos for the softening trivialities of their charming life. Accepting their creed in business and in politics, he had no great use and scant respect for the way they lived.

After the police strike while he was a Presidential candidate, Calvin Coolidge became a national figure. He was somebody in the eyes of State Street and in the financial district of Boston—somebody politically to be reckoned with—a power in Massachusetts politics, scarcely comparable to Lodge and Crane—certainly not their equal. Thus it happened that occasionally and always without his seeking their intimacy, Calvin Coolidge became the confidant of the rich and the great, a listening man in a world whispering gallery. What did men tell him? What did they really know—these mag-

[1] A phrase used by Federalist writers, quoted by Fisher Ames, "Works" Vol. 1, pp. 310–316. He says: "The federalists must entrench themselves in the State governments, and endeavor to make State justice and State power a shelter of the wise, and good, and rich, from the wild destroying rage of the southern Jacobins."

nates who sought his counsel? Probably they really knew little. Also most likely he could understand less than they told him. But the facts which he might have learned were these: We were going into an inflationary period. American banks had to finance the flow of goods to Europe, to expand credit very heavily in financing the twenty-five billion increase in government debt which, however (happily), was placed chiefly with investors rather than with banks, while the banks had to finance the transformation of our industry from a peace time basis to a war time basis, from America to Europe. Production in the United States was greatly overstimulated by the war. Production lagged behind consumption and war destruction. By the spring of 1916, labor was fully employed. The accustomed immigration stopped with the beginning of the war. Between 1916 and 1920 we had a labor shortage. Rising wages increased the percentage of the American national income going to labor, and probably labor had an absolute gain—housing excepted.

People on fixed incomes, of course, suffered—bondholders, school teachers, salaried people generally. Hence the high cost of living pestered Coolidge and other politicians in those days. From 1915, for five years, as farm prices rose rapidly and land speculation was buoyant, agriculture prospered.

As orders for war goods poured in, Massachusetts and all New England felt the warm hand of prosperity and the boom times. The belligerents paid for what goods they could with their balances in this country. But when the foreign exchanges were disrupted, and the allies' balance was exhausted here, they adopted the simple expedient of borrowing from us to pay for what they bought from us—and never paid us!

Coolidge was probably not enough of an economist to understand that the allies were unable to pay with their own money and credit for the goods we were making in those world war days.

So great was the European demand for commodities that even the return of four million soldiers to industry in 1919 meant very little shock to the labor market. An odd thing is revealed by index numbers of production. It shows that in 1919, Americans actually produced less than they did in 1918, despite the return of four million soldiers. As labor efficiency went down, labor discipline went down. There again the dial of destiny was pointing Coolidge's way to glory. The Boston police strike came out of that economic background. Who could have foreseen it? Certainly not Frank Stearns who in the midst of these alarms tried to make Coolidge President in 1916. And to pretend that Editor Barron of the *Wall Street Journal* read the crystal ball of Coolidge's fate in 1917, is to believe in wizards. Yet the rise of Calvin Coolidge was not luck. His real luck rose out of the inefficient management of American industry. Business being bedevilled by many problems, became overconfident because it was receiving more orders than could be filled. High wages and strikes were inevitable.

2

Late in his first term, Governor Coolidge decided to push to a complete fulfillment the reorganization of the state which was necessary under a Constitutional amendment adopted by the people during the McCall administration. Vigorously he pressed a group of measures through the legislature which called for a radical overhauling of the state institutions. The amended Constitution consolidated one hundred forty-four departments of state into twenty, and the Governor had to depose one hundred twenty-two heads of departments, selecting twenty new heads out of the original one hundred forty-four. Some of the men deprived of office by the consolidation of boards were Coolidge's old friends. He had to reshuffle all the names in the various state departments, making them fit places provided under the new Constitution. A few new appointees were called, but in appointing practically a new state service, he had to use most of the men who were on the payroll.

For the State House was Crane's fortress. Apparently Senator Crane and his vicegerent, Mr. Butler, decided to let Governor Coolidge take his own way. His secretary remembers [2] that Coolidge "asked little or no advice but that the job was on his mind and greatly depressed him." During the period when he was preparing the list, he saw few callers.

In reorganizing the State patronage, Coolidge ignored Lodge as far as possible. Crane was the kind of boss who picks a man carefully and lets him alone, knowing that his man will serve better without orders. When Governor Coolidge was making his plum melange with one hundred twenty-two appointments to make, if none of the Crane crowd wanted a plum, he passed it to Lodge. Those who were about Coolidge in that day remember that Lodge asked the Crane forces only for what he knew he could get. And Lodge knew the Crane men kept books on him. He would have to give where he took. But when and where he knew begging would bring results, he begged and would bow fairly low, being a better beggar for being born with a silver spoon in his mouth.[3] But at that time and for several years after, Lodge paid no heed to Coolidge.. To understand Lodge's contemptuous unconcern, we may well recall how Frank Stearns' *faux pas* in the Chicago Convention of 1916 choked Lodge with scorn.

So far as Lodge was concerned, Coolidge was just another short term governor who would be gone tomorrow or soon thereafter. The procession had been filing by for twenty-five years while Lodge was in Congress. In

[2] Interview with Henry F. Long, member of the Tax Commission.
[3] *Ibid.*

this attitude Lodge typified Back Bay and the Brahmins. They did not know that Calvin Coolidge existed. For four cloudy days of the strike they were tremendously conscious of him. But after that Lodge and Back Bay and the Brahmins forgot Coolidge and sunk sedately, solemnly into their urbanity, staring at some invisible esoteric navel as oblivious to the little stranger on Beacon Hill as though he were a new wart on the face of the moon. It was in the midst of Coolidge's perturbation over these appointments when he had denied himself to the friends of scores of candidates and was working out his problems, that he sent for Judge Henry Field, of Northampton, in whose office he had studied as a law clerk. With scarcely a word, he and his old friend left the Governor's office, passed a long line of waiting candidates and politicians in the outer office. Entering an automobile they motored in absolute silence for more than an hour. They rode out through Cambridge to Watertown and turned in at a little long forgotten cemetery. The Governor led his friend, still without a word, to two crumbling brown headstones that bore the names of John and Mary Coolidge who had migrated with the Puritans to the wilderness of Massachusetts in 1630. Judge Field relates [4] that he stood "at the feet of his forefathers lost in contemplation." Finally the Governor turned to his friend and simply said:

"Those are my first ancestors in this country." Nothing more then or all the way back. Together they climbed the State House steps, together they went through the line of anxious politicians. Reporters clamored at Judge Field. It was impossible to tell them that for two hours Coolidge had been enjoying one of his amiable and necessary cerebrating silences. He went back to his desk and plunged into his problem refreshed.

Such sentimentality did not check certain other impulses in his heart—toward certain displays of sinister teasing. Not long after his wistful mourning at his ancestors' graves, and after fondling "Dear Newt" with an affectionate dehydrated billet-doux, he slipped out of Boston on a street car to a public meeting, avoiding the pomp of guards and the honorable escort of the local committee; and after having the crowd and the sponsors of his meeting wild with worry, he suddenly bobbed up serenely explaining: "No I didn't have an auto, and I have no aides. Thought I'd take a night off and come on a street car!" and was impishly satisfied with the stir he had created. He never really weaned his vanity—just hid it!

It is easy to fancy Calvin Coolidge sitting alone in his big office in that grim, silent brown study which gripped him with a sort of epileptic thrall hours and hours, days and days, all alone feeling "sort of naked" as he played providence with the lives of that host of politicians. With what acute intellection he pieced together his cardboard universe; building, re-

[4] *Good Housekeeping*, March, 1935.

casting, rejecting men to fit the tapestry of some pattern of a need that was at best the hallucination of a season. Is it strange that finally strained to a sour milkshake of hysteria by gazing upon his mother's mute patient photographed face, always on his desk, he summoned Frank Stearns and Henry Field to sit by him? On and on he played silently with the phantasmagoria of his problem! Is it odd that in his perplexity he let the friends he summoned sit by him in silence? A man was needed to look at now and then —otherwise one might begin to jabber. But he refused to discuss the appointments with the regular run of office seekers. He kept the big bosses out of his office. When he had the list made up, he rose from the nest and cackled mildly to his secretary:

"I am glad it is done! It is the worst job I ever had to do! I shall never have to do anything like that again." [5]

After he had agonized for weeks over his long list of appointments, he did a wise thing. He could have trolled that list of nominees along the top of the waters for six months or more—one or two names a day for the state senate's consideration and confirmation, luring all the big fish in Massachusetts to jump at the glittering bait. But he presented the entire list of nominees one morning and the whole list was confirmed.

Three years later, Coolidge commenting upon some chance remark of his friend, Frank Stearns, to the effect that he had acted courageously in the Boston strike, broke out with this, one of his few apostrophes to his friend: [6]

"You say it took courage to act in the Boston police strike. Well, maybe a little—but where it took courage was when the Massachusetts Constitutional Convention provided that the state government reduce the number of state bureaus from one hundred eighteen to about twenty, and gave the legislature three years to do it; and I insisted that they do it in my administration so that I could be sure it was started right—that took courage."

And when he had done it with all his courage, of course he played practical politics. By sending the list to the senate for confirmation in one lot he had in effect compelled the friends of one hundred twenty-two possible office-holders to organize to see that no one candidate was denied confirmation by the senate.

Of course if a fight had started, Governor Coolidge knew that back of him stood Murray Crane and his vicegerent, William M. Butler. But no

[5] Interview with Henry F. Long, member of the Tax Commission.

[6] Interview with Frank Stearns, 1933. Stearns recalls that he had been seeing Coolidge every day during the weeks before the list of these prospective office holders had been completed. Just before announcing it, he handed Stearns the list. Stearns, before opening it, put down his guesses. On the names the Governor had listed, Stearns remembers that "except for three or four obvious reappointments, he guessed wrong on the whole list, and got three of the four in other offices than Coolidge gave them." Coolidge was that tight-mouthed with his dearest friend.

fight started. Lodge and his friends had scant satisfaction in Governor Coolidge, then or ever.

About this time the exigencies of Massachusetts politics required the deposit of state funds in a bank which later failed. Governor Coolidge had just appointed James Jackson, state treasurer. The deposit in the bank was suggested by a man with great power in the state senate who could help or hinder the Coolidge legislative program. Also the state deposit might have saved the bank and so saved losses to certain powerful supporters of the bank, men in the Sanhedrin of Republican politics. Governor Coolidge called James Jackson, his recent appointee, into the gubernatorial office. The Governor explained the situation, asked the state treasurer who was beholden to Coolidge for his job to make the deposit. The state treasurer explained to the Governor that the bank was doomed to fail,[7] and refused with decent tact to make the deposit. The Governor again called attention to his own obligation to the men who were bolstering up the bank and to his political obligation to a state senate group interested in the deposit. The Governor was between the millstone of his political godfathers above and a powerful senatorial clique as the nether grindstone. He elaborated the reasons for his insistence and repeated what he thoroughly believed, that the deposit would tide the bank over without loss to the state. Still the treasurer refused and when the Governor again insisted with something like a demand, Treasurer Jackson said:

"Have you a fountain pen, Governor?"

The Governor had. Taking a pad of blank paper from the desk, the state treasurer wrote out a brief, unequivocal resignation and pushed the pad across the desk. Whereupon the Governor blinked, took the resignation, sat in silence. Treasurer Jackson walked out. Coolidge knew that if the treasurer resigned under those circumstances he would talk and talk publicly. The Governor also knew that if the treasurer talked the bank would fail and the senate group would be on his back. A few days later, three days before the primary election, in which Treasurer Jackson was a candidate, the treasurer received this letter from the Governor:

> The Commonwealth of Massachusetts
> Executive Department
> State House, Boston
> September 27, 1920

Hon. James Jackson,
Treasurer and Receiver General,
State House, Boston.
My dear Mr. Jackson:

I wish to compliment you on the way in which you are handling a very difficult situation. I take great satisfaction in having appointed you and

[7] Letter from James Jackson to the author.

feel sure that you are doing all that anyone could do to relieve the situation. In all your actions I want you to know you have my support and approval.

Very truly yours,

(Signed) CALVIN COOLIDGE

The episode seems to have a place in this narrative to illustrate (a)—something of the character of Massachusetts politics of that day, (b)—the intimacy between the Republican organization in Massachusetts and the ruling class in high walks of finance, (c)—and finally, it illuminates the mental and political processes of Governor Coolidge, later to become President. He played the game of politics but with his cards against his vest. In his heart probably he was happy that he could take Treasurer Jackson's resignation to his Beacon Street and Back Bay friends and to his senatorial supporters as a good and sufficient reason why he had failed. They would know, as he knew, that to force a Treasurer to resign and talk would flush the covey. And when the Governor wrote those lyric lines of praise to his political protégé without referring to the resignation, his gratitude and appreciation were gushing from a full and happy heart.

He played the game according to the rules. About that time he called the head of an important department of the state government and told him to give a certain influential Republican a job. The department head said there was no vacancy. Looking out of the window, Coolidge said laconically: "Make one!", and went on with his work.[8] Thus spake the partisan from Ward Two, the county Republican chairman, the organization Republican trained to "take program," finally entitled to give orders.

After the strike came the felicitous epilogue and the basis of Coolidge's campaign for reelection as Governor of Massachusetts. The spotlight did not leave him even when he was in bed with influenza. He made but three or four speeches and won by one hundred and twenty-five thousand votes. President Wilson sent him a telegram of congratulation. The boy who started from Plymouth in the gray, cold dawn of that September morning, thirty-three years before to ride to Ludlow, had arrived in Boston. New England, the mother [9] of Vice Presidents, was about to be delivered again.

[8] Conversation with Thomas Green, State Civil Service Commissioner and, although a Democrat, an intimate associate of Mr. Coolidge, as reported to the author in a letter from F. W. Buxton, Editor of the Boston Herald.

[9] John Adams who was Washington's first Vice President came from Massachusetts; Elbridge Gerry who served with Madison in his second term, Hannibal Hamlin who served with Lincoln, and Henry Wilson who served with Grant—all were New Englanders.

CHAPTER XVII

He Pussyfoots the Path to Glory

"I HAVE known Calvin Coolidge," Senator Henry Cabot Lodge is quoted as saying in the apocrypha of politics in 1920, "only as long as it has been necessary to know him."

Probably Lodge did not say exactly that, but National Republican leaders were speculative in their early attitude toward the new Massachusetts hero. To Lodge and his Massachusetts followers, the presence of Coolidge, under the spotlight as a Presidential possibility, was disquieting. Moreover in Washington the senatorial managers of the Republican party were the Massachusetts Senator's friends. They controlled the National Republican Central Committee and so dominated the National Republican Convention that was to assemble in June, 1920. Clearly they regarded as an upstart this new figure from Massachusetts, disporting himself with rather obvious modesty under the spotlight. The plans of the Republican organization were not, of course, definitely settled. Harry Daugherty, of Ohio, one of the Republican Temple high priests, was being widely quoted in the winter and spring of 1920 as uttering a prophecy that when the Republican Convention had been in session for two or three days, a small group of leaders would meet in a smoke-filled hotel bedroom to choose the man to be nominated about 2:11 o'clock Saturday morning and that some time Saturday the nomination would be made. And Daugherty further proclaimed that Senator Warren G. Harding, of Ohio, would be the man.[1] So came Harding into his first fame as a Presidential candidate; a fame that became a byword linked to a boast about the power of the manipulators in that

---

[1] As a matter of cold fact the prophecy to all intents and purposes was Daugherty's. The actual language save for the time, 2:11, was a political reporter's colloquialism. See Mark Sullivan's "Our Times: The Twenties."

"smoke-filled bedroom." Harry Daugherty was an Ohio state boss. He had
one political passion—to be kingmaker to a President. And Harding was
the apple of Daugherty's eye. He had befriended the Senator politically for
a dozen years, spoonfed him and reared Harding by hand. Looking over the
parapet of national fame, they were brother cherubim. Daugherty was the
stronger, the sharper and on the whole possibly the more scrupulous. Harding yearned nobly at times. Daugherty got results. He had to drag Harding
into the Presidential race by the ears, protesting. Other leaders than Daugherty felt that the Convention's range of choice would be limited to General
Leonard Wood and former Governor Lowden, of Illinois. Daugherty's
prophecy of Harding and the "smoke-filled bedroom" seemed grotesque to
the wise ones in politics.

When Daugherty spoke thus, it is likely, knowing his political acumen,
that he relied upon something more than the friendship of the senatorial
group to control the Republican Convention. But his high hopes for Harding came out of a lifetime affection. Daugherty had been a legislative representative for several leaders of American industry. He knew his way around
in the labyrinthian bypaths which led to the backstage control of politics by
business—"the rich and the wise and the good." Powerful industrial captains knew of Daugherty and his power. He could not have escaped the
knowledge that certain oil interests, indeed the organized oil industry,
itched to get the federal oil reserves back into private hands. The attempt
during the Wilson administration was known to men like Daugherty. He
knew what steel wanted, what railroads were eager to obtain from government, what protection lumber needed, and so on down the line through
the whole litany of privilege. It is a reasonable assumption that knowing
these things, having years of useful political association with the leaders of
the great commodity industries, he would try to enlist their help for Harding. He knew what industries might influence certain Republican Senators.
Knowing how badly the Republican party needed campaign funds and
knowing where to get them, he could persuade the Senators, or thought
he could, "in the smoke-filled bedroom." So when he blabbed with a
politician's bravado, of Harding's coming fortunes in a day when Harding's
name was scarcely mentioned, Daugherty must not be blamed for indiscretion. The blabbing helped Harding more than it hurt him. It was Harding's
one asset. Politicians said in the secret places of their hearts when they
thought of Harding, "there is the money side of our contest."

Governor Calvin Coolidge, serving his second term, who still held his
place as a minor national hero, realized, being a smart politician even
though he was a provincial New Englander, that his Presidential chances in
1920 were remote. Yet he must have known he was not entirely impossible.
Despite his respect for his party organization, he realized that sometimes

organization is impotent. Events refute the logic of situations occasionally. Calvin Coolidge watching the game with an appraising eye, knew that the way he could help his own fortunes was to avoid offending the powers that be. His earthly inheritance could come only through meekness. His cue was to get in no one's way, to be as modest as possible without overplaying his humility. For if he incurred enmities among Republican leaders, they would smite him in his day of opportunity if it ever came.

His friends, alas, did not encourage his modesty. They [2] opened a Coolidge headquarters in Washington. Coolidge vetoed their endeavors and the headquarters closed. They started to get a Coolidge delegation at the Massachusetts Republican Presidential primaries in the spring of 1920, but Coolidge stopped that movement. His friend, Frank Stearns, soon after the Boston strike, arranged a meeting of a few newspapermen and a political manager of national weight and authority. Stearns asked Coolidge to come to the meeting without telling him of its purpose. Coolidge came. After dinner when they were smoking, the talk began to slant toward the Coolidge Presidential candidacy. Coolidge picked up his hat and stalked out. He did not say: "All right, you boys talk it over. I'll finish my stogy in the next room!" He left without a word, indicating a powerful sincerity in his desire to smother talk of his Presidential candidacy. The friends of General Leonard Wood came to the state and were fairly successful in capturing a handful of the Massachusetts delegation to the National Republican Convention. In the meantime, during the spring of 1920, as delegates to the National Convention were being elected all over America, Frank Stearns sent copies of the Coolidge book, "Have Faith in Massachusetts," to every delegate or alternate as soon as he was elected. With it was a pleasant note from Frank Stearns. The note did not emphasize the Coolidge Presidential candidacy. It was a friendly note—in good taste—calling the attention of delegates to a man who might possibly interest them later on. Frank Stearns' relation with Coolidge was always circumspect.

In the meantime, Senator Lodge was friendly to General Leonard Wood. Senator Lodge, General Wood, and Theodore Roosevelt had been cronies for thirty years, three Harvard boys who had made good. Lodge in the early days of General Wood's candidacy before the Boston police strike, had suavely promised to present Wood's name to the National Republican Convention in such a nominating speech as only Lodge could make, but when the Massachusetts delegation finally came out of the primaries, it was evident that Lodge had lost his grip upon his delegates. Senators Lodge and Crane were delegates at large as also was Speaker Gillett, of the House of Representatives, who was his own man. It is interesting to look back upon the poll of that delegation to see how Coolidge without announcing

[2] No evidence can be found in 1936 that Guy Currier was a party to the folly.

his Presidential candidacy, indeed after denouncing his candidacy and making no overt effort to secure the nomination, was still the strongest element in the delegation, in spite of the fight of the Wood men for the delegation. This strength of Coolidge among the rank and file Republicans of Massachusetts and his weakness among the leaders is a significant political phenomenon. Charles Willis Thompson in his book, "Presidents I've Known," writing about his visit in Massachusetts that year, declares that he "was staggered to find that men who were not for Coolidge hated him bitterly. . . . [Among] these men were Republicans, and Republicans of importance and standing. When they pounded their fists on the table and got raucous over Coolidge, it was not with contempt but with that hate which is compounded partly of fear." In the Massachusetts State Convention, the strong Lodge men and the thick and thin supporters of Crane seem to have harbored no intention to nominate Coolidge for President. The Lodge and Crane followers who did not regard Coolidge's candidacy as a joke quietly were on guard against it.[3]

But even before Senator Lodge saw this delegation rolling like dice out of the ballot box, he realized that he could not make the nominating speech at the Republican National Convention for General Wood. Probably Wood's friends felt that Lodge had deserted him. He did not desert. He was stuck in the mud. Perfunctorily he offered to make the nominating speech for Governor Coolidge. Politely but grinning in his heart the Governor declined Lodge's courtesy gesture.[4]

In the spring of 1920, liberal Republicans hoped to make a showing and to produce a decent minority, perhaps a majority in the Republican National Convention. When Theodore Roosevelt died in January, 1919, most of the Republican leaders who had opposed him so bitterly in 1912 had become reconciled to his leadership. He felt and he told his friends that he would be nominated by the Republican convention the next year. But his death left the liberal wing of the party without leadership. Many of Roosevelt's friends, particularly his personal friends who did not share his economic and social liberalism, tried to rally the liberal strength to General Leonard Wood. In a measure they succeeded. In certain matters—foreign policy and national defense, for instance—Wood and Roosevelt were in agreement. The West was for Wood, for he was identified with the West in many ways. He had been an army doctor during the Indian border troubles of the early eighties. He trained men during the World War at Camp Funston, in Kansas, and the West felt that Wood had been

[3] "Autobiography," pp. 143, 146, 147.
[4] Letter from Henry F. Long. In the diary of Henry F. Long, Governor Coolidge's secretary, is this entry: "November 26, Governor Coolidge told me Lodge had asked if he could present Coolidge's name to the convention as the Massachusetts candidate for President."

badly treated by President Wilson.[5] He was their martyr. Wood also was a Roosevelt Roughrider in Cuba which was more western in its western kind of noise than in its per capita of western enlistment.

On the other hand, Governor Frank O. Lowden was of the Middle West, a real mid-westerner by birth, by education, by environment. He had served in Congress and was governor of Illinois. In the matter of farm relief he was radical compared with Wood. Many Rooseveltian liberal Republicans went to Lowden, because of his attitude on the agrarian issue. He was a vice president of the Pullman Company and had married a Pullman daughter. He was rich. In that primary campaign, money was easy to get and was lavishly spent for advertising and for such honestly branded services as politicians can sell, which are few and not important, but conspicuous. No question could be raised about Lowden's money. What Lowden's candidacy did was to divide the liberal Republicans in the West and blur the issue. Moreover, Senator Hiram Johnson, of California, a Progressive of Progressives, who had run with Roosevelt for Vice President in 1912, was in the Convention as a Presidential candidate with a respectable showing. No one could question his liberalism. He was an irreconcilable Republican Senator in the matter of the League of Nations, one of those whom Wilson had denounced. This weakened him with certain Wilsonian Republicans like Taft, Nicholas Murray Butler, Hughes, Hoover, Root and Wickersham. No definite line of cleavage like that of 1912 could be drawn across the Convention. It was a zigzag line, the old crack which had broken the Republican party.

No one knew this better than the reactionary Old Guard statesmen who still controlled the National Republican Central Committee and so could organize the National Republican Convention. It was this sense of liberal dismay and defeat that Harry Daugherty caught. It gave him confidence to predict that the Republican nomination would be made off the floor in a hotel bedroom, Friday night before the nomination was achieved in the Convention. Governor Calvin Coolidge in the spring of 1920 also must have sensed this liberal weakness.

Here is how Coolidge stood in the race for the Republican Presidential nomination in the spring of 1920: Coolidge kept his partisanship regular in 1912. His record in the state senate was in line with his party in Massachusetts. Yet for half a decade after the passing of the Progressive party, the virus of Rooseveltism, like an antitoxin, pulsed in Coolidge's blood, and for that matter in the Republican platforms of Massachusetts. Generally it lived as a suppression.

Probably Senator Lodge, who was certainly never touched or tainted

---

[5] President Wilson recalled General Wood from the boat as he was about to embark for Europe with the 89th Division which he had trained.

politically by his relation with Colonel Roosevelt, instinctively must have been irked when he was not contemptuously amused by Coolidge's Rooseveltian symptoms. Coolidge's labor record for a dozen years was satisfactory to the A.F. of L. and quite out of line with the Lodge ideals. Coolidge also had greeted President Wilson in a speech which was tinctured with internationalism. This speech gave Lodge weariness and pause. The Governor had signed a law imposing a forty-eight hour week on certain Massachusetts industries. Even as the delegates to the National Convention were being elected, a hue and cry arose against the high cost of living. Massachusetts under Coolidge leadership was trying to reduce rents. Also Massachusetts under Governor Coolidge was leading the nation in state legislation, cramping the power of landlords, checking rent profiteering, authorizing cities and towns to take property by eminent domain in order to provide dwellings for the people. Coolidge himself sponsored laws giving the courts power to stay eviction proceedings in certain cases, laws prohibiting rental increases of more than twenty-five per cent a year, laws penalizing landlords who failed to keep agreements regarding heat, light and other services, laws requiring thirty days' notice to vacate a tenancy.[6] Governor Coolidge appointed a commission to study the whole matter of housing. He established the office of fuel administrator and signed a bill which became a law to define areas in cities and towns in which the municipality might operate public street railway lines. He enlarged the power and usefulness of cooperative banks and generally was regarded as an open-minded, forward-looking Governor for his day and time. He did not try to rally the liberal Republicans of the country to his banner. For he was not then, or ever a liberal. William Howard Taft defined Coolidge's political faith after nearly a decade of association with him on terms of something approaching intimacy. To a friend,[7] the Chief Justice wrote: "He is a conservative, and willing to stand for certain principles that appeal to the whole public, and is able to show by what he has done in the past that such principles for him are rules of action and not mere planks in a party platform." In 1919 and 1920 "the whole public" in New England was disturbed by the economic maladjustment that was unsettling the world. It seemed at that time to men of vision and character that government should broaden its base of endeavor in the effort to "promote the common welfare." All governments, federal, state and municipal, were trying, for the most part futilely, to restore the world that was before the war. Hence this Coolidge program to establish peace between landlord and tenant, between the mine operator and his customers, between the banker and his creditors, between

6 The High Cost of Living became America's first alphabetical abbreviation as the H.C.L.!
7 Col. J. M. Ulman, Rockwood, Maine, April 10, 1927.

the grocer, the clothier, the baker and the hardware dealer and their customers. Governor Coolidge became interested in these problems before the Boston strike. He did not get unduly excited about these problems afterwards. The reduction of the high cost of living was part of his day's work, not an incident in a Presidential contest. He took these measures as they came. When campaigning for the election of delegates to the Republican National Convention of 1920 was progressing, Coolidge made no attempt to advertise his achievements. It would have been easy to play the demagogue. He had a record to rouse the rabble. He made no speeches, wrote no letters to the press; he permitted no committee to organize and preen for him. But as he saw the returns come in, which gave him and his friends rather than Lodge and his supporters control of the Massachusetts delegation to the National Republican Convention, his satisfaction was obvious. "Coolidge—my god!" reechoing from Lodge's room at the Convention four years before, no longer rankled in his heart. That he could happily wave aside Lodge's punctilious offer to present the name of Coolidge to the Convention, must have been unction to his heart.

So the spring of 1920 passed. Coolidge's name had appeared in several state primary contests for Convention delegates. He was a decent fourth rater in the contest, but only that. But that was much to a man who half a decade before was glad to be Lieutenant Governor for the $2,000 salary that he could earn!

CHAPTER XVIII

He Waits Off Stage for His Cue

BUSINESS leaders generally scorn politicians. But business and politics are parts of an interacting whole. To know the politics of an era intelligently, one must study its economics. Kings and bankers, demagogues and industrial captains are blood brothers born of one womb. This biography of Calvin Coolidge, consummate craftsman of politics, is the story of a stage in the economic development of his country. The man is but a wraith, giving human features and evanescent flesh and blood to a dramatic moment in American history. Politics being the human shadow of the economic reality makes it necessary to review business conditions both before and after the Boston police strike. The period from 1914 to 1920 was a war period in our economic history, dominated by a gigantic outpouring of goods to Europe on credit. That credit affected the politics and the banks of Massachusetts, but neither the bankers nor Calvin Coolidge realized how this bank credit later would affect his life.[1] The absorption of foreign securities from 1914 to April of 1917, was greatly facilitated by

[1] The credit was rather easily given because Europe was sending us gold, increasing our bank reserves, that made possible a multiple expansion of bank credit, which could be used to finance the European securities that were coming over. This does not mean that the banks were buying these European securities, but it means that they were lending money to facilitate their issue and to release funds which would not otherwise have been available for the purchases of these European securities.

Various things were happening as this immense demand grew. The first impact of this war demand was a quickening of American industry. Commodity prices rose only about five per cent in 1915 as compared with 1914, but by early 1916 industry was fully employed and labor was fully employed. Then commodity prices began to shoot up rapidly. The rises in commodity prices were so rapid that country retailers were caught utterly unaware. Further, labor was gaining disproportionately in these years, because of the cessation of immigration. The war stopped immigration. It was during these years that wages were rising fast.

an expansion of bank credit based on the large inflow of gold from Europe. This expansion of bank credit increased our bank reserves and made still more bank credit. The newly organized Federal Reserve System helped to lower reserve requirements of banks which came with the Federal Reserve Act. And so we lent practically ten billion dollars to our allies in Europe, somewhat before they were our allies! These loans met Calvin Coolidge in a few years on a larger stage.

As the government issued its famous Liberty Loan bonds many persons, who had never saved a dime, bought government bonds. In the financial pages of the *Boston Herald*, Coolidge's political Bible in his Boston days, he read of bond purchasers becoming what the high-pressure salesmen of a later day would call "conscious"—"security conscious." In every Boston paper which he read at the Adams House before breakfast, he found evidence that persons of moderate means could save through investment in securities. Glib stock salesmen began "swapping" Liberty Loan investors out of their government securities into stocks, some of doubtful merit. Scores of stories of swindling went to the Boston State House while Governor Coolidge sat under the sign of the sacred cod. The field was fertile. But the War Baby stocks of Wall Street "wa'n't his stove." He started no prosecutions. This mad speculation was nobody's stove. So politics and business walked together—twins in cherubic beauty!

When our government loans to Europe stopped, credit continued going to Europe, regal in its prodigality, in unfunded form: (1) exports on open account and on long credits; (2) wholesale speculation in marks, francs, sterling and other European currencies; (3) loans were made by American banks to strong European banks, particularly British banks.

Then in the winter and spring of 1919–20, railroad congestion manifested itself. In spite of what the technocrats have often told us about producing incredibly more than we now produce if only there were fuller demand for our products, the railroads met an overwhelming demand with congestion in the last half of 1919 and the first half of 1920. That economic setup in America, the money spent in Massachusetts before and after we entered the war, forced Calvin Coolidge into American history. Owing to the compactness of Massachusetts and New England, the Allies' spending had a more dynamic effect there than in the West, where the high prices of grain relieved the pressure. The cost of living in New England quickly responded to high wages and a scarcity of labor. So labor grew cocky and capital grew mean. The esoteric unity between business and politics was, for a tragic hour, revealed!

2

In the period prior to 1920, a shoddy bank failure [2] in Boston shook New England. But none of the old Boston banks was even dented in the financial orgy that marked the closing year. While shoddy collapse did not break in Boston and New England until the summer of 1920, just before Coolidge was easing out of the state picture into Washington and into national politics, national politics began to reflect the tawdry national economy. Flashy trust magnates swindled the poor, while others, higher placed, who were dumbly swindling the rich and well-to-do scorned the more palpable swindlers. Bucketshop failures and mushroom installment houses crashed. The old-timers of Boston's State Street, who had seen the tinhorn financier rise and fall without a sign of restraint by the state authorities, were ashamed. Governor Coolidge knew, however, what all informed citizens knew in Boston, that these bucketshops were symptoms of a speculative disease. The bucketshop swindlers were reproduced in a tribe of noisy politicians who at that golden moment in 1919–20 were organizing the Republican National Convention [3] which was to make Calvin Coolidge the Crown Prince who should be King of the big Bull market in another, gaudier day. Oil was revealing its power in American politics. In early 1920, Governor Calvin Coolidge in the State House was dourly enduring the presence of the swindlers of State Street, and Harry Daugherty, Harding's political manager, was palavering with Jake Hamon, of Oklahoma, political pal and associate of the Western oil magnates. Even if Coolidge and Lodge and Crane scorned Ponzi and his kind and kept virtuously on their side of the seam of the sheet, out West politics and oil were thicker than three in a bed!

By the middle of 1920 when our credit limits were clearly reached, when the foreign exchange rates had broken, when sterling reached a low level of $3.18 in February 1920 and other European exchanges fell much lower, when Europe was buying without selling and no backflow of goods returned to us, and when money was a drug in Wall Street, the political side of the picture showed that both major political parties in the campaign of 1920 had more cash than was good for them.[4]

[2] Ponzi and his kind. Read "The Big Money," by John Dos Passos.
[3] And the Democratic politicians were cut off the same piece of shoddy.
[4] Commodity prices, which, in May of 1920, while Coolidge was garnering the Massachusetts delegation, stood at 247 per cent of 1913 prices on the index number of the United States Department of Labor dropped dangerously, reaching 141 by August of 1921. Wages fell violently, though not nearly so much. Prices at retail fell, though again not so much. Six million or more men were unemployed in the summer of 1921. The banking structure generally held, though many banks in the agricultural parts of the country failed. But, on the whole, American banking came through with flying

A labor shortage during the war was due partly to labor-saving machinery, partly to checked immigration, and finally to the raising of an army of four million men. Governor Coolidge, whenever he spoke of it publicly, was enthusiastic about the new labor-saving machinery. As he rode to and fro fortnightly between Boston and Northampton, he saw proudly that machinery in his district was doing the work of more and more men. The manufacturing cost of electricity distributed over a wide area, began to decrease. But living costs mounted. Americans had no time to build houses, and living space became scarce. So looking on the political side of the picture we see why the Coolidge gubernatorial Administration tackled the rent problem. But inevitably as we now know, Coolidge's programme for housing, so well-intentioned and so well conceived in its legislative terms, failed to help the situation. The labor troubles which had swept across the land from Seattle to Boston in 1919 met the reactionary answer of red-baiting and strikes and lockouts. Labor grew militant. Witness the threat of a general strike in Boston, when the police went out. Also observe how automatically Coolidge, the carnal embodiment of Hamiltonian Republicanism, reacted to the situation. The Crane machine began to stir; Coolidge moved. It was not luck. Coolidge was, in a way, the inevitable political epitome of the economic picture. That picture, in 1920, is an essential part of the story of the political Puritan who was to rule the economic Babylon of a later day.

Calvin Coolidge probably sensed what he could not define. As he took a degree of Doctor of Law at Williams College he declared:

Those who seek for a sign merely in greatly increased material prosperity, however worthy they may be, disappointed through all the ages, will be disappointed now. Men find their true satisfaction in something higher, finer and nobler than all that. We sought no spoils from war; let us seek no spoils from peace. Let us remember Babylon and Carthage and that city which her people, flushed with pride, dared call eternal.

In those lyric lines the politician spoke who subconsciously may have realized, as he saw the industrial changes in Massachusetts without de-

colors. In previous crises we had often had money panics due to our inflexible monetary system. The elastic Federal Reserve system worked and worked well in the 1920–21 crisis and depression. There was no shortage of hand-to-hand cash. Banks that had commercial paper, agricultural paper or loans against government securities, or that owned government securities could borrow at the Federal Reserve Banks, and consequently were able to hand to their solvent customers the sums that were needed to protect their solvency. The crisis did not degenerate into a panic. There were many bankruptcies, but, with a few exceptions, there were no unnecessary bankruptcies. The general system held intact, and, with the completion of the violent readjustment, a strong upward move started. With six million or more men unemployed in the summer of 1921—to look ahead a little—we had labor shortages in many lines in the spring of 1923.

fining what they meant, the disturbance and maladjustment of the times. It was to reach its nadir in the years when Coolidge passed out of national life. But he could not time that tide. In that, the first year of the third decade of the century, when the tide was slowly moving in, Calvin Coolidge was surely moving up with it.

3

So much for the business of that day. Now for its alter ego—politics. After the Massachusetts primaries, it was evident that Governor Coolidge had a majority of his home state delegates genuinely for him. Senator Lodge, when it was decided that he should present neither Coolidge's name nor Leonard Wood's to the Convention, was chosen by the powers of the National party organization to be permanent chairman of the Convention. Speaker Gillett from Coolidge's home congressional district had declared publicly as Lodge had, for Coolidge, and Senator Crane had also. But division in the Massachusetts delegation—upon the surface between Coolidge and Wood, but somewhat between Senators Crane and Lodge— was obvious to those politicians over the nation who were sufficiently sophisticated to become delegates to a National Republican Convention. So the Coolidge Presidential candidacy became one of the lesser candidacies in the pre-Convention estimates. This political subsidence of Coolidge was one of the minor evidences of the swing of the Republican party to the reactionary right. Theodore Roosevelt was dead, his followers divided. A fear psychology infected by the discontent of labor was in the heart of the propertied class, who controlled the Republican party at that time. The party, with the passing of Theodore Roosevelt, was without a great leader. So the economic reaction of the time was producing political reaction and in late May, 1920, it was evident that the ultra-conservative Republican United States Senators would dominate the Republican National Convention.

The Republican party, which boasted proudly in the sixties and seventies that it had "wrested the shackles from four million slaves," in the eighties and nineties and the first three decades of the new century became by the strange alchemy of politics, the party of property and privilege. But some slight pinking of a millennial dawn glowed in the Republican party in June, 1920.[5] And oddly enough considering his Wilsonian bias for

[5] Theodore Roosevelt's challenge to privilege, his valiant endeavor to put property in its place, had failed to achieve all that he would have done. His failure left a residue of Republicans who hoped to liberalize the party. Under the leadership of Will Hays, who was in 1912 an Indiana Republican Rooseveltian, and of Ogden Mills, who in 1918 and 1920 was in polite revolt against the domination of Penrose, Barnes, Cannon and the reactionaries in the Republican party, an effort was made to liberalize the Republican platform. A pre-Convention draft of a liberal platform was worked out under Hays's

the League of Nations, probably Governor Calvin Coolidge was on the whole the most vital candidate of the Progressive hope who had any chance in the Republican National Convention.

4

Consider the party national convention as a peculiar institution in the United States, unique among the political institutions of the earth. Under the two-party system in America, executive government generally comes out of the November ballot box in quadrennial years, to abide for four years with one of two party groups. The party which wins at the election of the Presidency generally carries the Congress with it. And also it has the power to fill all vacancies in the Supreme Court.

The fate of America more or less impinges upon these National party conventions. For during the sessions of party conventions the American government actually is in a fluid state. There, for nearly a century, rather than at the ballot box, the people surrender for a few hours their rights as free men. No other country knows such stupendous dramatization of its politics as America has witnessed in the conventions of the two major parties since 1860. Calvin Coolidge, a politician by instinct, an intelligent student of politics if the country ever bred one, watched from Boston the gathering hosts in the Republican convention at Chicago, and of course realized that the inheritance of the meek had not arrived. But he knew that he had moved into the national king row.

The conduct of these conventions as well as their distinct institutional quality amazed visitors from abroad. For in these convention battles between Goliath and Cyclops, the warriors seemed mad. Yet the clash of arms followed a pattern—like a joust. Possibly the deliberately planned mass appeal of the party convention arose out of American soil. This mass appeal in American conventions, political and non-political, is a complete surrender to mob psychology. The hooting, the howling, the braying, the blatting, the bellowing of the tens of thousands who sit in the galleries, a vast bowl of lunatic spectators, is directed at the stolid, immobile delegates in the pit below to infect them with the contagion of paranoia from the galleries. It is all like an Indian pow-wow. It is medicine making; savage, reasonless, uncanny. Yet out of this cauldron, this bowl wherein the incanting mob jabbers and yammers at the delegates in the pit below, slowly is formed by some psychological miracle, the purpose and direction which translates this into ordered events in our institutional life. The mania of the pow-wow spreads across the land. Through the newspaper, the radio, the swift telepathy of the insanity envelops millions who are

leadership with Ogden Mills as chairman of a tentative platform committee with representatives in every region.

watching the struggle at the party convention. If the struggle is prolonged the wires buzz, telephone wires, telegraph wires with messages of indignation or delight. While a party convention sits, half the people of America surrender to that party their political birthright.

5

The Republican National Convention which met June 3, 1920, at Chicago, was typical of its kind. Also it was in a sense unique. For at that convention the Republican party was bidding farewell to an era. Sixty years before at Chicago, a similar convention nominated Abraham Lincoln and opened an epoch. Between the nominations of Lincoln and Harding the Republican party had represented what it boastfully called the brains of the nation, meaning those possessive Americans who one way or another had acquired wealth and were able to hold it legally. The organizing talent of America in the huge enterprise of colonizing the Trans-Mississippi country was more or less a Republican enterprise. Republican states were carved out of the prairie, the desert and the mountains. Under the genesis of Republican federal leadership the major enterprises which welded those states into a homogeneous civilization were conceived and established. The railroads, the telegraph wires, the pipelines, even the newspapers and most of the western statesmen were Republicans. Protesting Democrats were political second fiddlers colonizing the trans-Mississippi country. But these migratory Democrats were strangers in a strange land which was on the whole a happy, diluted Republican plutocracy in its ideals and in its achievements. Yet the western Mississippi basin and the mountain states beyond, being the backbone of the Republican party, was in a sense agrarian despite its New England leadership. Even for a decade and a half before 1920, the West was chafing at the leadership of the East—the leadership of Pennsylvania, of New York, of New England.

Theodore Roosevelt's insurgent Progressives were for the most part from the Middle West and from beyond the Mississippi country. From 1904 until 1916 [6] they had kicked up a ruinous Republican row. Theodore Roosevelt took his party by the ears and led it to the consideration of problems arising from the distribution of wealth. The party had small talent for considering such problems. Its brains once creative had become possessive, being interested chiefly in amassing wealth.

In 1920, however, liberalism in the Republican party was a negligible minority and during those hot June days in Chicago the Republican party

[6] In that year a fairly liberal Republican platform written by Nicholas Murray Butler and a group of liberal conservatives was thrown out of the platform committee by Senator Lodge in one of his flashes of arrogant pride. See "Across the Busy Years" by Nicholas Murray Butler.

paid a long farewell to the liberals. But also it drifted down the time stream and away from its old leaders, the stiff-necked, selfconscious plutocrats of the seventies, eighties and nineties and also the Grand Army of the Republic. Gone were the blue-coated veterans of the sixties, wavers of the bloody shirt. The party of 1920 turned to the business of gorging itself in the opportunities for wealth that followed the World War. What was happening was that the economic phases of our life were affective politics. The million dollar precampaign funds of convention candidates reflected the loose and easy money of the period. Republican leaders in official life were personally honest. Yet for all their money honesty many Republican leaders in the party management formed a malignant group— a minority, not generally but too often interested in bond issues, in vast mergers, in tax dodging, in the chicane side of banking, ravenous speculation, gouging the public domain, plundering the gullible in many pious ways.[7]

And so as the crowd milled in the hot hotel lobbies on Michigan Avenue in Chicago, waiting for the party of wealth and brains to seek wisdom through the savage diabolism of its ancient ritual, in Boston, in the sepulchral serenity of the State House sat Calvin Coolidge, a fourth-rate candidate, stewing serenely in his own obscurity. How could the witches foresee that out of the alchemy of their hotchpot would rise this honest, forthright, repressed Puritan Slim Jim—Thane of Cawdor, who would be King thereafter?

[7] Calvin Coolidge's explanation of the complex economic troubles of 1920–21 (outlined in detail a few pages previously in this chapter) centered around the one point of debt, which he abhorred. Looking back at that time for a decade he remembered that he was wise in his day and generation. Quoth he, as he took that degree at Williams College: "The whole country from the government down, had been living on borrowed money. Payday had come, and it was found our capital had been much impaired. In an address at Philadelphia I contended that the only sure method of relieving this distress was for the country to follow the advice of Benjamin Franklin and begin to work and save. . . . Within a year the country had adopted that course which brought an era of great prosperity!"—This quotation is taken from his "Autobiography," p. 153.

BOOK THREE

PREFACE

In prefaces to the other Books of this volume we have looked at the outside world as background for the American scene in which Coolidge moved. In the period following 1923–24, however, it seems well to present developments in the United States as background for the whole world picture.

We had retained our economic strength and financial stability, and above all, our sound gold currency. The free gold of the world was coming to us, because the outside world knew the gold would be returned without loss, from the United States and from no other great country. On our financial and economic policies, the rest of the world pivoted, even though we refused to take international political responsibility. We had refused to enter the League of Nations. Neither Harding nor Coolidge had the remotest conception of the immensity of our economic and financial power, or of the extent of our responsibilities. We did accomplish a naval limitation treaty in 1921. Our government did timidly and indirectly assist in the making of the Dawes Plan in 1924. But the iniquities of the Versailles Treaty, which we might have helped to ameliorate in time had we been in the League of Nations, remained to poison the future.

The German inflation and general chaos continued progressively worse until the Dawes Plan of 1924. Under this Plan, Germany and the governments to whom she owed reparations undertook to restore German currency and credit, and to put Germany in a position to pay reparations. The Plan had intrinsic merit. Its basic error was that scheduled payments were much too high. But safeguards in the Plan under which transfers of payments should not be made provided that if the German currency or her economic life were to be endangered thereby, then the accumulations of reparation money in Germany were to be discontinued unless they could be safely transferred. The fatal error came in the Young Plan which superseded the Dawes Plan in 1929, under which these safeguards were removed.

The nations of Europe which had borrowed from the United States during the war undertook to refund their debts to us. Most of them obtained heavy concessions in the settlements that were made, though Britain proudly refrained from asking for any concession, except that the rate of interest be reduced to three and one-half per cent which she felt her historic credit entitled her to. Also Britain asked that for the first ten years, in view of the strain of the war, she should be allowed to pay only three per cent.

These settlements were made, however, under the influence of fantastic financial illusions. These illusions grew out of the ease with which foreign loans were floated, and the ease with which great sums of money were moved from nation to nation, as a result of the cheap money policies of the Bank of England and of the Federal Reserve System. High tariffs all over the world, and ridiculously high tariffs in America, the world's greatest creditor, meant that grave difficulties would arise when we stopped our foreign lending, or even if loans were sharply reduced, and when the debtor countries had to pay in goods and services.

Financial disorder in France continued until 1926 when the franc, overwhelmed by the gigantic deficits of the French government, took a nose dive to two cents in June. At this point, the Paris mob swarmed about the Palais Bourbon, threatening physical violence to the Deputies unless they should get behind the conservative old lawyer Poincaré, who instituted drastic economies, increased taxes, and pulled the franc up to four cents again.

In England in 1926, the general strike came, raising for a few months grave doubts in the minds of many as to the stability of the British political system. One consequence of the doubts of the middle of the third decade was a further expansion of bank credit in England; much of it went to the continent of Europe where England bought commodities including coal, which England should have produced herself, or at least should have paid for these commodities with her current production of goods. Much of this sterling subsequently came to France, and had to be bought by the Bank of France as a means of holding down the rise in the franc. England became dangerously over-extended financially in 1926, though the day of reckoning did not come until 1931. The temporary effect was, however, to put new liquid funds in the form of sterling at the disposal of the Continent, making it easier for the great boom of 1924–28 to go on.

Russia had come through her revolution. The extreme form of communism demonstrated its hopelessness in 1919 in a breakdown of production, from which Lenin rescued his country by his "New Economic Policy," based on a partial revival of private trade and enterprise with relatively stable money. By 1926 or 1927, Russia realized that she had been chiefly living on the capital accumulated in the pre-Bolshevik period, and turned in 1928 to the Five-year Plan, under which the government forcibly diverted a large part of the labor and resources of the country from the production of consumers' goods to the production of capital goods. The internationalist communism of Trotsky gave way to the realistic nationalist communism of Stalin. Stalin exiled Trotsky, and sent his followers to Siberia, shooting a few by way of diverting example.

The German middle class had been wiped out by inflation. But, following the Dawes Plan, an extraordinary revival of commercial and industrial activity appeared in Germany, accompanied by a growing determination to meet the obligations of the Dawes Plan. An era of good feeling followed the meeting of Stresemann and Briand at Locarno. Frenchmen and Germans tried to be friends, and succeeded surprisingly well until the heavy pressure of the great depression came.

Mussolini had marched on Rome, and the first tyrant in western Europe had established his peace of force in a world to be made safe for democracy.

Even in Asia, rumors of war were shaking the land. The hate and greed bred in Versailles which had checked the free play of commerce in its mission of healing the world, were starting the whirling wheels in a dozen self-determined nations that had no business on the planet—little nations whose very existence was conceived in malice against the defeated Central Powers.

Meanwhile, credit was expanding, which is to say that debt was increasing. The real name for credit is debt. And finally, there came so vast a use of credit in speculation, centering in the United States, that the price of credit, in other words the interest rate, which had been abnormally low, rose to abnormal heights. In 1928–29, the whole outside world was drawn into the maelstrom of Wall Street speculation. Banks in Cairo were sending money to be loaned on call in Wall Street at high rates of interest, and speculative orders for radio stocks and United States steel were cabled from Algiers.

In Washington, with no sense of what the world was really doing, sat Calvin Coolidge, declaring that "the business of America is business"—giving the blessing of his vinegary silence to the witches' brew that bubbled in Wall Street—the Coolidge Bull market.

CHAPTER XIX

Curtain for the Big Second Act

THE band played gaily in the galleries of the Chicago Coliseum, packed to the dome surrounding the pit where a thousand delegates were to assemble for the Republican National Convention of 1920. Delegation after delegation came straggling in. New faces appeared in the leadership, young soldiers fresh from the war. Now and then an old leader toddled down the aisle: Senators Watson and Beveridge, from Indiana; Senators Lodge and Crane, from Massachusetts; Mulvane, of Kansas; George W. Perkins, of New York, vicegerent of the late Theodore Roosevelt; and William "Boss" Barnes, of New York, Theodore's ancient enemy; Governor Herrick, of Ohio; and Walter Brown, the Progressive kingmaker; Senator Philander Knox; Andrew Mellon, who attracted scant applause; William Vare, enemy of Penrose and his friend Joseph Grundy, of Pennsylvania; Redfield Proctor, of Vermont; Bascom Slemp, of Virginia—men who represented the Old Guard in the old party victory over the agrarian rebels who had followed Colonel Roosevelt through the futile charge of 1912 and the surrender of 1916. Senator Boies Penrose, of Pennsylvania, a field marshal of that campaign, still remained the most powerful figure in his party. He did not appear in the opening parade that day in Chicago. He was ill at home in Philadelphia. Here and there a feeble soldier of the Civil War shuffled in with that strange bewilderment in his eyes which the aged have, blinking in fear at the new world. The band that greeted the delegates played new tunes. Silent was "John Brown's Body." Forgotten were the stirring war songs of the sixties: "Marching Through Georgia," "Tramp, Tramp, Tramp the Boys Are Marching," "Tenting on the Old Camp Ground"; the sprightly, martial tunes which had tickled the heels of the soldiers from Vicksburg to Appomattox. For thirty years those ritual hymns of bloody-shirt Republicanism had enlivened the Union veterans' hurrying feet as they came into the Republican conventions to nominate Grant,

Hayes, Garfield, Harrison and McKinley, their Civil War comrades, and Blaine their dear defender.

But the delegates assembling on the floor in the Convention of 1920 heard the songs of another war. Young Fiorello La Guardia and young Ogden Mills, poles apart in their political ideals, marched together with the New York delegation as it took its seat on the floor, while the new street songs flared out from the brass horns in the galleries, trumpeting a new day. With the new day came new issues and the rise of a new dementia in the world whose fever chart was beginning to point even then toward the danger line. But no one could read the chart. Men thought the fever was a glamorous rejuvenation—that they were drinking at some new fountain of perpetual economic youth. When the delegates all were in and the Chaplain had made his prayer and the band had played the "Star Spangled Banner" and the hosts had sung it, up rose Albert Edmund Brown, Director of Community Singing of the Republican League of Massachusetts, who cried:

"Now give three cheers and a tiger for the greatest country on earth— the United States of America!"

And three mighty hosannas arose: "Hurrah for the United States! And a long-tailed T-I-G-E-R!"

Whereupon Chairman Will H. Hays, of Indiana, of the Republican National Committee declared:

"The next order of business is the reading of the call for the Convention."

And so the new era opened. It was one of those coincidences which sometimes seem more significant than they are, that the man who really was to typify that era, and in a certain sense to dominate its expression in the United States, was at the moment nearly a thousand miles from the Convention, sitting in an old-fashioned, mid-Victorian room in the Governor's suite in the State House in Boston, busily engaged in the trivia of the day's routine. He was as remote from the mind and heart of the Convention as though he lived on another star.

The National Republican Committee, according to tradition and rules, names the temporary chairman of all Republican conventions. In 1920, the Committee had decided upon Senator Lodge as temporary chairman. The National Committee in these matters traditionally follows the suggestion of the leaders who are in control of the Convention, generally Congressional leaders. But in the thirty years following Harrison's day senatorial leaders generally had been stronger than leaders in the House of Representatives. In 1920 the leaders in the Senate over-shadowed the Republicans in the House of Representatives and it was fairly well understood before the Convention assembled that those Senators would control who had led the fight on President Wilson's League of Nations Covenant. Sometimes they were referred to as the Senate cabal. President Wilson called them a group of "wilful men." Senator Henry Cabot Lodge was their ideal, but not their leader. He merely gave them facade. As facade he was chosen to be temporary chairman and chairman of the Convention. Among his senatorial associates were Senators Edge and Frelinghuysen, of New Jersey; Wadsworth and Calder, of New York; Watson and New, of Indiana; Curtis, of Kansas, who was to be official party leader in the Senate a few years later; Fall, of New Mexico; McCormick, of Illinois; Phipps, of Colorado; Knox, of Pennsylvania. Wherever they went in the corridors of the political hotels of Chicago during the few days before the Convention met, these men walked with distinction. In their hands was the power, in their presence the glory of the Republican party at its seventeenth national convention.

When he came to the chairman's dais, properly escorted by a committee of distinguished delegates, Senator Henry Cabot Lodge was at the peak of his fame and influence. Many times he had been chairman or temporary chairman, or chairman of the Committee on Resolutions of Republican National Conventions for thirty years. His name had been connected with many party policies of national scope. He had been the friend and adviser of Theodore Roosevelt, Taft, and McKinley. Five times he had been elected Senator from Massachusetts and he had served with distinction in the House. His party and country had showered him with honors. Yet he was held back from higher honors by his own qualities, vindictiveness and grudge bearing, the deep unconscious arrogance of conscious class, an intellectual snobbery which was essentially offensive to certain Republican leaders fundamentally democratic, and to certain one-gallused aristocrats who dominated business and Republican politics during the forty years of Lodge's political life. Governor Coolidge basked under the shadow of Lodge's unconcern through Coolidge's entire political career.

Nothing was more remote from Henry Cabot Lodge, when he rose there in the Republican National Convention, June 8, 1920, meticulously clad in gray tweeds to match his gray fox beard, his gray mop of woolly hair, esoterically matching also his bleak gray heart and his gray granite mind, than the disturbing thought that possibly the wizened-faced, dry-voiced son of the Vermont constable and justice of the peace, then governing in the Massachusetts State House, might some day cross his path and bring him low from his high estate.

Lodge stood there that hot sunny summer day in the stuffy barn-like auditorium at Chicago and for an hour cried anathema upon Wilson, upon the Democratic party and upon the Covenant of the League of Nations. It was a monument of denunciation, that speech of Lodge's, echoing

scores of senatorial speeches equally implacable which had kept America
out of Wilson's League of Nations. Lodge was a hero.[1] They made him
permanent chairman when they organized the Convention regularly. No
great economic or social issue was tightening the nerves of the thousand
delegates who sat on the floor before him. The audience, eager for clamor,
for noise, for pandemonium, but without a serious sense of drama, howled
and hooted easily through the purely rhetorical exhibitions that were staged
to while away the hours during which committees were doing the work of
the Convention. Uncle Joe Cannon came and sang his swan song; Uncle
Joe Cannon, who had helped to nominate Lincoln, and had been a delegate to every Convention but two in sixty years. Chauncey Depew also
came, turning back the clock for a moment to Lincoln's day. The old era
of Blaine, of Harrison, of Quay, of McKinley and Hanna, of Taft and
Aldrich was saying adieu. The new era of the boomers was making its bow.
When Lodge, who entered politics in the days of Grant and Blaine and
Conkling, laid down his gavel that day, the scepter had passed into new
hands. But it was the same old scepter!

2

The Committee on Resolutions wrangled three days and two nights
trying to find some politely confusing phraseology for its declaration on
foreign policy. Finally, Thursday afternoon, June 10, the Committee agreed
upon an innocuous paragraph written by Elihu Root—a paragraph formu-
lated at the suggestion of Senator Lodge and his friends just as Root was
leaving for Europe. He sent it down to the senatorial sponsors of the Convention. Lodge and McCormick brought it to the Resolutions Committee.[2]
It meant nothing. It was a straddle. The two days' accouchement of the
mountain brought forth a pale and bloodless mouse. But it held up the
Convention until Thursday night. The balloting on President did not begin

---

[1] A contemporary estimate of Lodge is recorded in "Across the Busy Years," by
Nicholas Murray Butler, who writes: "The figure that made the least appeal throughout
all these years was that of Henry Cabot Lodge. He was able, vain, intensely egotistical,
narrow-minded, dogmatic, and provincial. For him Pittsfield, Massachusetts, represented
the Farthest West except on the quadrennial occasions when he was willing to cross the
state boundary to attend a Republican National Convention at Cleveland, at Chicago,
or at St. Louis. One would hardly have suspected his background of education and literary
work. Lodge came first to the Convention of 1880 and was present again in 1884, but
he was not a member of the Conventions of 1888 or 1892. He was a delegate in 1896,
1900, 1904, 1908, 1916, 1920, and 1924. His persistence in asking for honors for himself
made him permanent chairman of the Convention of 1900 and 1908, and both temporary and permanent chairman of the Convention of 1920. He was chairman of the
Committee on Resolutions in 1904 and 1916, but in 1924 he cut a sorry figure. So much
out of touch was he with the controlling forces of that Convention, that both he and
Colonel George Harvey wandered together about the corridors of the Cleveland
Hotel asking the newspaper men what the news might be."
[2] Of which the writer hereof was a member.

until Friday, June 11. All day long the rafters reverberated with the nominating speeches presenting and seconding the nominations of General Wood, of Governor Lowden, of Senator Hiram Johnson, of Governor Sproul, of Pennsylvania, of Governor Coolidge who was presented briefly but with meticulously measured enthusiasm by Congressman Frederick Gillett, Speaker of the House of Representatives, who declared Coolidge "patient as Lincoln, silent as Grant, diplomatic as McKinley, with the political instinct of [Theodore] Roosevelt," whose "head is above the clouds standing unshaken amid the tumult and the storm."

The demonstration that greeted the name of Coolidge when Speaker Gillett presented it, was hardly a flutter, lasting less than a minute. The Convention had devoted forty-two minutes of its time, baying, bawling, squealing, screeching its lungs out for Wood, and forty-six minutes of the same mechanical lung-rasping for Lowden. Then came the "also rans," those who received few votes and little applause and who for one reason or another desired the empty honor of a nominating speech. Marion Butler, of North Carolina, presented Judge Jeter Connelly Pritchard. Mr. Ogden Mills nominated Dr. Nicholas Murray Butler. Judge Nathan Miller, of New York, gave the Convention the name of Herbert Hoover. Then as an afterthought, Senator Willis, of Ohio, nominated Senator Warren G. Harding in a short unimportant speech. But one of the really gay moments of the day came when Senator Willis leaned over the railing of the Speaker's runway and said: "Say, boys—and girls too—why not name as the party candidate—." Whereupon there was a spontaneous outburst of laughter and applause and voices: "That's right, we're all boys and girls!" "Why not name the man whose record is the party platform?" [3] But the Convention gave Harding only a two-minute rouser, when Willis closed. By way of diversion, the Washington delegates presented the name of Senator Miles Poindexter, who had no rouser, and when with some grandiloquence a West Virginia delegate took ten minutes to present the name of Senator Howard Sutherland, the Convention was weary.

It was nearly five o'clock when the balloting began. General Leonard Wood led with 287 votes, Governor Lowden with 211, Senator Johnson with 113, Senator Harding with 65, Governor Sproul with 84, Governor Coolidge with 34. Coolidge's votes on the first ballot came 28 from Massachusetts, one from Kentucky, two from New York, two from South Carolina, one from Tennessee. Herbert Hoover had five and a half scattering votes, three from New York. Senator La Follette had 24 votes all from Wisconsin. Nicholas Murray Butler had 68 votes. On the second ballot, Wood and Lowden gained, one having 289 votes, the other 259 votes. Coolidge lost two votes and Butler nearly 30. The afternoon was old. The

[3] Proceedings of the Seventeenth Republican Convention—Official Report.

sun had set. In the twilight another ballot was taken. Wood climbed to 303, Lowden to 282; Coolidge dropped to 27. The Convention tried to adjourn but the senatorial plan demanded another ballot. Lowden was stopped at 289; Wood had climbed to 314; Harding to 61. Coolidge stuck at 25.

Then the Convention was adjourned summarily by Chairman Lodge after a vote to adjourn had been obviously defeated. He and Senator Smoot had decided adjournment must come or the Convention would go into a night deadlock which might have been bad for the Republican party. The two had no authority for their decision, and Lodge, the scholar in politics, arbitrarily took the authority which he knew his position gave him morally if not under parliamentary law, and sent the delegates home.

While the Convention stands adjourned Friday night, it is necessary to go behind the scenes where the real drama of the hour was playing. A thousand delegates, another thousand alternates and ten thousand of their friends milled up and down Michigan Avenue that night, after the adjournment, bobbing in and out of the great political hotels, packing the elevators, roaming through the hotel corridors, wandering with busy vacuity into the headquarters of Wood, of Lowden, of Johnson, of Coolidge, of Nicholas Murray Butler, hunting for crumbs of gossip which they munched with zest. Little did they know or realize the forces that were directing their political destinies and that of their country.

It is hard to describe with any sort of accurate intelligence what forces moved the men behind the scenes. It is easy to call it a plot. It was not a conscious conspiracy directed by greedy and unsocial forces. It was the tendency of the hour. The greedy and unsocial forces of that day had just gorged themselves upon the profits of a war, the most horrible war with the most outrageous profits that mankind had ever known. These forces of amalgamated wealth were not dominated by keen foresighted men who could intrigue with a plan and a purpose. It is true that in most cases the Senators and the powerful members of the national House of Representatives who manipulated the Republican convention were more or less friendly to the great organized commodity interests of the country. Each one in addition to being a Senator from his state was likely to take the color of his record from the leading industrial interests or commercial organizations in his state. So steel had its Senators, and oil its Senatorial leaders; textiles sometimes had three or four New England Senators; railways had their favorites, coal its pets, the food industries their special champions, and so through the list. But these industries did not direct their Senators behind the scenes at Chicago. Rather the class feeling of the Senators motivated their action. A movement rather than a plan, or plot, or definite intrigue controlled the puppets backstage while the delegates were milling through the halls.

Perhaps to understand why Coolidge came to the White House, one must look for a moment at the cast of the drama staged behind the scenes that Friday night in June, at Chicago. Few of the men there had a remote notion that Coolidge might arise out of their rather aimless but always subconsciously class-controlled palaver that night. First of the dramatis personae of the folk play let us consider Senator Boies Penrose, of Pennsylvania, rich in his own right with a Harvard background, a bachelor who played with politics as men play solitaire. For ten years he had been the will of the Republican party. He was not in the Convention. Yet even though from his sickroom in Philadelphia, he haunted the Convention, a wistful ghost, he could not direct its manipulations. He was a specter in absentia. He did not know it but he was beginning to be a tradition. Yet the tradition had its power. A few leaders visited him before the Convention, got his ideas, and because he had dominated the Senate until the day they took the train for Chicago, his ideas and suggestions were powerful. And though sick and five hundred miles from the Convention,[4] Penrose, as a sort of time spirit of the occasion, more than any other personality influenced those into whose hands the delegates unwittingly gave the power to direct the course of the Republican party. He was a large man—perhaps huge is a better word, with a Neanderthal jaw, a good cortex between his skull and his brain, a long memory that forgot grudges when it was expedient, a heavy hand, a scorn for idealists because he had been disillusioned, and a capacity for detail in that labyrinth where sometimes even honest Senators have to trade public welfare for their private interests. Penrose was the black beast of the progressives. Theodore Roosevelt used Penrose and William Barnes, the New York boss at Albany, to typify all that he scorned in American politics. Yet in the Republican party they and their kind always held genuine leadership. The party did not move faster than the moral sensibility of its bosses travelled, generally Senatorial bosses. Penrose succeeded Aldrich, who succeeded Quay, who succeeded Hanna, and so on back to Thaddeus Stevens and Thurlow Weed.[5] The surrender of the Republican party of the United States at its Convention every four years frequently placed all the delegated power of millions of partisans in

[4] Nevertheless, according to a letter from Leighton C. Taylor, secretary to Senator Penrose, to Senator Grundy, August 7, 1936, he had decided definitely to go to the Chicago Convention. Arrangements had been made for a special train for the Senator and his party, and Marshall Field had offered his house in Chicago to Senator Penrose for a week. Note, in passing, Marshall Field, one of the richest men in America, welcoming Penrose (a part of the instinctive bund of American plutocracy). At the last minute, however, the Senator was beset with another disease entirely foreign to his major illness and had to remain at home in Philadelphia, at 1331 Spruce Street.

[5] Grandfather of William Barnes, of Albany, state boss from 1905 to 1915.

the hands of men like Penrose. These leader-bosses were calculating without personal money greed,[6] patriotic for their party's success at any cost.
They weighed issues and men from a party standpoint. Probably that day
in June, 1920, at Chicago, they weighed them more wisely than men and
issues would have been weighed by the Convention uncontrolled by the
bosses. As for the country, that is another matter.

Penrose had conceived the idea during January, 1920, and the spring,
that Harding might be a presidential possibility. The candidacy of Governor Sproul, of Pennsylvania, was merely a feint. "Of course Penrose was
never sincerely for Sproul at any stage of the game." [7] Penrose preceded
Daugherty as the original Harding discoverer. Writes his secretary: [8]

"In the summer of 1919, on a hot afternoon, the Senator did not go to
the Capitol because he wanted to clean up some correspondence and do
a lot of long distance telephoning. He asked me to call Harding. When
Harding came in, Penrose said: 'Take off your coat, Senator, and sit down.'
The next thing from Penrose was: 'Harding, how would you like to be
President?' I don't think anyone could have registered more surprise than
Harding. He said: 'Why Penrose, I haven't any money and I have my
own troubles in Ohio. In fact, I will be mighty glad if I can come back to
the Senate.' Penrose said: 'You don't need any money. I'll look out after
that. You will make the McKinley type of candidate. You look the part.
You can make a front porch campaign like McKinley's and we'll do the
rest.' From then on Penrose began to talk Harding, until Harding made a
speech before the Manufacturers' Association in Philadelphia. That night
I took Harding's speech down in shorthand, and as soon as he finished
speaking I hurried down to 1331 Spruce Street and read it to the Senator.
One of his comments was: 'Harding isn't as big a man as I thought he was.
He should have talked more about the tariff and not so much about playing cymbals in the Marion brass band.' "

Penrose was looking for a winner. He rejected Harding. Later on Senator
McCormick tried to interest Senator Penrose in Senator Knox, of Pennsylvania, and a few weeks before the Convention on a Sunday afternoon while
Knox was living at Valley Forge,[9] "the Senator had me telephone Knox to
come to 1331 Spruce. The Senator put the same question to Knox that he
put to Harding: 'Knox, how would you like to be President?'

[6] They found a quarter of a million dollars in cash in Penrose's strong box at his death.
Probably it was a party fund, money needed for swift and sometimes emergency use.
The fund came possibly from sources that would not bear inspection and possibly was to
go for ends equally devious, but not Penrose's personal ends.
[7] Letter from Leighton C. Taylor, secretary to Senator Penrose, to Senator Grundy,
August 7, 1936.
[8] *Ibid.*
[9] Letter from Leighton C. Taylor, secretary to Senator Penrose, to Senator Grundy,
August 7, 1936.

"Knox was sitting on one of those large high Penrose chairs." For a man of Penrose's size, huge chairs of the Italian renaissance, stiff and high backed, were fairly comfortable. Mr. Taylor adds: "Knox's feet didn't touch the floor. He almost fell off from the shock. Knox once told Penrose before that he wouldn't come to see him if he didn't get smaller chairs. He looked like a boy in one of those chairs." Penrose always enjoyed Knox's discomfiture. For Knox worked the other side of the street. He was the great lawyer, the intellectual leader of those Senators who made speeches. Knox had been in the Cabinet. He was used to the spotlight. Penrose worked in the shadow. "At the Senator's query, Knox began to plead age, talked of his bad health; said that he had all the honors he wanted, and on and on. Penrose coaxed with Knox: 'You won't need to do much work. We'll get a good hard-working Cabinet for you and you can sit at the head of the table and run the show.'"

Knox continued coy and when he had left, Penrose turned to Taylor and said: "'That little devil would like to have the job handed to him on a silver platter.' Every day for ten days the Senator would ask me: 'Have you heard from Knox?' We never did hear from Knox, and no consideration was given to his candidacy at Chicago." [10] Penrose had said thumbs down. Mr. Taylor remembers that Senator Penrose had a private telephone wire into a suite in the Auditorium Hotel, in Chicago, probably Senator McCormick's, or the late John T. King's. Mr. Taylor writes:

"I have before me the shorthand notebook I used at 1331 Spruce Street during the week of the Convention. In it I find, among other things, the verbatim notes of three conversations held over our private wire. Two conversations were held by Senator Penrose with Senator McCormick and John T. King; [11] another a day later with Senator Brandegee, of Connecticut."

There can be no doubt that for all his illness, Penrose kept his hand on the Senatorial leaders of the Convention. His mind directed the preliminary proceedings. Mr. Taylor recalls that Senator Penrose followed the organization and the work of the Resolutions Committee, the Credentials Committee, the Southern delegate situation, Lodge's speech as permanent chairman; that he took a hand in drafting the plank on the ratification of the Versailles Treaty and that he kept close tab on the status of the various candidates, and watched the Pennsylvania and New York delegations to see that they did not get out of hand. [12]

[10] Letter from Leighton C. Taylor, secretary to Senator Penrose, to Senator Grundy, August 7, 1936.

[11] Republican National Committeeman from Connecticut.

[12] In his testimony before the Senate Committee in Part 2 of the document entitled "Lobby Investigation" consisting of hearings before a subcommittee of the Committee on the Judiciary of the Seventy-first Congress, Senator Joseph Grundy testified "that

But in the actual direction of the final work of the Convention, Penrose had but little influence. He was a receding figure standing under the shadow of death, and everyone knew it.

3

Lodge, Smoot and George Harvey dined together that Friday evening after the summary adjournment of the Convention. They dined in George Harvey's suite at the Blackstone Hotel. It was obvious that if the Convention was not to be wrecked upon the rock of bitter deadlock a compromise candidate must be found. They telephoned to various Senators and politicians controlling delegations who belonged to the Republican Sanhedrin, the inner council of the pharisees. These men curiously were not rich men. They served rich men or they served riches, the thing that Theodore Roosevelt called amalgamated wealth. They were honest, competent servants personally. Politically they were without fear and without reproach, according to their lights and standards. They summoned to Harvey's room, Joseph R. Grundy, the Pennsylvania leader, Senator Charles Curtis, of Kansas, who was nimble and tireless, and Senator Brandegee, of Connecticut. Senator Wadsworth was in and out of the room casually at various times during the long conference. This informal gathering took over the powers of a thousand delegates who milled about the hotels, rode or walked up and down Michigan Avenue, and in a carefree way enjoyed themselves, not realizing how they were being moved upon the board by the chess players in Harvey's room. Curiously enough—Harry Daugherty, Harding's Warwick, was not there. Apparently he knew little about the slow accumulation of Senatorial sentiment for Harding. Certainly he did nothing that night to accelerate it.

No unanimous decision, not even a majority decision, was reached during Friday night for Harding. But the current ran his way. The more active, eager, decisive members of the Senate cabal were for Harding, and went to their delegations, passing the word down the line that in due time the gods would nod and Harding would be named. Curtis was of this group; also Brandegee, McCormick, Lodge and Smoot. Will Hays had some following as a candidate; Knox only a little. And it was in those hours when Hiram Johnson, being asked to take second place with Knox who was known to have an afflicted heart, answered impetuously: "You would put a heart beat between me and the White House!"

Senator Penrose was so seriously ill that nobody talked to him during that period." Senator Grundy's statement is corroborated by Nicholas Murray Butler and contradicts the statement above quoted which being supported by stenographic notes is probably technically correct. But also probably at the last Penrose had little to do with the nomination of Harding.

Later the Senatorial leaders went to Johnson and asked him to take second place with Harding. Again he refused.[13]

Harding, himself, was in Chicago for a few days during the Convention. Friday he appeared on the fringe of the crowd in hotel lobbies, rather a battered figure; slouchily clad, with a two days' beard, with weary eyes and a figure slumped, dejected. For he was not confident about what Harry Daugherty kept telling him of the inner workings of the Convention. His case seemed hopeless. Harding knew little of the informal milling caucus that gathered about the table in Harvey's room that night, where his Senatorial friends were pressing Harding's candidacy.

This Senatorial conspiracy developed in the Republican Convention of 1920 because Republican Senatorial leadership wished to submerge the Presidential office. Under Theodore Roosevelt and Woodrow Wilson for fifteen years of the twenty years in the new century, the President had taken leadership, and Congress, particularly the Senate, was losing its prestige. The phrase, "the Roosevelt policies," and the flare of the Wilsonian doctrines convinced many Senators, and particularly Republican Senatorial leaders, that a hero in the White House was a bad thing for their party and their country. Naturally they desired no hero. They distrusted the strength of General Wood. They questioned the strength of Governor Lowden. Coolidge certainly was not of their kind. But Harding was one of their number. Harry Daugherty's prophecy might easily have been made by one who understood the determination of the Senatorial leaders not again to be deflated by a strong man in the White House.[14]

4

So Friday night wore away and the activity in George Harvey's room literally petered out. It was not the smoke-filled room that Daugherty had prophesied, and at 2:11, only a few dawdlers sat around snoozing, hanging on like old dogs dreaming with their paws on the bone, hungry to be insiders and actually remote from the situation even there. George Harvey himself was not sure how the day would go when he went to bed just before the dawn. This story is elaborated that a fairly authentic picture may be made of the way national conventions go. There is direction; there is purpose; there is also chance. The mixture of chance and planning achieves the

[13] Mark Sullivan, "Our Times: The Twenties," page 77.

[14] In his memoirs, "Across the Busy Years," Nicholas Murray Butler who was a delegate to the 1920 Republican convention and an insider of insiders writes: "The carefully planned and assiduous work done on Harding's behalf by Harry M. Daugherty was of greatest importance, but even that would not have won the battle had not a group of Senators put their heads together to nominate a man who, as one of them cynically said, would, when elected, sign whatever bills the Senate sent him and not send bills to the Senate to pass."

end. But in this Convention, of 1920, the planning was all done back of the scenery. It was done by the faithful and sometimes intelligent servants of the ruling plutocracy which in that year quietly directed America's destiny. Most of these leaders in Harvey's room wore unwittingly the collar of some commodity unit; steel, coal, oil, textiles, banking, copper, thus proving the theory that the possession of economic advantage carries with it political power even in a democracy.

By common consent and agreement—and with scarcely more plot or direction—early Saturday morning a score of state leaders, chiefly Harding's supporters in the straggling conference in Harvey's room, went to the morning meetings of their state delegations and announced the program. Wood and Lowden would be given four ballots in which to develop their strength or prove to the delegates that the Convention was deadlocked. Lowden's strength was destined to disintegrate. Harding was to be gradually pushed into third place during those four ballots. Then the band wagon would start—so quoth the prophets. Of course, the Wood and Lowden strategists knew the program as soon as the delegates. But knowing the program did not help them. If they could have united they might have stopped Harding. But Wood was objectionable to many Lowden men who thought Wood was too much of a militarist. A scandal had developed in the Missouri delegation for which personally Lowden seems to have been in no way responsible, but which revealed the use of money for Lowden in choosing the Missouri delegation. This convinced the friends of Wood that Lowden could not win at the election even if the Wood delegates nominated Lowden. The inevitable deadlock approached.

Among the rank and file of the delegates, no one knew much about Harding, good or ill. He was "a man made of a cheeseparing" by the Senate leaders who had followed Penrose for a decade.

While the Senators were pondering and parleying in George Harvey's room, those who were near the inside but not of it, resented the arrogance of the Senatorial leadership. Will Hays, Chairman of the Republican National Committee, held a press conference for no particular reason and introduced Governor Beeckman, of Rhode Island, to a group of newspapermen. Suddenly to Hays' dismay, the Governor broke out with a speech of protest "demanding to know if this was a Republican convention or just a Senatorial caucus." He cried in irony: "Are mere Governors going to have anything to say or are they going to vote just as the Senators tell them to?" [15] Outcries similar to that might have been heard that night all through the Convention crowd. Governor Henry J. Allen, of Kansas, was vocal, as were the genuine friends of both Governor Lowden and General Wood. Excepting Governor Allen, those men, indignant men, of the sec-

[15] Charles Willis Thompson, in "Presidents I've Known."

ond line were not progressive followers of Theodore Roosevelt. They believed in the divine right of the party of "wealth and brains" to rule. They did not question the infallibility "of the rich and the wise and the good." They were men who thought they controlled delegations until they came to Chicago. Also to these Governors and to others who were in the second line, Penrose was a symbol. Men said: "Penrose in his sickbed at Philadelphia is the real power behind the throne." Reporters printed the story Saturday morning that Penrose had the final veto on the nomination.[16] Probably Penrose did not give the signal that closed the contest. Yet the thing that Penrose stood for had power in that night's conference of the high command. Penrose, the incarnation and epitome of plutocratic power in a democracy, gave purpose to the Convention and Penrose on his death-bed by some ironic coincidence was the ominous shadow that hovered over the last phase of a dying era. He died in less than six months. The era collapsed in a dozen years. It was never able to conjure into flesh and blood the strength to give him a successor.

5

At 10:30 Saturday morning, a fetid day in the Convention Hall, the fifth ballot for the presidential nomination occurred. Ohio's vote on the third ballot had been 9 for Wood, 39 for Harding. Apparently, the 9 votes for Wood were expected to leave him on that ballot. The delegation was polled. Wood's men stuck. But Harding on that first morning ballot jumped from 58 to 78. Wood began to collapse. His vote dropped to 299. Lowden came up to 303. Coolidge dropped to 23. Nicholas Murray Butler's strength was down to four. Harding picked up a few from New York, Missouri, Texas, West Virginia, and from odd corners of the country he acquired small gains. Without pause the Convention proceeded to the sixth ballot, a heroic attempt to rally the Convention against the Harding plot. Wood and Lowden rallied, each to 311 votes. Harding dropped to 89. Coolidge gained half a dozen. The delegations were restless, apparently annoyed, certainly confused by Harding orders. Georgia, Michigan and Ohio were polled without important changes. And the Convention hastened into the seventh ballot. There again the deadlock between Wood and Lowden was maintained with only a half a vote difference between them. But Harding had 105 votes. Coolidge's vote was 28. In the eighth ballot signs of cracking began. Wood dropped to 299, Lowden to 307, Harding rose to 133, gaining his votes from Indiana, Missouri, Texas, Tennessee and West Virginia. The Lowden men in Missouri tried to break to Harding. A demonstration for Harding occurred, led by Ohio delegates. It was long past noon. A Kentucky delegate apparently fearing the worst tried

16 I printed several stories myself stemming out of the revolt against the Senators.

to force a recess. Ohio on behalf of Harding demanded a rollcall. But the Convention would not grant a rollcall and recessed on a standing vote until four o'clock.

In recess the deed was done. Lowden collapsed. On the ninth ballot Harding led with 374 votes. Wood held only 249. Coolidge's 22 Massachusetts delegates stood fast. Curtis brought in half of Wood's Kansas votes for Harding, Roraback, the Connecticut boss, brought in his state. Senator Watson brought in Indiana. Kentucky and the South joined the Harding stampede. Senator Spencer brought in Missouri. Wadsworth brought in New York. Grundy held Pennsylvania for Sproul. Senator Smoot stayed by Wood. As the rollcall proceeded, one demonstration after another started marching and cheering through the aisles, taking the standards of the state to the Ohio delegation and paying obeisance. The galleries caterwauled with delight. Here was drama but not climax. The Senatorial and the state bosses were preparing for the final blow. Senator Reed Smoot had been called to preside, a fitting imperator, to watch the gladiatorial slaughter of the Christians. Kansas led the real break. According to the official report of the Convention "one of the Kansas delegates [17] took the state standard or marker, raised it high in the air, having fastened a picture of Harding and a flag thereon, and started marching around the Hall followed by the Kansas delegation bearing pictures of Harding and flags, and joined by numerous delegates from other delegations."

But these cold words do not convey the deafening tumult, the maniacal clamor that arose in the galleries and spread to the Convention as Kansas marched to Ohio. Shrieking, whooping, noise-making machines zoomed, zipped, yowled and ripped the hot air of the Convention. Then all the world knew that the United States Senate had taken leadership in the party of "wealth and brains." Presidential leadership had received its death blow. Kansas would not move without Curtis. Curtis, Republican Senatorial whip, would not move except under the impulse of the Republican Senatorial organization. Lodge was Republican leader of the Senate, the high priest of Republican conservatism. It was an hour of triumph for the great god status quo, and the Senate was his prophet. Here was the ritual of his glorification. Hosannas to the Senate in the highest were these thunderous roars of applause torn from ten thousand throats. The presidential office in the American republic was put in its proper place. But alas for the vanity of man!

After twenty minutes of clamor, the Ohio delegation requested the other delegates to stop the pandemonium. Kentucky was called and 26 votes plunked for Harding. The South controlled by the Republican organization began falling in line for Harding. Wadsworth brought New York into the

[17] Former Governor Bailey, head of the Kansas delegation.

Harding cavalcade with 66 votes. When the nine Wood votes in Ohio were cast for their losing leader—the last strong man to menace the Senatorial cabal, the gallery hissed. Oklahoma announced twenty votes for Harding, but upon a poll of the delegation two voted against him.

Michigan, which had been polled on every roll call, finally voted 27 for Harding and 3 for Johnson. The vote on the ninth ballot stood: Harding 374, Wood 249, Lowden 121, Johnson 82, Coolidge 28, Hoover 6. The end was obvious. No one in the multitude doubted that the Senate would have its way.

The tenth ballot proceeded swiftly and in comparative quiet. It disclosed 692 for Harding, 156 for Wood, 11 for Lowden, 80 for Johnson, 5 for Coolidge (one from Massachusetts, 4 from New York).[18] Senator Lodge and Senator Crane controlled the Massachusetts delegation. Lodge of course had no regrets in leaving Coolidge. He was one of the conspirators in Harvey's room. But Crane left Coolidge because it was the sensible thing for a Senator to do when the tide was running so powerfully to Harding. Curiously enough Speaker Gillett, from Coolidge's district, stood hitched. As Chairman of the Massachusetts delegation he did not vote for Harding but refrained from voting until Harding had won. When the leaders decided that the coup de grâce should be given to Wood at the close of the tenth ballot, the chairman of the Pennsylvania delegation announced sixty of its sixty-seven votes for Harding. On the last ballot Herbert Hoover[19] had nine votes—the extreme protest against Harding.

It was then shortly after six o'clock. Again the raucous bark of the mob

[18] Wrote Nicholas Murray Butler a dozen years later, "Across the Busy Years," 1936, Scribners:

"On Friday morning, June 9, while the nomination speeches were making at the Convention Hall, I sat alone with Warren Harding in his rooms at the Auditorium Hotel. The weather was very hot and he had taken off his coat and waistcoat and was fanning himself vigorously. In the course of our conversation he said: 'I cannot afford to keep these rooms any longer and I have sent word downstairs to say that I am giving them up this evening. This convention will never nominate me. I do not propose to go back to the Senate. I am going to quit politics and devote myself to my newspaper.' These words were spoken between eleven and twelve o'clock on Friday morning. At about six-fifteen on the following afternoon Warren Harding had been nominated for President of the United States. At the moment when his nomination was made, Frank Lowden and I were sitting with Harding in one of the small rooms back of the platform of the Convention Hall. We three were alone. The roll was being called on the tenth ballot. Suddenly, there was a tremendous roar from the Convention Hall. In an instant, the door of the room in which we three were sitting burst open and Charles B. Warren, of Michigan, leapt into the room, shouting: 'Pennsylvania has voted for you, Harding, and you are nominated!' Harding rose, and with one hand in Lowden's and one in mine, he said with choking voice: 'If the great honor of the Presidency is to come to me, I shall need all the help that you two friends can give me.' In another instant Harry Daugherty arrived, seized Harding, and took him back to the Auditorium Hotel before a crowd could assemble."

[19] One of the nine Hoover votes came from Kansas and was cast by the author of this chronicle.

echoed through the barn-like rafters of the Hall. Around and around the floor of the pit marched the howling delegates led by Ohio. For ten minutes the demonstration proceeded. North Dakota, Arizona, Washington, New Mexico, Colorado, Illinois, all changed their votes. A motion to make the nomination unanimous was offered. A loud chorus of "aye" arose from the delegates and the few weak Wisconsin "no's" which followed when the Chairman put the opposing questions were "greeted by a storm of hisses from the other delegates and the gallery," says the official reporter, and adds, "The chair declares Senator Warren G. Harding, of Ohio, unanimously nominated for President. (Applause loud and prolonged.)" [20]

Coolidge [21] had only 34 first ballot votes, 28 on the ninth ballot, and only 5 who stayed with him against the pressure of the Harding stampede. And only one of these was from Massachusetts. On that last ballot Coolidge polled less than Johnson, who quit with 80 votes, and Hoover, who quit with 9. La Follette held his 24 Wisconsin stalwarts who had voted for him in national conventions for a dozen years.

The deed was done. The Senate leaders had won a glorious victory. Those who were intelligent like Lodge and Penrose felt that they had changed a tendency in American political life, the growing exaltation of the Presidency. They had named for the Presidential office one of their own men, a man of their ideals and background, in a year when victory seemed certain. They felt that eight years of Senatorial peace and prosperity lay ahead of them, a tranquil and placid course. So when the gavel fell and the "applause loud and prolonged greeted the declaration of the Chairman that Senator Warren G. Harding, of Ohio had been unanimously nominated for President," [22] the Senators stretched their legs, put on their hats and walked with due Senatorial dignity from the scene, victors in no mean battle. And fate, shaking the dice in the wings, smiled and rolled the bones.

Saturday morning, in the gray dawn of that June day, one thing was decided in conference in George Harvey's room, to nominate Senator Irvine Lenroot, of Wisconsin, for Vice President. After Harding's nomination late Saturday afternoon, Senator Medill McCormick took the floor to nominate Senator Lenroot and in his prefatory remarks declared (though he knew better) that the Convention had "nominated for the Presidency a man of ripe experience, of deep learning and of great power." Also the Convention instinctively knew better. The Convention had swallowed with grimaces the straddling platform plank on foreign relations which the Senate had prepared. It had taken the

[20] Official report, p. 224.
[21] From tabulation of the tenth ballot in the "Official Report of the Proceedings."
[22] Official proceedings.

Senators' candidate for President. Lowden men and Wood men knew that they had been thwarted by a cabal which in its heart was a nega- tion of representative government. For all the noise, for all the tramping of thousands of feet in the tenth ballot parade, the delegates were unhappy. But orders were orders. The Senators, who saw their delegations in the morning and passed out the Harding program, put Lenroot's name upon it. When Senator McCormick had finished his two minute speech nominating Lenroot, Hurt, of Kentucky, seconded the Lenroot nomination as did Remmel, of Arkansas, Senator Calder, of New York, Governor Herrick, of Ohio. Orders were obvious—Senatorial orders. Then the revolt of the mob came quickly and with amazing directness.

6

Hundreds of delegates had left the Hall. The show in the big tent was all over. More than half of the Massachusetts delegates were strolling down Michigan Boulevard toward their hotels. In the barn-like hall the roaring tramp, tramp, tramp and clatter of falling seats was like a rifle charge. The presiding officer of the dwindling Convention could scarcely hear the pro- ceedings on the floor. He assumed that program was being followed. When suddenly far over on the north side of the floor of the Convention a man rose on a chair and began to cry out with a fog horn voice for recognition. He was not on the chairman's list, but the chairman was an easy boss and recognized Mr. Wallace McCamant,[23] of Oregon. What he said no one heard clearly, but the delegates above the rattle of crashing seats and the shuffle of departing feet realized that he was nominating Calvin Coolidge for Vice President. Here is what he said:

"When the Oregon delegation came here instructed by the people of our state to present to this Convention as Oregon candidate for the office of Vice President a distinguished son of Massachusetts, he requested that we refrain from presenting his name. But there is another son of Massachusetts who has been in the public eye in the last year, a man who is sterling in his Americanism and stands for all that the Republican party holds dear; and on behalf of the Oregon delegation I name for the exalted office of Vice President, Governor Calvin Coolidge, of Massachusetts." [24]

The official stenographer of the Convention records that "this nomina- tion received an outburst of applause of short duration but of great power."

---

[23] Charles Willis Thompson writes in "Presidents I've Known": "Portland, Oregon's McCamant turned out to be an ex-Judge and a man of sterling worth. As soon as Calvin Coolidge became President he appointed him to Federal office, but though the years had beaten down much of the 'Soviet's' [meaning the "wilful man"] power, those who remained fanned a little spark of revenge and evened their score with the rebel Mc- Camant by refusing to confirm the nomination."

[24] Official report, p. 226.

The "great power" was the hysterical voice of revolt. The Senate bosses were gone. For once in four long days the Convention was to have its own way. The Michigan delegation seconded the Coolidge nomination and quickly the delegates from Maryland, North Dakota, Arkansas, Connecticut, Pennsylvania—all delegations that were supposed to be controlled— rose to their feet. A Colorado delegate tried to close the nominations but yielded for a moment while Kansas placed in nomination Governor Henry J. Allen. Then a delegate from Illinois, another from Nebraska, a woman from Nevada, a Vermont delegate, and a New York delegate began clamoring for the nomination of Governor Coolidge. The names of all the nominees for Vice President were read and the name of Coolidge produced something decently approaching an ovation. The Convention leaders were appalled but powerless. The thing came out of the air like lightning. The resulting ballot gave 674 for Coolidge with the Convention bosses able to assemble only 146 for Lenroot. The nomination was made unanimous. No response came when the negative vote was called.

All that afternoon in Boston, Governor Coolidge had been in contact with the Chicago Convention through the newspapers, through telegrams that came from friends. The failure of the Governor to make an impression in the contest for the Presidential nomination was reflected in Coolidge by a grievous silence. It lasted half an hour. But Frank Stearns, ever faithful —"grief could not drive him away"—clung to the wire. When it was six o'clock in Chicago and the nomination of Harding had been achieved, in Boston it was seven o'clock. The Governor and Mrs. Coolidge were alone in their two-room suite in the Adams House. With wifely anxiety, she had come down from Northampton to be with him while the Convention at Chicago was deciding his fate.

His interest in the Convention was waning. He sat in his room at seven o'clock when suddenly the news came of the demonstration for him following the Oregonian's speech. Newspapers were calling him. Mrs. Coolidge knew the signs, realized that her husband was interested. He told her the story of McCamant's surprising speech. It was time to go to dinner. After they had chatted a moment, the telephone rang. He answered it. He put it down, gazed at his wife a puzzled second, and then quacked: "Nominated."

Out of that one lean word he sucked the brackish heady wine of his triumph.

"You aren't going to take it are you?" quizzed Mrs. Coolidge, soothing his vanity.

"Well—I suppose I'll have to," he replied. The episode closed.

Twice again the telephone rang as two friends notified him from different phones—one in Chicago, one in Boston—of his nomination. Being wise and kind, he let each friend feel that he was the bearer of the first glad tidings, after the way of politicians. In a few minutes friends were knocking at the door to tell him in person of the news.[25] There they found him, in the shabby little sitting room—a bedroom plus a table and two chairs and minus a bed—sitting cross-legged, sphinx-like, repressing his satisfaction, smoking a big stogy cheroot. He had himself in hand. He let them know that if McCamant had not named him others were about to do so, and who and whence and how. Mrs. Coolidge soon retired to the other room of the suite as the crowd thickened. Reporters came, politicians, supporters, friends. He managed to get on his feet, to shake hands, to bark in monosyllabic satisfaction and in the excitement of the hour to light another stogy. Mrs. Coolidge, looking back on the exultant hours through which her husband had passed, wrote fifteen years later: "Mr. Coolidge was always sentimental!" [26] But in that hour he gloated in his heart and turned a sour serious adamant face to the world.

Cinderella, riding home in the pumpkin behind the white mice, was not half so pleased with herself.

[25] Letter from Henry F. Long, Mr. Coolidge's gubernatorial secretary to the author.
[26] Good Housekeeping, February, 1935.

CHAPTER XX

"Then Cuffing Cinderella, the Stepsisters Went Upstairs to Bed"

AN HOUR after Judge McCamant had mounted his chair that mid-June evening, 1920, and called the name of Calvin Coolidge across the emptying Hall at Chicago, the bell in the tower of the Edwards Congregational church at Northampton was gaily proclaiming the nomination of Calvin Coolidge for Vice President. The town was delighted, as what small town would not be pleased to have its first citizen nominated for Vice President in a year when his election seemed sure? In the twilight, the main street blackened with citizens, everyone buzzing, all eager with the news. Four days later New England joined Northampton in a celebration. When Governor Coolidge left Boston for the twenty-fifth reunion of his class at Amherst, as he crossed the state along the route upon which he had travelled so many years, the towns turned out to greet him and at Northampton a great multitude surrounded him as he paused on his journey. Buildings were decorated with bunting. His picture flashed out everywhere. Bands and fife and drum corps blared and squeaked and boomed. The College glee club sang. A parade thumped the pavement. And when Governor Coolidge came to his journey's end at Amherst, it was an ovation. It was then and there that they took him to see the memorial cottage where Emily Dickinson wrote her lovely lyrics. He stared at the manuscript with his granite eyes and azoic face and ventured a clowning quip: "She writes with her hands. I dictate."

Which among the dumb passed for dumbness, as often his humor was mistaken.[1] It did not seem to bother him. Probably he figured, Life Is Like That.

[1] But the best comment of all on the subject is Coolidge's plaintive remark: "Whenever I do indulge my sense of humor, it always gets me into trouble."—"The Quick and the Dead," p. 246.

A few weeks later he went to Plymouth for Old Home Week, he and Mrs. Coolidge and the two boys, John and Calvin. It was a great day at Plymouth Notch. In ancient Capua, Lentulus returning with victorious legions had no more sense of triumph. The hills roundabout shed all their human habitants who crowded into the little village street and surrounded the Home Town Boy who had Made Good, a typical American scene, noisy, happy, with its venders and fakers, its little country band, and its pretty girls and boys in their Sunday best, all doing homage to their neighbor. The governor of Vermont was there, and three ex-governors, and the leading Republicans, and when they had all made speeches, the Vice Presidential candidate declared:

"Vermont is my birthright; it is a high and noble birthright. Rising to it entails a great and high obligation."

Of which there were five hundred or a thousand words more. Whereupon the crowds, who had walked down the valleys and up the hills to hear their old friend whom they called "Cal" behind his back, and then circumspectly addressed as Governor, went their various ways. And still the tide did not begin to recede. The State University of Vermont gave the Governor an honorary degree. When he had received it and made appropriate remarks, he journeyed back to Northampton where all New England and delegates from each of the forty-eight states appointed by the Republican National Convention assembled to hear Governor Morrow, of Kentucky, make the notification speech.[2] Governor Morrow's speech as befitted the eulogy of a good Kentuckian, ended in poetry, Longfellow's "The Building of the Ship":

> Sail on, nor fear to breast the sea!
> Our hearts, our hopes, our prayers, our tears,
> Are all with thee,—are all with thee.

The Senatorial associates who had named Harding were in charge of the Republican National Committee. It soon became evident to these stepsisters that they had a problem child on their hands in Calvin Coolidge. He lacked above everything that double-breasted, balloon-jawed, Senatorial dignity which Harding assumed so easily. The Yankee's voice quacked and his platform manner had no oratorical appeal. After the notification of the Governor at Northampton, the National Republican Committee sent Governor Coolidge to Minneapolis, in September, to make a speech at a

---

[2] On page 234 of the "Official Proceedings of the 17th Republican National Convention," the reader may see that the writer of these lines was "appointed Chairman of a Committee to notify Governor Calvin Coolidge of his nomination for Vice President." He did not qualify. He was loath to support Harding. He desired to be free during the campaign, and the notification of Coolidge also required a trip across the Continent, and he declined the honor.

State Fair. The audience stared at him for a few minutes as they might have gazed at a prize bull or a blue-ribboned boar. Then the crowd turned away cruelly and left Governor Coolidge talking to empty benches. They did not like his Yankee drawl. They did not like his compact matter of fact English style nor his emotionless oratory. So the Speakers' Bureau of the National Committee bundled up the New Englander and sent him into the South where they knew he could do no harm and where there was a chance that his rural charms might win some votes. They did. He made good. In Tennessee and in other southern communities where he spoke, the Republican ticket won. Kentucky and Tennessee for the first time in a generation returned the Republican electors. The back woods of the Green Mountains could speak the language of the back woods of the Appalachians. The simple direct appeal that Coolidge made to the mountain folk won them. It was a typical Coolidge performance. The singed cat of the legislature was repeating his old tricks, the everlasting American success story—conquest by the humble. And then again perhaps Coolidge had little to do with these Southern gains. The South had no taste for Governor Cox, the Democratic candidate for President.

In the midst of Coolidge's campaign for election to the Vice Presidency in October, 1920, Senator Murray Crane died—Calvin Coolidge's political sponsor, his guide, philosopher and friend. It was the Crane organization with Guy Currier's benediction, plus J. Otis Wardwell, plus Charles Innes, plus Arthur Russell, representing the business interests of Massachusetts, that gave Coolidge a leg up and took him out of the infantry into the cavalry of Massachusetts politics when they made him president of the State Senate in 1914. It was the Crane organization in western Massachusetts that rounded up the squirearchy of the mills in the Connecticut Valley and made Coolidge first lieutenant governor, and then governor. He was loyal to Crane always, indeed loyalty was one of Coolidge's fundamental virtues, the first evidence that he was a sentimental man, not too sordid a realist. True he was a competent judge of where his loyalty lay. Perhaps at odd times he rationalized his judgment to his own self interest, but not always, nor ever conspicuously. In cases of divided loyalty he sometimes nobly chose the public interest rather than his own advantage. Which is by way of saying that the Garman faith in some sort of moral government of the universe guided him. On the other hand Lodge never even remotely feared that the little hickory stick of a man from the Connecticut Valley would pry Lodge out of his place at a Washington dinner table, a place Lodge held in proud esteem as chairman of the Committee on Foreign Relations, and leader of the Republican Senate.

When Crane died, Coolidge dropped every duty to attend his funeral. There, about the grave, gathered the nobility and gentry of the Connecticut

Valley, the barons, the earls of the nation. And there in his ceremonial silk hat, with his ceremonial morning coat and appropriate gray trousers, with his face flinted for the formal function of the hour, Governor Coolidge stood near Senator Lodge at the open grave of his friend. Being what he was, Calvin Coolidge was internally deeply moved. A thousand kindnesses from the hand of the man in the coffin flooded his memory and stirred him to grief. As he turned from the grave and was walking toward his carriage, a press photographer appeared. Beckoning Lodge, the news man would have made a picture of Massachusetts' two chief mourners. Lodge was not averse to standing in that hour beside the Republican candidate for Vice President. Visions of that seat beneath him at Washington dinner tables may have mellowed Lodge's pride. He stepped toward Coolidge. The suppressed emotion of the hour rose in the Governor's rasping voice as he drawled harshly: "I came to bury my friend. It is no time for photographs!" and turning his back moved circumspectly to his carriage.

After Crane's funeral, gossiping with his friend, Robert Brady, of the *Boston Post*, about the new alignment, Coolidge revealed his attitude toward Lodge. The two men were discussing the inevitable political readjustments, the break-up of factional lines that would follow the death of Crane. After Coolidge had indicated that Weeks was impossible as the successor to Crane, and after expressing the belief that William M. Butler would take over the Crane command, the Governor leaned back in his chair, lighted his cigar and drew this picture of Lodge: [3]

"Now you ask about Lodge's friends. I don't think Lodge has many friends. He has a host of admirers. But there is a big difference between admirers and friends. Crane had friends and those friends will stick to Crane dead in state politics. Lodge's admirers will stick to him until he gets his first setback. When that comes you won't see many people sitting on the mourner's bench."

Upon the basis of that estimate, Coolidge governed his attitude toward Lodge. Four years later his prophecy came true. And Coolidge gave Lodge his first setback. When Lodge came to realize the truth of what Coolidge had prophesied, he died of the revelation.

The election of 1920 was a Republican landslide as everyone knew it would be. America had turned from Wilson, the idealist, and all his visions. America was in the throes of a crisis and facing a minor depression. The high cost of living still was worrying the householders. High wages were blamed. Labor being restless and envious of the war profiteers, was making trouble. Employers were using the government in a red baiting campaign. And all hands, labor, capital and harried consumers turned upon the Democratic party which had made the League of Nations a vital issue and had

---

[3] Letter from Robert Brady to author, July, 1936, from notes made at the time.

nominated a candidate for President who did not attract the electorate. So
Calvin Coolidge, tied by a revolting Republican Convention to the Harding kite, rose to the Vice Presidency, an office which he had not sought but
which he proudly accepted.

On his last day as Governor in January, 1921, he folded up his mother's
photograph, cleaned his desk with great care and then in his own hand-
writing slipped under the blotter on the desk where his successor would
find it after the pomp and ceremony of the inauguration, a letter releasing
his mingled emotions of gratitude to Massachusetts and to the world in
general in two or three cordial lines quite carefully directed to his successor.
Then he and Mrs. Coolidge took the old trek for Northampton as the guns
on Boston Common were booming their first salute to the new Governor.
Years after he recalled that afternoon train "which I had used so much
before I was Governor. It had only day coaches and no parlor cars. But we
were accustomed to travel that way and only anxious to go home. For nine
years I had been in public life in Boston." [4] It was a familiar sight, that
train for Northampton, puffing across Massachusetts into the twilight over
the hills and down into the Connecticut Valley. Across Middlesex he went,
over the uneven hills and valleys with the bottoms opening upon snowy
fields and orchards, with thousands of stone walls peeping out of the snow
and the marsh willows bending over the frozen brooks. The color had gone
from the landscape. The houses of the aristocracy setting in wide lawns
built in the rococo of the seventies and eighties, were sentinels of respectability. He went through Wayland, South Sudbury, out of Middlesex
County into the high tableland, into the hills covered with second growth
timber and on into Hudson. Berlin and Clinton passed with their mills
and factories. They passed West Boylston, Holden and Rutland. Men were
cutting ice on the ponds near Holden, and the old white church near the
cemetery on the road glittered in the snow. The modern mill as he passed
through the town looked strangely incongruous. On they went over the old
familiar route. At Coldbrook Springs they struck the Ware River and followed it to Gilbertville and its mills down to Ware with its dam and Main
Street paved with brick; and so to Amherst and a dozen little towns, mill
towns.

During the nine years the Coolidges had been going up and down this
railway, a new world had been built under the roofs of the mills. Speeding looms and new machinery were everywhere. And also, significantly,
in all the savings banks of these towns deposits were rising rapidly; more
rapidly indeed than the loans and discounts, the cash reserves were growing, steadily, perhaps unhealthily. Literally millions of dollars were
stored in these savings banks in Massachusetts from Boston west to

[4] "Autobiography," p. 155.

the Connecticut Valley. These savings Calvin Coolidge hailed with deep satisfaction. They were the substance of things hoped for in his faith. "Work and save; save and work," the alpha and omega of his philosophy. It did not occur to him remotely that in this cosmos anything untoward could happen to a world where men worked and saved. So as he rode through this fat little kingdom toward his own homeland, toward the hearth in the little duplex apartment on Massasoit Street in Northampton, knowing how fast the looms were spinning, knowing of the tons of bright new machinery that had been moved into these mills and factories, and seeing the incense of work rise in the form of savings that were mounting in these little country banks, his soul was at peace, his god was sanctified.

Before he left for Washington to be inaugurated in March, 1921, he spent an hour on the Main Street of Northampton saying goodbye to his friends. "Jim" Lucey, the shoemaker, remembers Coolidge's visit. He stood in the doorway and when the shoemaker looked up said: "Well, I've come to say goodbye!," shook hands, turned around and walked out. He repeated that performance in a dozen places. This knack of holding friendships was a lifetime habit. "I never thought he was cold or fishy," said Lucey to a reporter a dozen years later.[5]

2

March 4, 1921, inauguration day, in Washington, was a bright crisp day and after the short inauguration parade, the Vice President in the Senate made his inaugural address. It had no distinction save brevity; attracted no attention. The Coolidges, who at first expected to rent a Washington house and live simply as they had lived in Northampton, putting their boys in the public schools where Theodore Roosevelt's children had gone a decade and a half before, found that rents were too high. A decent house could not be found within the Vice Presidential salary. At that time Calvin Coolidge's entire wealth was probably less than twenty-five thousand dollars. He had been in public office practically all his life. He had managed to save a little every year but only a little, a few hundred dollars in the first years, a few thousand dollars in the latter years. These sums had accumulated in the savings bank of which he was once a trustee and attorney, later vice president. When he came to Washington he believed that it was his solemn duty still to live within his income, still to save a little every year. So the idea of a house was abandoned and the Coolidges took an apartment in the New Willard Hotel. It was a modest suite vacated by Vice President Marshall. The New Willard at that time was many cuts above the Adams House in Boston. It was a proper and fashionable hostelry.

For two years Calvin Coolidge for the first time in his life had no major

[5] From the *Boston Post*, interview with James Lucey, September 14, 1934.

political worries. His duties required him to dine out frequently. As a diner out he was a rare bird and he knew it. A myth grew up about him that he was dumb. A counter-myth appeared that he was a sly, keen fellow. Both myths were based upon the mistaken premise that in the heart of him he was dry and cold and hard. Alice Roosevelt Longworth denied that she had said he was weaned on a pickle. But it was smart enough for her rapier tongue. Only the reverse was true. Coolidge was never weaned. He carried to his grave a mother complex. Her photograph he moved carefully from the governor's desk to his desk in the Vice President's office. It was always near him in Washington. He was his father's idol. And always in his heart he sought the protecting shadow of his father's approval even to the last. Because his table manners needed Grace Coolidge's care, because he was everlastingly nibbling nuts, crackers and oddments, a sincere, competent but discriminating feeder who sat at the table to eat and not to talk, dinner partners in Washington first found him hopeless and then sought to make him do parlor tricks. They provoked him to Yankee aphorisms and he knew what they were up to. So he clowned a little for his own delight, played the dumb man, impersonated the yokel and probably despised his tormentors in his heart.

The political apocrypha of the day contains a story that certain Senators, amused at the dumb, silent Vermonter, spiked the punch heavily at a stag party where Coolidge was a guest and plied him with it, thinking to release the playful goat beneath his skin. The more he drank the quieter he got and finally with his mouth consciously puckered, yet yearning for release, he walked carefully down a floor crack toward the leader of the band and asked deliberately: "How are you enjoying Washington?" returned on the floor crack and said no more.[6]

In those Vice Presidential years Coolidge seems to have been lying fallow, studying the game, accumulating strength. He sat in the Senate owlishly watching the show, but comprehending it. Few Vice Presidents had come to the Speaker's desk in the Senate with so much parliamentary experience and with so wide a political background as he. The parliamentarian who sat at his elbow had little better equipment than Coolidge.[7] His service in both houses of his state legislature, his work as president of the state senate, and as lieutenant governor had made him competent as a parliamentarian. He was intimate with the legislative end of a first class state. As governor for two terms he had watched legislative procedure with a discerning eye, anxious for the advancement of his own bills and measures. The United States Senate was the Massachusetts senate dealing with

[6] George Moses told me this story.
[7] Conversation with reporters in the Senate Press Gallery—also personal observation.

larger problems but in the same old way that legislatures from time immemorial have handled public matters.

In the final years of the third decade of the century, the difference between Boston and Washington made the new life of Calvin Coolidge seem to be a break with his past. Washington was the center of the world vortex. Here business and politics turned upon their common axis. Self seekers from all over the United States jostled and elbowed with their kind from every quarter of the earth, all seeking special privileges from our government, the most stable and the most powerful of human institutions. The air was full of clamor and uproar. Reformers and cranks appeared with political and social panaceas, adding strident notes to the din of the roaring voice of an unbridled plutocracy. But Calvin Coolidge kept his head. To all outward appearances he was living in the shabby purlieus of the Adams House or in Massasoit Street. He remained the son of Plymouth Notch.

So Coolidge, silently dining and meekly clowning his quiet way through official Washington society by night and watching the Senate by day with no responsibilities, no anxieties except to send a part of his pay-check every month back to his Northampton bank to watch the twenty-five thousand grow and grow—symbolizing the doctrine, work and save—he was, as it were, politically embalmed.

CHAPTER XXI

He Stands Before Kings

PRESIDENT HARDING, whom the Coolidges had visited after the election in the autumn of 1920, had invited Vice President Coolidge to join the regular Cabinet meetings.[1] This he did, but always in character. He saw much and said little. Probably he sensed something of the shame and disgrace that the managers of the Veterans Bureau were putting upon the administration in their swindling operations there. The Vice President could not have failed to get some hint of the scandal in the bureau where the interned property of German citizens was administered. Whiffs of the stench from the Ohio gang must have offended his political nostrils. Coolidge's real heritage from the Harding administration was to be the oil scandal. A brief chronicle of the affair is necessary here. The oil scandal arose out of the secret sale of certain United States oil holdings in three tracts known as United States Naval Oil Reserves No. 1, No. 2, and No. 3. Naval Petroleum Reserve No. 1 was created by an executive order of President Taft, in September, 1912; the Elk Hills, California, Reserve, more than 37,000 acres, of which, probably a sixth, was in private ownership of the Standard Oil of California and of the Southern Pacific Railroad. Three months later United States Naval Oil Reserve No. 2 was created at Buena Vista Hills, California, also upon executive order. It contained 30,000 acres, of which one third was privately owned. Nearly three years later, President Wilson, in April, 1915, created United States Naval Oil Reserve No. 3, Teapot Dome, Natrona County, Wyoming. It consisted of more than 90,000 acres.[2]

In June 1920, still in the Wilson administration, Congress provided for

[1] At the suggestion of Harry Daugherty. Letter from Harry Daugherty to Cyril Clemens, June, 1937.

[2] And by way of an interesting item, it may be set down that Harry Sinclair who

the leasing of the United States Naval Reserves by the Secretary of the Navy, and in November 1920, two weeks after Harding's election, but before he was inaugurated, Edward L. Doheny, an oil baron, and J. Leo Stack, an oil promoter, of Denver, contracted to negotiate with the United States for a lease to develop offset wells along the boundary of Teapot Dome Reserve. It was highly important that a Secretary of the Interior should be appointed who would initiate for Doheny the preliminaries necessary to the leasing of those oil reserves. Albert Bacon Fall, of New Mexico, was selected by President Harding as Secretary of the Interior.

And two months after Albert Fall was sworn in as Secretary of the Interior, he sent to Edwin Denby, Secretary of the Navy, the outline of a letter for Denby to sign requesting the President to transfer all the Naval Reserve oil lands from the Navy Department to the Interior Department. Denby, two weeks later, wrote to President Harding requesting the transfer of those oil reserve lands from the Navy to the Interior Department. And less than a week later, the President signed the executive order transferring the control of the Naval Petroleum Reserves from the Navy Department to Fall's Department of the Interior. Fast work this.[3]

Two months later a lease and contract for drilling offset wells in Reserve No. 1, at Elk Hills, California, were entered into between the government and the Pan American Petroleum & Transport Company. The lease was signed by Acting Secretary of the Interior Finney and by Vice President Danziger of Pan American. So much for Naval Oil Reserve No. 1.

Now before considering Reserve No. 2, we must detour to pick up a loose thread in the narrative but an important thread—a golden thread in truth! In November, 1921, arrangements were made between Harry F. Sinclair, representing the Sinclair Consolidated Oil Company, H. M. Blackmer of the Midwest Oil Company, Robert W. Stewart of the Standard Oil Company of Indiana, and associates, to purchase 33,333,333 barrels of oil in the name of the Continental Trading Company, from A. E. Humphreys, in Texas, and his associates, at $1.50 per barrel. Sinclair and his group were to purchase the same amount of oil for the Continental Trading Company for $1.75 per barrel. Observe the profit of twenty-five cents a barrel. When the profits on this deal had amounted to three million dollars, the contract which had been signed and delivered in November, 1921, was closed out. The Continental Trading Company, chartered at Toronto under the laws of Canada, seems to have been organized for the purpose of accumulating a fund for political purposes. So much for the detour and that deal.

But Doheny, Sinclair, Stewart and Blackmer were fairly busy with other co-

had an eye on Teapot Dome, in 1920 in the administration of President Wilson contributed five thousand dollars to the Democratic National Committee.
[3] Fall and Daugherty confirm the authenticity of Denby's letter.

ordinated matters in those days. The last of November, 1921, a little less than nine months after Harding's inauguration and just a week after the Continental Trading Company had been organized, a little more than four months after Fall had leased Naval Oil Reserve No. 1 at Elk Hills, California, to Doheny, Doheny loaned Fall $100,000 and sent the money to Fall in a satchel which became famous later as "the little black bag." Doheny's son carried this apparently, surreptitiously to Fall. A few days less than a month after the "little black bag" had passed from Doheny to Fall, Harry Sinclair and his attorney, J. W. Zeveley, visited Secretary Fall at his ranch at Three Rivers, New Mexico, and negotiated for the Teapot Dome lease. And a little more than a month after Sinclair and his attorney had visited Fall at Three Rivers and had negotiated for the Teapot Dome Reserve, Harry Sinclair filed public application for the Teapot Dome Naval Reserve No. 3, with the Secretary of the Interior. While the lease was still pending, February 28, 1922, the Mammoth Oil Company was organized under the laws of Delaware, to exploit Teapot Dome. Ten or a dozen days later this Company paid a million dollars to a subsidiary of the Midwest Oil Company, for the Teapot Dome contract. In March, a few weeks later, H. Foster Bain, Director of the Bureau of Mines, issued a memorandum showing the loss of government oil as a result of delayed government drilling and a week later Senator La Follette, having been told of what was going on, introduced a resolution to investigate the leasing of oil reserves. Captain John Halligan, of the United States Navy, wrote to the Secretary of the Navy, protesting against the private exploitation of Reserve No. 3, Teapot Dome. The newspapers were filling up with questions and revelations concerning the leases. The drift of public opinion was definitely against the negotiations. Less than a week after La Follette's resolution appeared, the contract for leasing Teapot Dome Naval Reserve No. 3 to the Mammoth Oil Company was executed and signed by Secretaries Denby and Fall, though the information was not made public at the time. The existence of the lease was even publicly denied. Three days after the contract was signed Director Bain,[4] of the Bureau of Mines, innocently wrote to Senator Kendrick, of Wyoming, who was inquiring about these oil leases in his own state that "as yet no definite contracts have been made," though the leases for Teapot Dome and Elk Hills had been signed and delivered for nine months.

In the meantime, the Continental Trading Company in Canada bought $30,000 worth of Liberty Bonds, easily negotiable securities, almost as fluid as cash, but alas not quite so fluid for they were numbered. These bonds were bought for purposes of political intrigue, and two days later the Trading Company bought $200,000 more of this high-powered financial

---

[4] Bain was ignorant of the lease at that time.

fluid to operate in politics. The next day the *Wall Street Journal* published the story of the lease of Teapot Dome and the Senate adopted a resolution offered by Senator Kendrick, asking Denby and Fall to advise the Senate if negotiations were in progress to lease national oil reserves. A few days later, the Canadian Continental Company's war chest was strengthened by another $50,000 in Liberty Bonds. And April 25, Assistant Secretary Finney of the Interior Department, and Secretary Denby, of the Navy Department, wrote a letter in answer to the Kendrick resolution, explaining the oil leases. Senator Robert M. La Follette, the elder, in a Senate speech, demanded an investigation of the whole oil lease conspiracy. The next day the Senate after a bitter and fiery debate adopted the La Follette resolution. On May 10, Fall's son-in-law, M. T. Everhart, received from H. F. Sinclair, on Sinclair's private car in Washington, $198,000 of the Continental Trading Company's Liberty Bonds. A week later he received $35,000 more in Liberty Bonds from Sinclair's office in New York. May 29, Everhart left $90,000 of those Liberty Bonds with the First National Bank of Pueblo, Colorado, declaring they were the property of Secretary Fall, and delivered $140,000 in Liberty Bonds to the M. D. Thatcher estate in Pueblo. These bonds which later came to Fall were identified by their numerals as those purchased for the Continental Trading Company.

Agitation was beginning to interest the people. The smell of oil was over Washington. To calm the public, President Harding, always a good fellow, addressed to the Senate a letter which declared that the oil land leasing policy was submitted to him before its adoption and that the policy and acts under it "have at all times my entire approval." Of course Harding did not know about the Continental Trading Company's activity in Liberty Bonds, nor about Fall's deals with Doheny and Sinclair.

By the summer of 1922, the stench of these oil scandals was beginning to rise and among informed people it was known that the Secretary of the Interior had formed some kind of alliance with Doheny, Sinclair and certain elements of the Standard Oil that were interested in the main chance. December 11 of that year, the second lease of Reserve No. 1, at Elk Hills, California, was turned over to the Pan-American Oil Company and the lease was signed by Fall and Doheny. A few weeks later in early January, 1923, Fall, realizing that his day of usefulness to the oil companies was over, announced that he would resign from President Harding's Cabinet on March 4, after serving two years. That day, or about that time, Harry Sinclair, in his room in Washington, at the Wardman Park Hotel, handed $25,000 in cash to Fall's son-in-law as a loan to Fall, taking no note. When Fall retired from the Cabinet, after two years of service, he gave as a reason for his resignation "the pressure of private business interests," and when he resigned, President Harding announced that he had once offered

Fall an appointment as Justice of the Supreme Court, a sadder commentary upon Harding than upon Fall. Fall had been out of the Cabinet less than a month when $70,000 of the $90,000 lot of Liberty Bonds turned over to the First National Bank of Pueblo in May the year before, were deposited with the Exchange Bank of Carrizozo, New Mexico to the credit of Albert Fall. After Fall's resignation Harry Sinclair gave Fall an additional $50,000 as a loan and $10,000 in cash for Fall's expenses for a trip to Russia on behalf of Sinclair and his oil interests.

Probably the men drifting in and out of George Harvey's "smoke-filled room," at the Chicago Convention in June, 1920, were as innocent as Warren Harding of conscious guilt in the plot to steal America's oil reserves. They were naming a President there—"playing the game." The stakes in the game were on another table in another room. Often in the alliance between business and politics, the politicians do not see the stakes.

But in retrospect one can see why Fall appeared in the Interior Department and not in the Supreme Court. Fall, in his autobiography, recounts with pride that he went into the Interior Department only under pressure.

Returning to Daugherty's prophecy about the "smoke-filled room," going back to Daugherty's knowledge of what the oil industry coveted in Elk Hills and in Teapot Dome, and what they tried to get from Secretary of the Interior Lane and Secretary of the Navy Josephus Daniels, under Wilson, one may construct at least a reasonable hypothesis that Daugherty's vision of power for Harding had in its background the need for Fall in the Interior Department. Into that hypothesis fits the motive for organizing the Continental Trading Company to pay the deficit in the Republican campaign expenses of 1920, and also to fill the "little black bag," and to spread cattle on a thousand hills of Fall's New Mexican ranch. Upon the hypothesis [5] that Daugherty, with something more than a pleasant prescience, knew exactly what he was doing, when he spoke in February, 1920, may be constructed a fairly stable, logical thesis that the ox knew its master's crib. But this hypothesis and all speculation about the connection between his party and the money-grabbers, never once fluttered across the consciousness of Vice President Calvin Coolidge. In his "Autobiography," it is curious and significant that he never mentioned the oil leases. Yet all of the facts just recounted were developed in the federal prosecution which followed the investigation by La Follette's committee in the Senate. This oil scandal just wa'n't Calvin Coolidge's stove. What he saw in the Senate and heard in

[5] It is, however, only a hypothesis, not a thesis. But this footnote is addressed to the posterity of the next twenty years to be on the outlook for odd unrelated bits of information from diaries and memoirs to be published and biographies to be written which may furnish the links to make this hypothetical chain a dependable thesis.

its corridors, what he learned in the Cabinet room at the White House, he
never revealed by even a hint.

In Washington, during the first two years of the Harding administration,
these things were but vaguely known. But even in the outer rim of inside
administration circles in Washington, by the late spring of 1923, the breath
of scandal was getting warm and rancid. The Vice President who was keen
of ear, quick of eye and delicate of nostril, certainly sensed these iniquities
as soon as the insiders themselves. His nine years in public life at Boston,
his training in Northampton which after all was political (for Northampton
was a little Boston—Boston a little Washington) makes it hard to imagine
Calvin Coolidge going about his daily walk even on the remote periphery
of Republican politics, without encountering some knowledge of the
truth about the Harding administration. Yet the unsavory character of the
Harding administration, the very men he met around the White House,
even in the Cabinet, men like Daugherty and Fall, the loose talk of graft
in half a dozen bureaus that one inevitably heard in the Senate lobbies, in the
Senate restaurant and on the streets of the Capital, would have disturbed
another man with a feverish imagination. But Coolidge—probably not!

Harding looked husky. The heart beat between Calvin Coolidge and the
White House was ticking regularly. It was obvious that the President was
keeping too late hours, running with loose and frivolous people. Being an
insatiable consumer of gossip, Coolidge also must have heard the stories
about the President that were going around among the elect in those days.
Chief Justice William H. Taft, who also delighted in the trivia of politics,
heard a most fantastic story in 1921, and on May 2, of that year, two years
before President Harding's death, the Chief Justice wrote to Gus Karger:

"By the way, just as a bit of gossip, there's a curious little story going the
rounds to the effect that Mrs. Harding is worrying a good deal. The tale
has it that she is a believer in 'High class' clairvoyance and that for a long
time it has been her custom to consult one of the Washington elect of the
esoteric circle. Some time before Harding was nominated this soothsayer,
grasping at the obvious, informed Mrs. Harding that her husband would be
nominated and elected President of the United States. That was very nice,
and easy and comfortable to believe. But it is reported that she has consulted the prophet again, since the inauguration, and that the oracle this
time indicated that the President would not see the end of his term. This
is said to be preying on Mrs. Harding's mind. I have no personal knowledge
as to the truth of the story, but got it from some one right close to the
throne." [6]

It required no cosmic intelligence in 1921 for a Washingtonian who was
a good listener to prophesy that a man who was burning the candle at both

[6] From the unpublished correspondence of William H. Taft in the Library of Congress.

ends was doomed to die! A keen-eyed soothsayer might easily have plucked this fear for her husband out of Mrs. Harding's own heart. Being free spoken, she could not hide the truth about Harding. So the gossips had the story and as for Coolidge—! Who knows? What he saw of this disreputable Ohio gang, what he heard of the scandals in various departments, what he felt about the President's hard living, he may have recorded indelibly in his memory, probably labelled his impressions with his own judgment. But he certainly let it go at that. So Calvin Coolidge, Vice President, walked in prissy serenity across the ugly sands at low tide in American politics, low tide for fifty years.

2

Harding had a two years' honeymoon in which newspapers generally treated him with elaborate courtesy. To understand this period in American history it is necessary to understand Warren Harding, the President, who in a way sponsored his short day and dramatized it. In 1923, after two years in the White House, President Harding was getting on well with the newsmen, hail fellow well met. They suspected more than they wrote, hoping that appearances were deceptive, that they were wrong in their surmises. No one could have been more genuinely earnest in the manifestations of his public desire to be a good President than Warren Harding. He was frank in confessing his limitations, disarming in his candor to his friends, and even to casual acquaintances and always to newspaper men who crowded into his press conferences every Friday. He stood before them bland, charming, even jovial at times, but with an actor's quick sense of dignity; a fine, well set up figure of a man, clearly of the emotional type with the eager wistful lineaments of a friendly pup written on every flexible feature, with the warmth of a woman's cordial glow in his eyes.[7]

Compared to the Vice President, the President was a political novice. Harding had served one term in the Ohio State Senate and a rather dull term in the United States Senate. He was defeated as a candidate for Lieutenant Governor. The faults of inexperience marked his Presidential course at every turn. A canny Yankee and low pulsed statesman like our Vermonter must have observed Harding's faults watching the President in Cabinet meetings. For Harding was expansive. He had the harlot's virtues —impulsive generosity, impeccable credulity and a love of palaver for its own insouciance. He made few actual promises but made many enemies who had an impression that he had promised them various favors. Harding was one of those smilers and political masseurs who paw and pat their way out of difficulties. Calvin Coolidge knew that type. For it he had the spinster's distrust of the courtesan's technique. All his life he took pride in his cold ex-

[7] Saw him myself—made these notes—W. A. W.

terior. "It took courage," wrote his secretary in 1935,[8] "to ask Mr. Coolidge for a favor!" The Vice President could look back at the twenty years of his own varied public career and, remembering his steady rise, he could realize how long this waxwork Adonis at the head of the Cabinet table in the President's chair had been lying fallow during most of Coolidge's fruitful years. President Harding had been a small town editor, a director in the local bank and in the country town lumber yard, a leader in the Baptist church. He was known in Ohio chiefly as a lodge orator—a silver tongue! He was chosen in 1912 to nominate President Taft in the Republican National Convention in the year of the Bull Moose revolt. Politically, Harding lined up with the Republican machine, was the apologist for George Cox, the Cincinnati Republican boss whom he had eulogized as "a peerless leader," [9] and he was the chairman of the National Republican Convention in 1916. He was chosen then by the National Republican Committee because chiefly he would denounce Theodore Roosevelt in his keynote speech in shameful vituperation! Yet in that Convention Harry Daugherty, the kingmaker, cherished hopes that Harding would be that dark horse who could beat Hughes; while Frank Stearns, the other kingmaker who backed Coolidge, was exciting the risibles and scorn of Lodge and his Massachusetts delegation by croaking gently the name of Coolidge in the same Convention! By a quick turn of time's kaleidoscope in four years Harry Daugherty and Frank Stearns saw their dreams realized when two country town provincials, unknown beyond their state lines in 1916, were named as leaders of a major party to guide the destinies of a nation that was driving the four horses of the economic apocalypse, inflation of credit, industrial folly, speculation and nationalism to the destruction of their world.

The pageantry incident to the Vice President's job fed a certain Coolidgean sense of dignity in his heart and he found it good. At Washington dinner tables where he sat sniffing, pretending to disdain the vainglory of it all, he sat a notch or two above Senator Lodge. Also Mrs. Coolidge had her social status in Washington. In Boston a Governor and his wife, in relation to the society of Back Bay and Beacon Street, have about the status that a country sheriff has when he comes up from the township to the county seat. He is known, recognized officially, sometimes rides in an open car with visiting dignitaries and has certain pompous duties in the opening of the court. But where soup and nuts and fish are concerned, he and his wife have a low place. So it was in Boston with the Coolidges, in Back Bay and Beacon Street. But in Washington, the Vice President ranks socially next to the President, and his wife is easily second lady of the land. Official

[8] March, *Good Housekeeping*, 1935.
[9] Which Democrats recounted in circulars in Harding's Presidential campaign.

society in Washington for all intents and purposes is the only society in Washington, even if official society is insignificant in Boston. And for the first time in her life, Grace Coolidge bloomed into all the beauty that her qualities of heart and mind had promised at Burlington, at Northampton and at Boston. For a husband so devoted and so intelligently appreciative of his treasure—even if he parleyed and palavered little—the triumph of his wife in Washington official society must have been one of the major satisfactions that he found in his office.

But he had his own high moments. The first of those came in November, 1921,[10] with the great national demonstration at the burial of the unknown soldier at the Arlington Military Cemetery. There in his high silk hat, heading the Senate, standing ahead of Lodge at the President's shoulder, amid the glitter of epaulets and the gleam of brass buttons and gilded sword sheaths, with the diplomats of the world standing by, the farm boy from Vermont must have felt that he had come a long way since he and the bull calf parted company at Ludlow a generation before. The sense of his official importance stiffened his neck [11] not too rigidly, but gave him a perspective from which he could look down the procession of his past and see how far and how steadily he had marched upward, taking a hand now and then from Lady Luck, but always after he had seen well ahead where he was going and had mapped his course to the next goal post.

Two days later in Memorial Hall where the world's ambassadors met at the first world Disarmament Conference, Coolidge sat with the Senators. Near him were the members of the United States Supreme Court. Beneath him were the Cabinet and the President. In the crowded room had gathered leaders from all over his country, from all over the earth to witness the opening of what everyone felt was to be a momentous show. But he, Calvin Coolidge, the constable's son, the scion of the governing class on his native heath, always a somebody in his village, in his country town, in his state, and now in his nation and the world, covered with the pleasant panoply

---

[10] A few weeks before, the Vice President had enjoyed another thrill which must have ignited a glow in his heart. In a letter to the author December 18, 1936, Bascom Slemp tells of hearing the President, Senator Lodge, and Vice President Coolidge from the same platform at Plymouth, Massachusetts, where they spoke at the Plymouth Tercentenary. After Harding spoke, Senator Lodge stepped forward and addressing the crowd in his best manner, declared that he had spoken before the Plymouth Society previously and added that while he spoke at some length he felt he had not "fully covered the subject, so I am coming back to discuss the further historical significance of Plymouth". Coolidge came next and Slemp relates that he said: "I likewise was present on the occasion the Senior Senator so eloquently refers to and I also spoke on that occasion, but unlike him, I exhausted the subject." Mr. Slemp declares that "the crowd caught this and there was a tumult of laughter and applause". A Vice President could tickle the ears of a Massachusetts Senator with a straw and the shy, Puckish little devil in Coolidge's heart prompted him to this gay absurdity.

[11] Saw him then myself. He looked most complacent!

of pomp—on that gorgeous day, looked down with rather cool, keen eyes upon the delegates at the green baize table and at the monocled Ambassadors. He was no longer a cat looking at a King. He was a ruler among equals. He saw the President when he stood and opened the Conference, heard him make his rather insignificant speech with its bass drum rhetoric, and saw him file out and leave the business of the hour in the hands of Charles E. Hughes, Secretary of State. Coolidge must have felt how Harding muffed the ball, and how he might have said what should have been said after the manner of Garman and in the philosophical language that had made Coolidge for a day the sage of Boston. Looking down his nose in that proud hour as though he were identifying some bad smell, he gave no sense of the serenity which must have soothed his soul as he realized the security of his position. He knew that he was ready for the next step when it came.

CHAPTER XXII

Tragedy Approaches with Her Spotlight

THUS President Harding's first year passed and his second year began, while Coolidge every month put by in the Northampton savings bank a considerable part of his salary, dined where he must, played the President's double when he could, found his speeches growing longer in Washington than they were in Boston, but had more time on his hands. So he read. He read Wells' "Outline of History" from cover to cover.[1] He always had in his office a sheaf of books from the Congressional Library, but not modern books. He was not a well read man in modern literature. He managed two or three current novels a year, mostly trash. He read the weekly magazines and every day systematically read the daily papers of the region, the New York Times for news, the New York Herald for editorials, the Baltimore Sun for politics, and the Washington Post for the local doings of the town.[2] But none of these papers, no newspapers and few popular books in America, were publishing the big news of the day. For that news was still largely subterranean. It was not news then in any newspaper sense. Ten years later, the story of these subterranean events became history.

The country during the two years of the Harding administration was trembling in the first light shock of the earthquake which was to topple the tall towers of Harding's utopia. Labor was wearing silk shirts, buying motorcars and moving into gaudy apartments. Also it was feeling its oats, and was striking. Industry had been speeded up to win the war. Wages had been raised to win the war. The war being over, two or three million young soldiers had returned from barracks to industrial and commercial jobs.

[1] March, Good Housekeeping, 1935.
[2] Observation of the author watching Mr. Coolidge at his morning newspaper diversions.

234

They were unhappy because while they had been living on subsistence wages, laborers in the shops, offices, and fields had been living high. Jobs were scarce, but not too hard to get. A red scare swept the land. It was partly an emanation of the employers' imagination and partly a vague yearning for a new era in the heart of labor. The red scare irritated labor and made employers nervous. Deflation came, possibly consciously to put labor in its place; but more likely to check a dangerous boom. But the upswing of real estate and housing kept industry afloat in spite of the deflation, though the called loans checked many a suburban real estate development and broke many a soap bubble boom. Instead of hitting labor, the first effect of the deflation was the wreckage of the cattle industry of the great plains. Large scale farm enterprises were also crippled on the high prairies from Texas to Canada within three hundred miles eastward from the Rocky Mountains. But the current of commerce in the world that Coolidge knew, moved forward sluggishly. Industrial New England and the great middle class across the land were not worried by what was happening to the American farmers. The farmer began to lose his purchasing power, not entirely, scarcely appreciably in the mass. But the first subsidence in what was later a general sinking of rural income occurred in those two years of the Harding regime. Harding was not responsible. His administration deserves little blame for the subsidence of rural prosperity. Harding's administration must answer for other faults than that.

Could anyone have known the truth in those days? Could any responsible statesman have foreseen the coming of calamity which started with the deflation of 1921? Certainly the middle class millions in their security and prosperity, had no way of facing economic reality. If wise men knew—cloistered economists—no one heeded the warning they were giving. And no one could look ahead far enough to feel the impact of approaching trouble. So Calvin Coolidge went his prim way—reading Wells' "Outline of History," meticulously picking out kernels of news, of comment, of political sagacity from his morning paper, enjoying in regal splendor the night life of a Vice President, listening to Senate gossip by day and beaming back of his sour face with delight at his wife's triumphs in the diversions of a gay Capital. On that sheltered path probably Calvin Coolidge knew nothing of submerged economic forces far below any plummet he could drop or read. Yet they began to gather—the forces that would bring the earthquake—the disaster that should cast down the shimmering baubles of prosperity. That prosperity seemed to Coolidge's class and kind all over the world, the substance of a new era. Men could still work, and save. So why should he be deeply disturbed when here or there an oil baron tried to cheat the government in a trade for oil reserves, or when a scoundrel sought to rob the government through the Veterans Bureau, or a pliant

Cabinet officer abetted by a careless colleague, let the public interest pass into private gains? Calvin Coolidge kept silent when a conspiracy among oil brokers to raise a questionable fund to pay the national party deficit began to be visible to insiders in Washington, even though a Cabinet officer was put in a tight place to explain it. These were not his kind of doings. He had gone through twenty years of public service, politically passing the time of day with unabashed lobbyists, even accepting their favors. Yet no scandal had touched him—no scandal had tainted him according to his standards or those of his day and place. The standards of real politics shock sensitive persons who stand over the game and occasionally catch glimpses of the cards and hear rumors of the rules. The old struggle between the haves and the have-nots—inevitable, ceaseless, eternal—generally is submerged. But neither combatant is moral according to higher ethical standards. Calvin Coolidge, being in politics, had to accept the subterranean rules or quit the fight. He stayed in. He was a shrewd judge of men who allowed no advertised crooks to stand too close to him. But while he did not go pharisaically up into the Temple to pray about his righteousness, he certainly must have realized that other men are different. Probably the corruption, the scandal, the vulgarity that enmeshed President Harding, who tolerated fools too gladly and accepted crooks too easily, annoyed his Vice President. But also the Vice President must have been almost equally ashamed of the Presidential bombast, of the all but meaningless pomposity of Harding's public utterances. In the Coolidgean wisdom, however, these were minor matters. Corruption, vulgarity, scandal, ineptitude in English and an addiction to trash in the White House seemed after all merely skin eruptions on the times. Also having considerable personal imagination, never forgetting that "only a heart beat" stood between him and the Presidency, Calvin Coolidge certainly must have been comparing himself, his powers, his record, his walks and ways with those of his superior officer. Probably his amour propre was being fed. Certainly it waxed strong.

In those early Vice Presidential Washington days this Massachusetts man who was always a conventional regular Republican came into his first direct contact with a really new kind of political bug: The irregular western Republican. Coolidge had of course heard of those Progressives—Theodore Roosevelt's legacy to his party. But in Washington where they assumed to hold a balance of power, these western mavericks must have shocked him at first and then annoyed him. Colonel E. M. House, President Wilson's political cataloguer of men, had said of Coolidge: "He is a timid man, frightened into conventionality." [3] And certainly the untamed Republicans, who were liable to run amuck any day in the Senate, must have scared the inner lights out of the Vice President when as a presiding officer

[3] Letter to R. M. Washburn, Coolidge's first biographer.

he tried to herd them. But his inexperience led him into a blunder that cost him dearly in after years. He permitted himself to be the instrument of a deliberate insult to Senator George Norris—the leader of the Progressives, second in command in 1921 only to La Follette. The trouble occurred over a motion in the Senate which substituted a less drastic bill for a farm bill drawn by the Progressive farm leaders and introduced by Senator Norris. The milder farm bill had the backing of the Harding administration. The play in the Senate probably followed Harding's orders. But Coolidge took them. The facts furnished in letters [4] by Senator Norris and former Senator Curtis follow: Vice President Coolidge who should have been presiding over the Senate, told Senator Curtis to take the chair on the morning of July 26, 1921, immediately after the routine business was done, and to recognize quickly Senator Frank Kellogg, of Minnesota. Kellogg was to introduce the weaker substitute bill. The recognition of Kellogg was a trick Coolidge did not care to play himself. But a Senatorial friend tipped off the plot to Norris. Probably knowing that Kellogg was ready to offer the Harding substitute bill, Norris asked Coolidge to recognize Senator Ransdell, who desired to address the Senate on the Norris bill. Norris claimed that Coolidge promised Norris to recognize Ransdell. And Norris primed Senator Ransdell to be on his feet demanding recognition. But at the juncture when recognition would have been in order Curtis, not Coolidge, was in the chair. Ransdell was on his feet clamoring at the top of his lungs for recognition. Curtis in the chair—where obviously Coolidge did not care to sit under the circumstances, kept saying with official composure: "The chair recognizes the Senator from Minnesota."

Frank Kellogg, "the Senator from Minnesota," did not rise. He was either for the moment absent-minded or confused by Ransdell's clamor. And only after Curtis as presiding officer had repeated his recognition three times, did Kellogg respond. As soon as Kellogg had done the damage to the Norris bill, Vice President Coolidge, who had remained outside the chamber while Curtis played his hand, coolly appeared like a cat with cream on his whiskers and sat him down as the Senate's presiding officer. This marked Coolidge for Progressive wrath. Coolidge, fresh from Boston, did not know how clannish the members of the Progressive wolf pack were, nor how or where they buried their bones! But in four months' watching them, he should have learned their ways.

In the summer of 1923, Calvin Coolidge said goodbye to Warren Harding. Coolidge went north to Vermont for a vacation in the scenes of his childhood; Harding went west on a speaking tour and for a vacation in Alaska. It was a simple and happy homecoming for the Coolidges, father, mother and sons, at Plymouth. There they entered the life of a mountain

[4] Letters to author, 1934-35.

village. Some friends or relatives resurrected or reconstructed a gray farmer's smock with some family history about it. Photographers came along, newspaper photographers and casual kodak fiends. Being a kindly man, off guard and relaxed, the Vice President posed for the picture makers in the smock. He worked in the hay fields. He put his boys decently to choring in the neighborhood. Calvin worked in the tobacco fields of the Connecticut Valley a few score miles away. Mrs. Coolidge helped with the kitchen work when she was needed. For it was a small family and help was scarce and servants were not known to the village in the sense that they thrived in the larger towns and cities of the land. Beyond a doubt, Calvin Coolidge was happy that summer. Being expansive in his happiness and also mischievous, slyly he measured his wife's hat band and slipping down to Bellows Falls, bought her a large Gainsborough model picture hat. She remembered a dozen years later that he insisted on her wearing it even when she left the hills at odd times to go down into the valley.[5] His taste in hats left something to be desired. But he was a good husband, and barring weary days and hours of perplexity, was as amiable as one could ask. Neighbors came in at twilight and sat on the edge of the porch to visit.

"Howdy, Cal!"

"Howdy, Newt!"

The man in the wagon passed on, waving his whip, "Cousin of mine," the Vice President explained to Bruce Barton. "Haven't seen him for twenty years."

Tourists sped by occasionally in the hot noontime and bought trinkets at the store next door to the Coolidge home—or women, seeking a special article and hearing of the smock and the work in the hayfield, came up for a Sunday story, and such were invited to take potluck with the family. It was a small table and the boys had to wait or eat in the kitchen. Frank Sibley,[6] of the *Boston Globe*, remembers that being politely urged to take a second helping of hash for lunch, he cleaned the platter. Sibley recounts: "Cal was unhappy, knowing the boys would go without." The ménage was that simple. It was all middle class, prosperous enough, happy enough, exactly typical of the millions of American homes which were the bulwark of our political and economic liberties.

During that summer while Vice President Coolidge was leading this bucolic life, President Harding, after bidding Vice President Coolidge goodbye, had gone to Alaska. His trip across the Continent was punctuated by addresses. The World Court was an issue. His advisors pulled and hauled him across both sides of the issue during his westward journey. At Kansas City, Mrs. Fall, wife of the Secretary of the Interior, was closeted

[5] Mrs. Coolidge in *Good Housekeeping*, March, 1935.
[6] Letter to author, 1935.

with him for an hour before his speech, and he went to the auditorium evidently perplexed and perturbed after the interview.[7] The next day he told an acquaintance [8] that his enemies gave him little trouble. His chief worry was his friends. He had denounced one friend, Forbes of the Veterans Bureau, and had expelled him from office. He knew that another friend, his Secretary of the Interior, had become deeply involved in a questionable transaction which was bound to be made public. For Senator La Follette was on Fall's trail. In March, 1923, Secretary Fall resigned. He quit while the quitting was good. The President knew that Daugherty and his Ohio henchmen—his old friends and social cronies—were compromising the administration in the Department of Justice. He knew that the Sinclair and Standard interests were involved with the government; and very likely he knew that oil money was being irregularly used to make up a deficit of the Republican National Committee, and that the tainted money was sure to affect his standing. To a man of Harding's volatile, emotional nature it must have seemed that his whole world, his whole administration was crashing about his shoulders. In Alaska a code message came to him by airplane which brought him near to collapse, his associates remembered. He talked often to his confidants about the hypothetical case of a President who had been betrayed by his friends.[9] Evidently the question in his heart was whether he should desert his friends and go frankly to the people confessing his mistakes, asking for their suspended judgment while he punished the offenders, or if he should stand by his friends and go down with the ship. For a man of his temperament—imaginative, emotional, given to impulse and always finding surcease in babble—the impending revelations must have seemed horrible. That kind of a man soon spreads the infection of his mind to his body. Returning from Alaska, at Seattle he was stricken. The bedside advisors declared it was ptomaine poisoning from eating crabmeat. An investigation of the steward's list,[10] showing no crabmeat, indicated that crabmeat had not been served to him. He was taken from Seattle to the Palace Hotel in San Francisco, where the physicians gave out the story that bronchial pneumonia had developed. He rested there several days.

Calvin Coolidge, at Plymouth in his shirt sleeves and work clothes was doctoring a sick maple tree [11] when the news came to him that Harding

---

[7] Statement of Senator Arthur Capper who was the Hardings' guest at dinner, to the author who was also present when the President and Mrs. Fall went into conference.

[8] The author of this book.

[9] Conversation with Judson Welliver, Harding's literary secretary who was on the Presidential trip.

[10] By a naval officer, a friend of the author of this book.

[11] "Calvin Coolidge: From a Green Mountain Farm to the White House," by M. E. Hennessy, G. P. Putnam's Sons, New York, 1924.

was ill. There also was a man of sensibility and not without imagination, whose manifestations he always held in check. When the heart throb between him and the Presidency began to tremble, his own heart may have skipped a beat.

Thursday, August 2, the Coolidges told the newspapermen who were writing of Coolidge's doings that summer, that they were starting Friday morning for a few days motoring trip to visit Frank Stearns at Swampscott, Massachusetts. They were expecting to spend Friday night enroute at the baronial estate of Guy Currier,[12] at Peterboro, New Hampshire. He will be recalled as that old friend of Coolidge's Boston days who as the bipartisan leader of the political forces of New England conservatism had given Coolidge a helping hand in many of his earlier upward journeys. It was a grand palace the Guy Curriers kept there at Peterboro, befitting a man who soon was to leave an estate of three million. The newspapermen noted the Currier visit, and left Ludlow in the afternoon. Nothing held them in Plymouth.

The nearest telegraph office to the Coolidges was at Ludlow twelve miles away on a rough mountain road. No telegrams came for them that day, and no news of President Harding later than Tuesday was available; but that news was most encouraging. Thursday night at the usual early bedtime for Plymouth, the Coolidge household went to bed. At half past ten in Plymouth the Vice President sank into a deep sleep. It was half past seven in San Francisco.

Then the doctor who had been at Harding's bedside all day left the sick room and meeting Raymond Benjamin, an old friend, in the lobby of the Palace Hotel, chatted with him awhile. The doctor explained that he was leaving his patient because he was so much better and the doctor was going for a walk. Benjamin suggested that he join him at dinner and when the doctor declined Benjamin strolled up Market Street to a little French restaurant where he dined leisurely, reading the evening papers, rejoicing that the tide had turned in the President's illness. As Benjamin came into the street at eight o'clock he found it in turmoil. Newsboys were crying the President's death. It had come that suddenly.[13]

Dr. Ray Lyman Wilbur, at that time President of Stanford University, a physician of the highest standing, declared that death had come to the President from an embolism. Because his death was sudden, because obviously those near him knew that he was not putting up a fight, many theories, some of them scandalous, were whispered about after his death. But if death came as a welcome release, how is not important. At 7:30

---

[12] Currier's obituary in the *Boston Herald*, 1930.
[13] Statement by Raymond Benjamin, Republican National Committeeman from California, to the author, in 1931.

Thursday evening his nurse coming into the room found him dying. A minute and it was over. It took an hour to arouse the nation.

After midnight into the graveyard quiet of the town of Plymouth, Vermont, came a chugging automobile. It stopped at the white roadside cottage where the Coolidges slept. A man dashed around the headlights and pounded on the door of the Coolidge house. A match flared out, a coal oil lamp was lit, and a head appeared at a bedroom window. Colonel John Coolidge asked sharply:

"What's wanted!"

Breathlessly the telegraph messenger who had come from Bridgewater, the nearest night station of the Western Union, cried:

"President Harding is dead and I have a telegram for the Vice President."

Colonel Coolidge aroused his son and daughter-in-law, dressed hurriedly, brought the messenger into the house, and taking the message to his son's room, read it in the lamp light—a communication from George B. Christian,[14] secretary to President Harding, informing the Vice President of the President's death. With this was a telegram from Attorney General Harry Daugherty advising the Vice President to qualify as President with as little delay as possible. The heart throb between Calvin Coolidge and the Presidency had stopped. His first thought in a flash came:

"I believe I can swing it." [15]

He had looked over the job for two years, had seen how Presidential failures come. He knew his own qualities. He was not afraid. There was exultation in whatever shock, or sorrow the moment may have brought. Here was dramatic, poetic justice: that Calvin Coolidge should have that first fine flare of exuberant self-respect, and have that eager zest in the very scene, indeed staged in the very home where for some sad reason in his childhood his spirit had been broken in shyness, and where either by blood or environing circumstances his whole nature had been turned inward behind a flinty mask!

[14] M. E. Hennessy, page 165.

[15] Statement of Dr. David S. Muzzey, official biographer of Calvin Coolidge, confirming Coolidge's own statement to an artist painting his picture.

CHAPTER XXIII

Our Hero Dwells in Marble Halls

HERE the story slows down. After that first self-reliant moment when the news of Harding's death flashed into the life of Calvin Coolidge, the tempo of his normal life began. He and Mrs. Coolidge dressed, and as they dressed, the Vice President decided what to do. His stenographer came up from Bridgewater in a car a few minutes behind the first telegraph messenger. The Vice President soon had a message on the way to Mrs. Harding. In an hour, Ludlow knew of Harding's death. The few reporters still lingering at Ludlow appeared about two o'clock. Telegraph linemen were tapping the telephone trunk line at Plymouth Union. At 2:30 the Vice President was talking to Secretary Hughes who advised him to come to Washington at once. It was Mr. Coolidge's idea—having a taste, if not for large drama at least for a homely cast of characters of the obvious sort—that his father, who was a Notary Public, should administer the oath which would make Calvin Coolidge President of the United States. So there in the little room, half living-room, half office, where Colonel John Coolidge kept his daily accounts and transacted his scant business, eight people saw a President inducted into office. What a beginning for the new President! How superbly he made his entrance into his role—the American classic—from poverty to the White House. The scene was so commonplace, so simple, that with one bizarre touch it might have been prepared as a travesty on democracy itself. Around the President and his father were Congressman Porter H. Dale, L. L. Lane, of Chester, President of the New England Division of the Railway Mail Association, Captain Daniel D. Barney, of Springfield, Vt., Herbert P. Thompson, Commander of the Springfield Post of the American Legion, Joseph H. Fountain, editor of the *Springfield Reporter*, Erwin C. Geisser, Mr. Coolidge's stenographer, and Joseph

McInerney, his chauffeur. It was 2:47 by the old fashioned clock on the mantel, a rococo clock with a pressed wood front, the worst of the seventies. A typewritten oath had been dictated by the Vice President which his father held. As the father read, his voice broke. His son repeated the words, phrase by phrase, after him. When the last phrase was spoken, the Vice President put his hand on the open Bible and with decent solemnity added: "So help me God!"

The new President turned to his wife, put his arm around her. She had been crying. The emotional impact of the news was too much for her. His father held out his hand. The son grasped it for a moment, let it drop, turned away wordless. Yet if that blood clot which closed Harding's life had only taken another turn in the artery—had delayed its course another day—the new President would have taken his oath under the glowing lights of Guy Currier's spacious parlors, surrounded by every bauble of sophisticated luxury that a millionaire's country palace could assemble, with one of the first score of America's veiled political prophets as the new President's host and sponsor.[1] Knowing in his heart of this day's narrow squeak that might have spoiled his humble entrance to power and glory, how this sharp self-centered Yankee must have wetted his lips to throw surreptitious kisses to his Lady Luck for this good turn!

As it was, after the oath was administered to the new President, he beckoned to his stenographer. The two went to the dining-room and President Coolidge dictated a brief statement to the country indicating he would carry through the Harding policies. He expressed a desire that the members of the Cabinet should remain in office to help him, and closed with the faith "that God will direct the destinies of our Nation."

While this ceremonial was progressing under the lamplight in the Coolidge home, hired motorcars from all over the region bearing newspapermen, shiny cars bringing public officials, rattletraps conveying townsfolk from Ludlow and Bridgewater, were coughing like a herd of paleolithic beasts as they came charging up the hill roads from all directions to Plymouth. Before dawn the village street was packed with autos.

Farm houses began to light up. But characteristically, at three o'clock the President blew out the lamps in the Coolidge home and went back to bed. When he got up three hours later, the street before the house was filled with townsfolk from Plymouth, from Ludlow, from Plymouth Notch, and the Union, from all over the adjacent hills. When he appeared at the doorway, he would not make a speech. He greeted his friends simply, with innate good taste. After breakfast he went across the field to the little family cemetery on Galusha Coolidge's farm, where he paused for a time in front of the marble headstone where his mother lay buried. There he

---

[1] Editorial in *Boston Herald*, statement confirmed by letter from Mrs. Currier, 1935.

stood bareheaded, and silent. Before seven he bade his father goodbye, and he and Mrs. Coolidge started across the hills to Rutland, followed by a long train of newspapermen, local politicians, old friends. That also was a simple, homely American scene. At no time did events get out of key or tempo with the slow, strong rhythm of the Coolidge spirit. He dominated it. At Rutland, Governor Redfield Proctor, surrounded by a crowd of Vermont statesmen, greeted the new President and went with him in a day coach to Albany where he changed cars for New York City. There a private pullman car appeared.

Yet in it no plumed knight was this new President. He was evidently nervous. He wore his second best suit and a golf cap and sat for a time on the divan with Mrs. Coolidge and one of the reporters.[2] Occasionally he would wander aimlessly, nervously through the group of politicians in the car, then duck into his stateroom for a few minutes. At every stop he was on the rear platform waving kindly greetings to the crowd. A photographer snapped him shaking hands with a boy.

Mrs. Coolidge was deeply moved. Her eyes were on the President watchfully. She could not have been unmindful that his nervousness was sometimes staging him awkwardly. She might have feared that the train of mourning by some unconscious gaucherie of the new President would be turned into a journey of triumph which she knew and he naturally realized would be most unfortunate. But their nerves were taut. For he was unused to pageantry and the arts of kings. Thus they came out of the mountains, down the Hudson Valley to New York City, crossed the town from the Grand Central to the Pennsylvania Station and arrived in Washington. They went to their old quarters at the New Willard.

The new President was not so important a figure as the body of the dead President. A spontaneous emotional upsurge of affection for a lovable man, an emotion which Warren Harding would have understood perhaps better than Calvin Coolidge, was moving the heart of the nation. As the funeral cortege spanned the continent, while the Vice President was coming down out of the Green Mountains to the Potomac, tens of thousands of people, perhaps ten times ten, were lining the railroads, crowding at the stations, to pay a last tribute to the dead President. The American people knew little of the scandal that was hovering about the White House, knew nothing of the tragedy of their President's life which made him a shining mark for death.

On the funeral train from Washington to Marion, Ohio, the new President and Mrs. Coolidge felt for the first time the power of the new office. They also began to know its distraction. They were the center of a vast polite pulling and hauling. The train was crowded with Republican statesmen

[2] Conversation with a reporter who was on the train.

who suddenly had to make their wants known to a new figure in the White House. After two years untroubled by patronage, Calvin Coolidge all at once found himself in the midst of the heaviest patronage pressure area in the world. Every man who approached him might well be under the suspicion of wanting something. When Chief Justice Taft stooped over and kissed Mrs. Coolidge's hand, just after the train left the Washington station, he felt it necessary to explain that "there could be no suspicion of the infusion of the spirit of royalty into the function because of that courtesy." [3] He explained to his wife that he had no office to seek and added of Mrs. Coolidge: "She is very nice." But he continued: "There was a good deal of political talk on the train, even talk of the nomination in 1924" and "Senator Spencer, of Missouri, was voluble on the subject." The Chief Justice makes it plain that when he could get a word in edgeways with the President, he warned the President that Mr. Spencer had the appointment of a federal judge up his sleeve. Every Senator, every Representative, had his little pet scheme to promote, and the Chief Justice noted, with mounting resentment, a Senator who "kept nudging up close to Coolidge to promote the appointment of his judge." There on the train, amidst the shoving and jostling of the new President, began Taft's four year campaign for a high class judiciary. He told the President that the Senator wanted a judge "so that he can appoint receivers or counsels to receivers in the courts." The new President promised the Chief Justice that he would not yield in such cases, and the Chief Justice, being wise and a bit sophisticated, in his letter to Mrs. Taft, expressed something more than a shadow of a doubt about the new President's high resolve.

Amid the whirlpool of self-interest that made the vortex around the new President, the Chief Justice, who was a superb old gossip, picked up and brought to the President on the train after the funeral in Marion this choice bit, that "Charles Evans Hughes had declared that he would not run for the Presidency and that was final." [4] Such a bit of information to take to Calvin Coolidge was worth something. Incidentally the Chief Justice found that in Marion they looked askance at the Coolidges because "they were the successors of the town's great man." They wanted to know "why it was necessary to bring the Stearnses of Boston to Marion, whom Harding did not like, nor they Harding."

The social vampires started to sink their suckers in the new President. A young military aide was fired and an order issued rescinding the discharge. In Washington, the Chief Justice notes that the Ned McLeans, rich social satellites with a newspaper back of them, who had led the gaiety in Harding's White House, began to woo the new President. The night

[3] Letter from Chief Justice Taft to Mrs. Taft, August 11, 1923.
[4] Letter from the Chief Justice to Mrs. Taft, August 11, 1923.

when Mrs. Harding started back from Washington to Marion with her husband's body, the McLeans entertained the Coolidges at their country place and advertised it extensively in the morning *Post*. And the *Post* also printed the story "that Mrs. Harding had to follow her husband's body from the station in the private car of a friend." [5] The Chief Justice explained that as a matter of fact Mrs. Harding had asked the White House people not to send a car for her because she was going in the McLean car. Whereupon the new President soured upon the McLeans. He said nothing at the time, but later to the White House staff, he wrote the word "Finis" at the end of the McLeans' social dominance of his administration.[6]

When Calvin Coolidge turned away from the Harding tomb, in Marion, Ohio, he was President with full powers. The white light beat upon him without a flicker. He and Mrs. Coolidge returning from Marion went again to the New Willard; Mrs. Harding to the White House. The Coolidges made it plain that she must not be hurried in assembling her belongings and leaving the White House.

While the new President waited in the New Willard, he wrote *billets-doux* to his old friends. One to a Northampton politician read:

Dear George—
    I know you will be thinking about me. I am all right.

Nothing more. Just that!

When at last, after ten days alone in the White House, Mrs. Harding had packed her last trunk, closed her last valise and had put all her personal belongings into the White House car which took her to her hotel in Washington, another White House car was sent immediately to the New Willard for President and Mrs. Coolidge. Their sons were still in New England. It was midafternoon when the new President and his wife alighted from their car and walked quickly across the portico to the doorway of the White House. He wore the proper afternoon garb. She was dressed in black, a color that he particularly disliked, but mourning for Harding was necessary.[7]

Just beyond the doorway in the spacious hall that divides the two wings of the mansion, they stopped for a moment, perhaps to gather themselves, to realize this new realm, the new life, the new magnificence that was upon them. They gazed possessively around for a moment, looked up at the portraits of their predecessors. Some hint, some intaking breath of the

[5] The Taft correspondence.
[6] The Taft correspondence with his wife at that time. It did not help matters when a few months later Ned McLean, hiding away from the importunities of reporters seeking to question him about the oil scandal, sent through mutual friends a gay ironic message "Love and kisses to Cal!", which message was duly publicized.
[7] See "Presidents and First Ladies," by Mary Randolph, page 94.

sheer power and glory of the greatest elective office on earth was thrilling through them. They walked a few steps into the hall and stood undecided. An usher approached them to guide them to the elevator if they preferred to avoid the stairs. Still shyly they stood for another second. The President turned to his wife and said:

"Now you run upstairs, mamma."

He turned into the little anteroom to greet the head usher, Irwin H. Hoover, known for forty years there as Ike Hoover. The President said:

"I understand how things are around here. I want you, yourself, to keep right on as you are. There will be no change so far as you are concerned. But one thing: I don't want the public in our family rooms on the second floor so much as they have been."

Briefly he indicated that Hoover should expect a change in the gaiety and informality that the Hardings had encouraged. He said:

"I want things as they used to be—before!"

Indicating that he meant the simple and dignified hospitality of the Wilsons which President Coolidge knew only by hearsay. That was his simple answer to a serious and sophisticated social problem, the problem that faces every President who is bedevilled by the rich residents of the Capital city to make the White House a rendezvous of fashion. No such problem had faced him in Boston. In Washington, he had only his common sense and invariable good taste to guide him.

Having delivered his oracular comment, he turned from the head usher and walked back into the hall. There he stood surveying his new domain with a comprehending eye, like an assessor setting down a list of goods and chattels. The head usher standing by him, caught almost a smile, and then slowly draining the moment to its dregs he walked to the elevator and went to his room.[8] In those nine words, "I want things as they used to be—before," the new President changed the whole social aspect of the White House. No more poker, no more Forbes and Fall and Daugherty, no more butterflies clustering around the younger Senate set and the rich idlers of the Capital, no more trash in the White House.

2

He arose at half past six the first morning and for a few months made it his White House rising hour. He went for a walk in the White House grounds before the servants were up. He looked over the whole ménage himself with the same critical eye which he used when as lieutenant governor he trotted over the penal and charitable institutions of Massachusetts. He had taken the Massachusetts post-graduate course in public kitchens. But that first morning, waiting for the day to begin, he installed his mother's portrait on his desk and sat down, the confirmed sentimental-

8 This episode comes by way of a conversation with Irwin H. Hoover, February, 1933.

ist, to write a love letter [9] to his old friend, the shoemaker philosopher of Northampton which was to pass through Shoemaker Lucey to posterity, guided by the President's far-seeing eye.

The White House,
Washington, D. C.
My dear Mr. Lucey:

Not often do I see you or write to you, but I want you to know that if it were not for you I should not be here. And I want to tell you how much I love you. Do not work too much now and try to enjoy yourself in your well earned hour of age.

Yours sincerely,
Calvin Coolidge

August 6, 1923.

The sentimental mood being upon him, he turned to his other and more powerful sponsor, Guy Currier, of Peterboro, New Hampshire, the grand vizier of New England politics, as it related to all business in New England. He remembered the engagement to visit Peterboro, the engagement broken by President Harding's death. All the business of his place of power did not crowd from his mind the recollection of that engagement, so he wrote with his own hand this note to Mrs. Currier: [10]

The White House
Washington, D. C.

August 7, 1923.

My dear Mrs. Currier:

We were very much disappointed at not coming to your house and seeing the presentation of the play.[11]

Tell Mr. Currier that Pres. Smith spoke most complimentary of him. With every good wish, I am,

Cordially,
(Signed) Calvin Coolidge

Mrs. Guy Currier,
Peterboro, N. H.

The note to James Lucey found its way into print. Being a rather perfunctory social billet and being politically of no consequence, the note to Mrs. Currier did not find its way into print for several years after his death.

Then, President Coolidge clicked off the current of his sentimentality but not until he had arranged that his first White House guest should be

[9] Washburn, p. 141.
[10] Printed here for the first time through courtesy of Mrs. Currier.
[11] The play was "Romeo and Juliet,"—staged al fresco in the Currier garden.

former State Senator Richard W. Irwin of Northampton who had given him "the singed cat" letter sixteen years before. Whereupon he turned to the business of the day.

His faithful friend, Frank Stearns, who had followed him to Washington became a sort of a spiritual buffer to the President who needed Stearns. He sent for Stearns five times during his first Presidential days; one time was a hurry-up call. And after each summons the President let his friend sit there while the President went about his work. Stearns figured it out, probably correctly, that in the new environment with the new duties and a new world pressing down upon him, the President wanted one familiar, dependable, immovable, restful object near him.[12] During all of Coolidge's White House years, Stearns left his business and became the President's silent monitor. Stearns said grace at the White House table. Otherwise it went unspoken.[13] Stearns and Mrs. Stearns often sat quietly with the Coolidges in the evening. And on occasion, he and the President retired to the Presidential study where Lincoln had sat, and Stearns, smoking his cigar, mutely sat by while the President babbled about his troubles and trials. Stearns offered no advice, rarely interjected a syllable of comment. Once in some moment of aberration Stearns told Coolidge that he had been asked to recommend a Judge, and named him. The Boston merchant mentioned the request casually and with no idea of pushing it. But the fact that Stearns had even been approached irritated Coolidge. He snapped out:

"What do you know about the qualifications a man should have to go on the bench?"

"Nothing!"

"What do you know of this man's qualifications?"

"Nothing!"

"Well then, I advise you in the future not to meddle in things you know nothing about."

It was that relation, a sort of father and son relation, one in which the son has a right to be as impolite as he pleases and the father does not care, a relation in which the generic father realizes that the boy takes after his mother and lets it go at that. Stearns who had no idea of meddling, was not peeved. He was merely amused at the younger man's jealousy and pique. There was no political bond between them. It was purely psychological, a desire for comradeship on the President's own terms. And no account of the Coolidge administration, and of Coolidge the inner man, may be written without some relation of this strange association made necessary by a repressed man's urgent need for a confessor. As Mrs. Coolidge has

[12] Conversation of Frank Stearns with the author.
[13] "Forty-two Years in the White House" by Irwin H. Hoover.

written, he did not respect her education, he never talked things over with her, probably felt that he could not, or should not for some vain reason. And because he knew human nature well, he dared not talk casually to the men he met officially. He was not the kind to make confidants of men who he felt were his inferiors, his office force for instance. Yet human nature could not stand the strain of utter isolation in the White House, so Stearns became the father confessor of the tight-mouthed Puritan with a constricted heart.[14] Stearns, living in the White House, found the new President took many of his problems from the executive offices to the residential section of the White House. He was forever studying questions, reading statistics, cramming for interviews with Congressmen like a student for examination. It was his technique to be ready for visitors before the visitors came. He had the figures. They had a general idea. He persuaded his visitors. He had a retentive mind for facts. Writing was a chore to him. Over and over he worked upon the verbiage of a speech to get the phrases balanced, even to make the syllables click with a metallic rattle trippingly on the tongue. But Stearns remained at the end as he was those first days, the psychological vessel of release. He asked Stearns for no advice, sent him on no errands, told him no secrets. After a thing had happened, he liked to gas about it to Stearns, but disliked questions. Stearns knew his heart, knew how kind he was, and how flinty of skin.

3

Calvin Coolidge in the White House was the same shy, imprisoned soul who had puzzled men in Boston, who had held them at arm's length in Northampton, who had baffled them in Amherst, who had let them alone in Ludlow, and amused but never warmed their hearts at Plymouth. At Washington he slapped no man on the back, he pawed no man's shoulder, he squeezed no man's hand, gave no man the glad come hither with his "pretty keen eyes." Yet in the secret places of his heart that almost puerile sentimentality waxed fat and warm.

The new President took over the Harding cabinet. Indeed, at no time in his administration, even when he was elected to serve as President in his own right, did he make any notable changes in his cabinet. Men came and went—Secretaries of State Hughes, for instance, and Kellogg, Attorneys General Stone and Sargent, Secretary Wilbur of the War Department, Secretaries Wallace, Gore and Jardine of Agriculture. Two major figures in the cabinet remained through his administration: Herbert Hoover, who represented, if anyone did, the liberal sentiment; and Andrew Mellon, who was the archetype of the complacent, convinced, class-conscious plutocrat. Mel-

---

[14] The foregoing paragraphs were written following a conversation with Frank Stearns in 1933, after Coolidge's death.

lon was advertised in politics as the richest man in the world, which he was not; and certainly the appellation "the greatest Secretary of the Treasury since Alexander Hamilton" was sheer adulation. It was characteristic of Calvin Coolidge that he kept two men so widely different in viewpoints in his cabinet through all the years. In times of crises they gave widely different advice. Instinctively, in those times, he would turn to the man who represented the militant power of concentrated wealth. Indeed, so completely did Andrew Mellon dominate the White House in the days when the Coolidge administration was at its zenith that it would be fair to call the administration the reign of Coolidge and Mellon.

Coolidge's first Cabinet meeting in the White House lasted just fifteen minutes. He cancelled three consecutive weekly Cabinet meetings and instead of discussing matters in the full Cabinet he had long conferences with various Cabinet members during the first month of his White House occupancy. He made Bascom Slemp his secretary early in his occupancy of the White House.

It is easy to understand why he called Bascom Slemp to the White House as his secretary. Slemp was a Virginia politician, a former Congressman, the son of a Congressman in a Republican district. He was familiar with Republican politics in the southern states. He had handled patronage there for the Republican National Committee. Political purists sniffed and did not like the odor of Slemp's transactions. But he knew his way around in Washington. His party status had not been questioned. His personal integrity was not at issue, however keenly his political activity had been criticized by those who had no great love for the Republican party. Slemp was the man whom President Coolidge needed, a liaison officer between the White House and the Republican organization in Congress and in the National Committee, a man "diligent in his business" who should stand before kings. From the Democratic press, from the independent press, from the Progressive group in Congress and out, a storm of protest rose over Slemp, but it beat vainly upon the White House. The new President knew exactly what he wanted and he had it. He specified in his preliminary arrangements with Slemp that he should keep no notes, copy no papers, write no intimate memoirs. Bascom Slemp lived up to the letter and spirit of his contract.[15]

Calvin Coolidge, of course, took color from his times. He felt that he was the head of the party of talent and wealth and that talent meant, of course, acquisitive talent which produced and owned the wealth of the land—the Hamiltonian idea made perfect. Bascom Slemp meant successful party manipulation in the interest of talent and wealth. After the World War the reaction in American social and political thinking snapped back

[15] Letter from Bascom Slemp.

beyond Wilsonian idealism, back even beyond Taft's chuckling *laissez-faire* complacence, back of Roosevelt to the Hanna period that followed the Spanish-American War and its rising imperialism.

On one of his earlier visits to the White House in September, the Chief Justice wrote to Mrs. Taft [16] that the President asked the Chief Justice what he thought should be done now. "I told him I thought the country was delighted to have a rest before Congress" . . . met, and that "with approaching and present prosperity the people wanted to be let alone." The next day when the Chief Justice took a group of Circuit Court Judges on a pilgrimage to the White House, he noted that one of the Democratic Judges declared that Coolidge was "the most insignificant looking man he had ever seen in the Presidential chair," but the Chief Justice thought he appeared to great advantage. And the day after that, Harry Daugherty dropped in for a chat with the Chief Justice and after talking over the appointment of certain federal judges in the lower courts, the Chief Justice, not unlikely inspired by his two recent visits to the White House, hinted that it might be wise for Daugherty to retire and found "that he has no intention of retiring. Indeed he is very sensitive on the suggestion." A day or so later, the Court "put on its silk hats and three button cutaways" and went to see the President. And Justice Taft felt "that it was a bit tepid to hang around the White House." He added with fine irony: "He was as interesting as usual!" No subconscious self-interest in Taft's heart could conceal the sad aridity of the Coolidge personality.[17]

Yet Coolidge, the politician, was the Yankee throw-back to McKinley's era. "Coolidge's political philosophy grew to be that of the mid-nineteenth century, the optimistic ideal of democracy as the final solution of all problems, the cure of all evils," wrote Gamaliel Bradford, who understood him,[18] and added, "there may still be something to be said for this theory but it can hardly be said to denote advanced thinking from the twentieth century point of view. It can hardly be said that Calvin Coolidge had much to do with the twentieth century." Which truth makes it more amazing that in this time of reaction democracy, obeying some deliberate instinct—or if you will by some almost unbelievable coincidence—carefully turned the clock in the White House around twenty-five years and then slammed down in the seats of the mighty a man to match the nation's beclouded mind.

Anyway the times called for a nineteenth century tolerance of official malfeasance in Washington in the autumn and winter of 1923. At that moment certain rich men, lumped by New England as "crude western-

16 In Pointe à Pic, Canada, September 27, 1923.
17 Taft correspondence, Library of Congress.
18 "The Quick and the Dead," page 288.

ers—" Sinclair, Doheny, Fall and the oil men, were beginning to appear in the newspaper headlines. The annoying shortcuts to the accumulation of wealth, which these oil magnates were taking, seemed to trample down formal plans of legal usage and ancient custom. The oil scandals were beginning to blacken the front pages of sensational newspapers. Instinctively the new President rejected this gossip. Among those near the President who contended that the oil scandal when it first appeared was a plot of the Democrats and Progressives to discredit a Republican administration, was Chief Justice Taft. His distrust of the Progressives who had caused his defeat in 1912, was excusably bitter. One could not blame the Chief Justice if he doubted the sincerity of the Progressives or the validity of their oil scandal. That he tried to bring the President to his way of thinking, his letters to his family seem to show. And it is also obvious that for several months, Taft's view was reflected in the White House attitude.

The Democrats and Progressives were, from the Coolidge viewpoint, discrediting the whole theory of immanent democracy—that brains are wealth and wealth is the chief end of man. Yet the ways of these western oil barons did not affect his philosophy. Before he had been five months in the White House he declared in a public speech that "the business of America is business." [19] So the activities of the oil mongers could hardly be explicitly condemned offhand even if for the moment they were discrediting the Hamiltonian ideal. But business men did those things more carefully in New England, more circumspectly, within due process of law and generally without crass bribery. There greed was deodorized. Certain Amherst, Williams, Dartmouth, Yale and Harvard alumni who wore braid on their morning coats, and who encased their legs in mouse-colored trousers, burned academic incense in charnel houses. So when Coolidge came to the White House, Sinclair and Doheny and Fall and the western oil men with sludge on their boots were for the moment barred, while the manicured associates of Kreuger and Toll were admitted. The bankers of New York and New England who were pyramiding shares and evidences of ownership and obligation of the railroads, of utilities, of mills, even of the oil wells, and of the mines were welcome. Had not the President said:

"For all the changes which they [the people] may desire, for all the grievances they may suffer, the ballot box furnishes a complete method and remedy!" [20]

4

In that first year of the Presidential administration of Calvin Coolidge, his countrymen were more interested in their new ruler than in the

[19] At the December meeting of the American Society of Newspaper Editors.
[20] "Foundations of the Republic," p. 58.

whims of the various markets. For he was a rare bird. His strange ways in the White House made news. America was passing into an urban civilization and here was the first farm boy to enter the White House for a generation, and barring McKinley, the first since Garfield's day. America smiled when Calvin Coolidge first came there and tried to sit in a rocking chair on the front porch under the great white columns. As Mayor, State Senator, Governor, even as Vice President, he could rock in his porch chair at Northampton and at Plymouth without exciting the multitude. But a crowd gathered in Washington whenever he brought his rocking chair out after dinner and he had to give up the country habit.[21] Irwin Hoover, head usher of the White House, recalls the President parading about the family quarters in his old fashioned nightshirt disporting his spindle shanks, in his rooms, much to the embarrassment of the White House entourage! [22]

How exactly he fitted into the mid-Victorian picture of political rectitude, how like a ghost he came back from Lincoln's day! The President liked to snoop about the kitchen and the pantry. Mrs. Jaffray the White House housekeeper [23] declares that the President liked to talk over the menus of meals with her and she sent the day's menus up to the President. He didn't like the custard pies made in the White House. Once he suggested that Mrs. Howard Chandler Christy, whose husband was painting the Presidential portrait, go to the kitchen and make some apple pies which he ate with a relish but without comment until Mr. Christy had finished his portrait when he declared that the portrait was almost as good as his wife's pies and let flattery stop at that.[24] Mrs. Jaffray made up a special recipe for his custard pies and his muffins. When the Prince of Wales came to lunch at the White House, the President worked out the menu which Mrs. Jaffray called one of the most extraordinary official luncheons "we ever had at the White House." It was composed of both fruit cup and clear soup, fairy toast, speckled trout, broiled chicken, a mixed salad, strawberries, ice cream, salted almonds and White Rock, and the Prince in one of the voluminous silences that adorned the meal remarked:

"What a marvelous chef you have, Mrs. Coolidge!"

Coolidge liked breakfast food but could not endure manufactured brands, so every few weeks the White House cook would buy a peck of wheat and a quarter of a peck of rye, mix it, and boil a portion of it for the President's breakfast, whole wheat and whole rye. Mrs. Jaffray remembers that once while she was in Mrs. Coolidge's bedroom just before an

[21] Mrs. Coolidge's reminiscences in Good Housekeeping, May, 1935.
[22] Unpublished notes of Irwin Hoover.
[23] Mrs. Elizabeth Jaffray, "Secrets of the White House."
[24] Reminiscences of Howard Chandler Christy, Good Housekeeping, May, 1935.

elaborate state dinner, the President popped in from downstairs. By way of making conversation the White House housekeeper asked him how he liked the dining room and he admitted it was all right. And then she asked him about the kitchen and he cackled:

"I don't see why we have to have six hams for one dinner. It seems like an awful lot of hams to me."

She explained there would be sixty people and that Virginia hams were so small that no more than ten people might be served with one ham. He reiterated:

"Well, six hams looks like an awful lot to me!"

So she closed the debate and left the room and shortly after quit her job. A sophisticated cook working for an old fashioned countryman would have her baffled moments!

Naturally the President slept in a double bed. Twin beds probably seemed new fangled and citified. He liked pets and being a man of routine he generally appeared in the White House from his day in the executive office at the same hour daily. A signal from the executive office sounding in the residential portion of the White House always precedes the President's departure from one section of the structure to another. His favorite cat, a big yellow tabby, hearing the gong sound which heralded his approach, used to stand waiting for him at the door. He would pick her up and drape her like a foxfur from his shoulders to her great delight and go carrying her into the family presence where there were always dogs and birds and various household livestock reminiscent of the barnyard of his early days. He even had a pet coon. And when some admirer sent him a moth-eaten bear, he put it into the Washington zoo and visited it with some regularity to see that it was well cared for.

The social life which the Coolidges instituted when they came to the White House was simple but dignified. No other mistress of the White House in the memory of men living in her day had more charm and brought more adequate training to her task than Mrs. Coolidge. Looking back at those first days of '23 and '24, she set down a dozen years later the fact that she was graduated from a co-educational State University and had taught a year in a deaf school and then had established her own home on Massasoit Street in Northampton where she remained until the Coolidges left for Washington and were installed in a suite of rooms at the New Willard.

When she came to the White House, it is well to remember that at the first White House reception she wore her best dress made by the town's dressmaker in Northampton,[25] and her utter lack of pretense was the secret of her charm.

[25] Reminiscences of Elizabeth Jaffray.

She had no official duties in Boston, no executive residence. At the New Willard when Coolidge was Vice President, the Coolidges entertained only formally under the culinary guidance of the Willard chef. But she remembers that when the appalling prospect of assuming her White House responsibilities faced her, she shrank from the task with fear and inner trembling. But she did it splendidly.

Writing to Mrs. Frederick J. Manning during the autumn which followed Harding's death, Chief Justice William Howard Taft commented upon Mrs. Coolidge's anxiety to please. He saw she was not quite certain of herself, yet he was sure that "when she gets entirely at home at the White House she will be a worthy First Lady." He looked forward to "a lovely social season in Washington this winter. The Coolidges expect to do all that is required of them."

However, Mrs. Coolidge was never a part of the President's political family. Over and over in her reminiscences in *Good Housekeeping*, which appeared in 1935, she reveals the fact that she scarcely knew or knew but casually, the members of the President's Cabinet; their wives—yes; Senators' wives, Congressmen's wives, diplomats' wives, of course; but the President drew her into no councils. She did her job. He did his. She spent the money. He went over her bills! She remained to the end the farm wife who looked after the chickens and pigs, butter and eggs, the garden and the kitchen and never fretted about the crops and the live stock.

Yet no more affectionate and dutiful husband ever came into the White House than Calvin Coolidge. And the President's pride in her was beautiful to see. In his family he was a disciplinarian and a strict one. This is revealed by John's reminiscences [26] though Mrs. Coolidge recalls that he only spanked the boys once and that was with a flat hairbrush. When they wrote for dressing gowns from their school, he asked why their nightgowns wouldn't do, and Mrs. Coolidge wangled the money for dressing gowns out of her private allowance. The President was always her problem child. Again and again in those curious comments upon Coolidge stories printed in *Good Housekeeping*, in 1935 she reveals the fact that he had a crusty temper which broke easily. Yet also it is obvious, looking at him through her eyes, that he carried no grudges, that he made up quickly. His son John, reminiscing about him, recalls that the President was a tease. He loved that gay hyperbole of language, part irony, part satire, which dances gaily, sometimes a bit puerilely, around its victim. The word "kidding" expresses it. He even liked to deceive himself and held up his lefthand virtues to mild derision. But he was not always so eager to be the butt of

[26] *Good Housekeeping*, April, 1935.

another's joke. It depended upon the joke, and who played it. He was what is known as a practical joker. He found and rang the front doorbell of the White House in his first few weeks. It had rarely been rung, for a footman or usher standing near the door usually greeted the caller before he could ring. A literal White House servant named Mayes, faithful but without humor, was the butt of many of Coolidge's practical jokes. When Judge and Mrs. Hughes were at the White House early in Coolidge's administration, the President rang for Mayes and told him to knock at the door of Judge Hughes's room and see if the Judge was ready for a shave and a haircut. Poor Mayes was worried. He submitted the matter to Mr. Stearns complaining that Hughes had never been shaved, and Stearns would not countermand the order. Mayes knocked gently and when Judge Hughes did not respond, hurried away. It was Mayes who explained to a friend of Taft that he noticed the newspapers were commenting on the fact that Mrs. Taft wore a skirt too short. When she came to the dinner for the Supreme Court, Mayes carefully measured her dress with his eye and found that it was nine inches from the floor. Then he added to Judge Taft's friend: "You and I know what a fine woman Mrs. Taft is and we know that when she wears short skirts, it is not because she is a sport but because she is in fashion." [27] With such a man in the White House, Coolidge had a source of continual delight.

At odd times the President used to press all the buttons on his desk at once and sound half a score of alarms over the White House and bring the servants to him for the sheer fun of it,[28] like the little boy playing at Plymouth, who never entirely died in his heart. Inside him that little boy —sentimental, mischievous, sometimes inconsiderate and cruel—never grew up!

Yet he did not trifle with serious matters. He thought them out as best he could. Bascom Slemp pictures the President in that day—"thinking and thinking and thinking." [29] That line shows Coolidge at his best. Certainly he talked loquaciously to his friends after he had thought. Now this kind of man, money-honest, old-fashioned, self-respecting, conscientious, in short a homely soul, is not the kind to gloss over the misdeeds of the rich vulgarians who were caught in the various scandals of the Harding days. We must remember also that he was sensitive to public opinion once it had set and hardened. By the late winter of 1923, American public opinion seemed to be definitely turned against Fall, Doheny, Sinclair and

[27] Letter from a friend of Taft to the Chief Justice found in his correspondence in the Congressional Library.
[28] Irwin H. Hoover's notes.
[29] In *Good Housekeeping*, May, 1935.

Daugherty and their associates in the oil scandals. So President Coolidge was ready to act.

But for the orgy of commercial speculation that was beginning to riot over the land he had no words of public rebuke, no dramatic warning. In the presence of thieves stealing the people's birthright in naval oil, even while he knew that in his cabinet men sat who had connived at the rascality, he was dumb. In Heine's study of Shakespeare's "Cleopatra," there is a brilliant portrayal of the contrast between the dark, solemn, austere, mysterious, dreary land of Egypt and the gay, frivolous, trifling Persian harlot who ruled over it. In a striking paragraph about Calvin Coolidge, Gamaliel Bradford declares that: "It would be possible to make an equally effective contrast between the mad, hurrying, chattering, extravagant, self-indulgent harlotry of twentieth-century America and the grave, silent, stern, narrow, uncomprehending New England Puritanism of Calvin Coolidge. And Heine caps his climax with the exquisite comment, 'Wie witzig ist Gott!' " [30]

[30] "The Quick and the Dead" by Gamaliel Bradford, p. 253.

CHAPTER XXIV

And Sits in the Seat of the Mighty

DURING his first months in the White House, when President Coolidge went across the land, he observed that national economic and industrial progress had run in a parallel course to the line of progress which he saw while he shuttled from Northampton to Boston and back during the two decades of his public service. Electrical invention had been knitting all industry in all America into a close class-conscious structure, into interlocking directorates, into industrial homogeneity. Here was a new fiduciary invention—produced by electrical financial welding, highly effective, highly possessive rather than creative. The savings mounting in the banks of the little Massachusetts towns were but replicas of mounting savings in tens of thousands of various institutions, trust companies, building and loans, insurance companies, savings banks, mortgage companies all over his America. These savings, by the legerdemain of new business methods and morals had become one lake of capital, placid on the surface, sustained in its waterline by constant springs in the hills while it was being drained slowly into Wall Street.[1] It was a lake of debt—bonds, mortgages, preferred stock,[2] all sorts of curious evidences of obligation were dumped into the lake replacing the fluid capital drained off the cash reserves. Thus as the machines to make machines had been creating industrial debt in plant expansion, the secondary machines themselves were speeding up consumption, piling up goods and chattels at the factory doors. The economic problem of perpetual motion seemed to be solved. Production had been hooked up to the golden machine of perennial credit. We were making

[1] "The Modern Corporation and Private Property" by Berle and Means.
[2] "Railroads: Finance and Organization," "Railroads: Rates and Regulation," by W. Z. Ripley.

wealth by making debt which made more wealth to make more debt. And Calvin Coolidge, President of the United States, with Garman mysticism in his heart, looking out from the portico of the White House across the lawn at Andrew Jackson's statue, which should have given him a shudder, viewed America to the westward and saw that it was good.

His daily routine was light and comparatively easy. He rose early, break-fasted at eight, went to work at nine, quit at half past twelve, lunched lightly, went back to work at half past two or three, quit in mid-afternoon. His schedule was set in steel. It clocked from minute to minute. In his office he was rarely ruffled. But he was not a hard worker.[3] He delegated his tasks. His relations with public men were much as they had been in Boston—wooden and graceless. But public men themselves were ordinarily not expansive—in Boston! A Yankee of the Coolidge type was unusual but understandable there and forgivable. In Washington, President Coolidge had succeeded five gracious men: Harding, whose one talent was his charm; Wilson, who was charming at times but grim on occasion; Taft, who was always amiable, who breathed frankincense and myrrh; Theodore Roose-velt, who was a gay and happy sprite when he was not one of the devil's own imps; and McKinley, suave, smooth, diplomatic, a trained politician, the second oldest profession in the world which had for ten thousand years learned the philandering arts of the oldest profession. But Coolidge had the sour manners of a stern and rockbound New England spinster. Washington hailed him with curiosity and delight, a new specimen of a rare species. There was no accounting for him by any of the rules of the game laid down in American politics. He was silent when it served his ends, but loquacious, even gabby when he wished to talk.

2

Three major matters lay on the desk of the new President which de-manded his attention between August, 1923, when he came to the White House, and December, when he would have to face the new Sixty-eighth Congress. They were, first in importance, the oil scandals, second, his first Presidential message, and third, a decision about his own Presidential candidacy at the Republican National Convention, in June, 1924, less than a year ahead. Let us take them up in reverse order:

Evidently the decision that he would seek the Presidential nomination of his party was made early. For before Congress met, in December, less than four months after the new President had crossed the portals of the White House, his nomination seemed assured. When he chose Bascom

---

[3] Irwin Hoover, head usher, in his book "Forty-two Years in the White House" declares that President Coolidge worked fewer hours and assumed fewer tasks than any other President the head usher had known.

Slemp as his Secretary, the choice was an announcement. Bascom Slemp, a former Congressman from Virginia, the son of a Republican Congressman, was a politician. He went to work on the Southern delegates. The New Englanders had rounded up the North Atlantic Seaboard. He called in John Adams, of Iowa, Chairman of the Republican National Committee, and his committeemen, working with the western Senators, who in 1920 had supported Harding, conservative westerners led by Senator Smoot. They went to work on the West as Slemp had taken over the South, and the New England Senators under Moses were in charge of New England. Chief Justice Taft, whose familiarity in the White House colored his views, wrote to his wife [4] that Lodge could be counted on to do "the Jingo thing," which was true of the Senate under his leadership, and the President built his plans for the Republican Presidential nomination largely ignoring the Republican Senatorial leaders. He gave Bascom Slemp plenary powers. The plan had its drawback. Justice Taft, gossiping in a letter to his wife, tells of a rich Republican who came to Taft worried about making his contribution through Slemp. The rich man wanted the President himself to know about it and be properly appreciative, because chuckled Taft, "Of course while he said he did not wish anything and was only anxious to save the country from the Reds, his wife had let out to me some months ago that she would like a diplomatic post!" [5]

But really Theodore Roosevelt in 1904 had made the same direct appeal to the people over the heads of politicians as Coolidge made in 1924.

Coolidge clashed with Lodge and the Senate jingoes. They favored Japanese exclusion, increased pensions and a soldiers' bonus. He opposed these things.

In spite of all his Yankee cunning, another Congressional group he could not control was the Progressive group. They also had a legislative program and were seeking a Presidential candidate; a score of Senators, or more, mostly westerners with at least eight leaders. The leaders included Senators La Follette, Borah, Johnson, of California, Kenyon, of Iowa, Capper, of Kansas, Ladd, of North Dakota, Norris, of Nebraska, and Smith W. Brookhart, of Iowa. They were forming what was known as the Farm Bloc. They were backing an agricultural bill known as the McNary-Haugen bill, which provided for an equalization fee to market surplus agriculture and to subsidize agriculture by stabilizing the home market. They were also fighting a rear guard action against the retreating railroads that for fifty years had been the particular devils of American reform politics. The railroads had flourished under the curses of the Grangers, of the Greenbackers, of the

[4] April 1924.
[5] So the Reds were thwarted, and then after all the rich patriot did not get the diplomatic post!

Populists, of the Rooseveltians and of La Follette. This progressive group which Vice President Coolidge had met in the Senate, was strong enough in December, 1923, in the House, to delay for three days, when the new Congress met, the election of Congressman Gillett, of Massachusetts, as Speaker, and Congressman Nicholas Longworth, of Ohio, as floor leader. This progressive group to all intents and purposes, was a third party deeply different in aim from the Republican party, able to ally itself with the Democratic party when the Democratic party saw the advantage of yielding to progressive principles and making a majority to harass the flank of the Republican phalanx. These progressives were blood-bred martyrs. They sought trouble. Turmoil was their meat and drink. The spotlight was ever their strongest weapon. Inevitably Coolidge, however earnestly he tried, would fail to tame them. For in the long run they would distrust his methods and suspect his aims. He had no personal quarrel with them. Yet the progressive congressional leaders did not long maintain cordial relations at the White House. At first, Hiram Johnson, George W. Norris and La Follette came and went and Brookhart when he pleased, but their days there were numbered. Between them and the President was an inevitable clash of temperaments. This was not true of progressive Senator Capper. When Mrs. Capper died, the President and Mrs. Coolidge asked Senator Capper to come to the White House after her funeral and live for a time. Yet for all his waving olive branches, the progressives sensed that Coolidge was a party man first and would only go with them so far as they could command a Republican majority. Capper had voted a score of times to defeat administration projects. Capper was the head of the Farm Bloc. When the President finally vetoed the McNary-Haugen bill, he used such emotional language that the message cracked with malicious static. Yet the President kept trying to placate the Progressives. During that first White House year, he appointed Governor Gifford Pinchot as coal mediator in a struggle in the bituminous coal region of Pennsylvania and western Virginia. He pardoned all of the prisoners convicted under the espionage act. Whatever he could do as a pious party stepmother for these impish party orphans—or worse—he did—everything but giving them the key to the pantry. He had his Presidential nomination bagged. But he had it with the sullen opposition of many Progressives and without much help from the regular Senatorial leaders, Harding's friends—the Senate cabal of 1920.

When it came to his second objective which was his first Presidential message to Congress, he was fairly sure of himself. He recommended adherence to the World Court. Many, probably, most of the Progressive Senators were internationalists. But that did not break their lines. Neither did his declaration in favor of recognizing Russia, nor his fair words for the farmer. A few days later the President let Secretary of State Hughes wipe out the implications of the President's fair words to Russia. That episode irritated the Progressives. Yet his message had many mellow spots, which should have attracted the Progressives. But they sensed his inner attitude of conservatism. They were intransigent—"agin the government," and so remained to the end of his term, a quarrelsome, militant, uncompromising group the like of which was never known on any New England land or sea. This group was on the whole without either distrust or enthusiasm for the nominee.

Perhaps the rise of this Progressive bloc in the Senate was rooted in the decay of Republican leadership. That decay began with the introduction of the direct primary and the direct election of United States Senators by the people. Senators elected by the state legislatures were beholden for their election to forces which controlled legislatures—generally those forces which would pay lobbyists to organize and direct legislatures—capital in its corporate forms. The Senate, which McKinley knew, had vanished. It is not fair to assume, indeed it is most unfair to presume that the Senators of the later day, coming directly from the people, were more nobly patriotic than those Senators coming from legislatures directed by the benevolent despotism of the last half of the nineteenth century. Probably the state legislators on the whole, discounting the scandals and the flagrant corruption of the times, were men of more intelligence, of more parts and consequence, than their successors in the days of Wilson, Harding, and Coolidge. It was not a question of patriotism, not a matter of erudition. In the days of Grant, Cleveland and McKinley, Americans of the highest vision were looking toward an expanding America. They were interested in problems of production. So were all the people. But when the primary came, and when the direct election of United States Senators was achieved, problems of production were secondary problems. The people—particularly the people of the agrarian West were interested in problems of distribution. So the complexion of the Senate changed—not in patriotism, not in honesty, not in nobility, but in aims, aspirations, purposes. It was a new world.

Calvin Coolidge, as Vice President, knew the Senate in the beginning of a decline. Senator Penrose, titular leader of the Senate, was dying. He had inherited his scepter from Nelson Aldrich and surrendered it to Henry Cabot Lodge. When Lodge came to power, he was sated with glory. The Lodge whom Vice President Coolidge knew had refreshed his spirit with a victory over President Woodrow Wilson in the contest for the American entrance into the League of Nations and the World Court. But Lodge was old and proud. The sublimated vanity of the conscious patrician gaited him in his public appearance. His hold slackened upon his Republican

Senatorial associates. Despite the diligence which had energized his earlier life, he seemed lazy. The Progressive Senators were younger men, chiefly westerners—outlanders to Lodge. Their creed, an inheritance from Theodore Roosevelt, revised by the elder La Follette, was gibberish to the Massachusetts Brahmin and the open liaison with the Democrats which the Republican Progressives flaunted in the Senate sickened him. They were living in open sin and Leader Lodge could do nothing more effective than to brand them with the Scarlet Letter. Which pleased and amused them and helped them at home in the hustings.

Often, perhaps generally, men of Lodge's caste of mind who appear in high places in politics are encrusted by a protective innocence. They put out of their minds the knowledge of how the mechanics of politics reinforces the shaky foundations of any jerry-built capital structure. But Calvin Coolidge was no innocent. He knew from the hour a bill was drafted in a lobbyist's office, how it was conceived, gestated, born and nourished into maturity. In the White House he played the game with the most astute regular Republican parliamentary leaders. He lost sometimes, but they knew that he knew their tricks and wiles. Often he had gone up and down the backstairs of Massachusetts politics, yet unsmudged with scandal.

Still this is important: His political advancement unquestionably came because he was willing to accept political favors from men who might much more conveniently have given him money for political attitudes and for political actions which from his own viewpoint were fundamentally right. He too often saw justice in the policy of those who would use government as a shield for economic privilege. He sometimes stood for the public interest. But the Coolidge quality that endeared him to the Congressional leaders of his party was his slant toward the dollar once it was legally galvanized as an "investment." Men of wealth, seeking to put the gold plate of authority upon legalized past deeds and deals which were built upon sand, found Calvin Coolidge in Washington, as in Massachusetts, generally willing to accept their claims as rights. He asked few questions. Generally speaking, property once it had gathered itself under a charter and was guarded by a silk hat, was sacrosanct to him.

On the whole therefore, the regular caucus Republicans in Congress were pleased with his message and accepted his leadership. He tried to lead where they delighted to follow. Moreover they realized he was not squeamish about this political alignment at home and in Washington. He was not offended by the knowledge that certain of the congressional leaders were beholden to representatives of the major commodity industries in their states—oil, steel, transportation, food, textiles, communication, banking, copper, coal, insurance, lumber. He declared in his speeches that "the business of America is business."[6] It was plain that in his first message he had consecrated his country under the moral government of an orderly universe, to profits.

So much for the first two items of business on the new President's desk: his presidential nomination and his message to Congress. The oil scandal was not so easily dismissed. No cry of alarm, no exclamation of wrath rose in public and little in private from the new President to rebuke those who were touched with the tar of the various scandals which the Harding administration had produced. To Coolidge, Harding was a Republican President who should be shielded in his mistakes. So at first, President Coolidge gave no encouragement to those who were fighting to uncover the scandals in the Interior and Navy Departments. It was evident from the start that he would "go along" with the party leadership.

3

The limitations of his provincialism may have charted his course in approaching this most vital matter that greeted the new President in the oil scandals. He could cope with a first-class political situation like a presidential candidacy. He could write a good message, but his temperament gave direction to his curious unemotional consideration of the oil episode. To him the Progressives seemed to be troublemakers in the oil scandals— deliberately stirring up a stink! Theodore Roosevelt [7] was moldering in the grave. He had helped to elect a Democratic President and here were his heirs and assigns trying to help the Democrats befoul a Republican President who was dead and defenseless. Evidently at first Calvin Coolidge had no zest in the political business of restoring the stolen goods in the oil reserves, or punishing the thieves. After weeks of reportorial prodding at his press conferences, the only evidence of personal interest—to say nothing of indignation at revelations that were shocking the country—was the dry phrase "let the guilty be punished" quacked at his hecklers with the stingy, gingerly reluctance of one surrendering to extortioners. Indeed he gave up that only after the country knew that Secretary Fall had surrendered Naval oil reserves worth millions to oil barons who in return for it had advanced to him a hundred thousand dollars in ill-gotten liberty bonds, and had used the questionable collateral to secure funds to pay the deficit of the Republican National Committee. By the time the new Congress was warm in its chairs, after President Coolidge had come to the White House, four members of the Harding cabinet, Fall, who had re-

---

[6] Speech before the American Society of Editors, Washington, Dec. 1924.
[7] "Whom Coolidge despised"—letter from Judson Welliver, Coolidge's literary secretary in the White House.

signed under Harding, and Hays [8] and Daugherty and Denby, were being criticized in Congress and in the press for their relations with various phases of the sale of the oil reserves. These men of course were President Coolidge's friends. He had sat with them in the meetings of President Harding's cabinet. He must have heard the sale of the oil reserves referred to at these meetings from its polite and respectable angles. When he saw the transaction revealed by the Senate investigation and blazoned in the press for what it was, his spinster sense of outrage must have been chilled with a terror that he who had graced the festal board also might be suspected of eating the tainted viands and drinking the blood of corruption. His righteous wrath was restrained by something deeper than maidenly reserve!

The oil scandal challenged him early in the autumn of 1923. When Senator Borah came back to Washington, he called at the White House rather perfunctorily. Borah, in August, immediately after President Coolidge had come to the White House, had issued a statement at Spokane calling on the country to support the President. Remember Borah was a bellwether of the Progressives. The President realized this and when Borah appeared at the White House his host saw to it that the perfunctory call should be something more serious. Characteristically the President did not refer to the Borah interview in Spokane. Borah who was not without perspicacity noted that.

But when the Senator first appeared, after the usual greetings, the new President asked if Borah had any suggestions to make. Borah assumed that Coolidge's remark was prompted by courtesy and ignored it. Later in the conversation, as Borah was about to go, again the President asked Borah for suggestions. Borah then declared: "I presume, having asked this the second time, that you mean it." And said: "My only suggestion is that you remove Daugherty from your Cabinet."

The President replied frankly: [9] "I am here to carry out the Harding policies. I am here as a Republican President. Daugherty was Harding's friend. He stands high with the Republican organization. I do not see well how I can do it."

Whereupon Borah told the President with unsparing candor what he might expect from the revelations which were sure to follow. "If you do not remove him now you will have to do so later." The President after that was pleasant but noncommittal. It became evident before the Christmas holidays in 1923 that the President was standing by his Attorney General. Republicans generally, excepting those of the highest caste, were disturbed.

[8] The personal honesty of Will Hays never was in question. None of the dirty money stuck to his fingers.

[9] "Borah of Idaho" by Claudius O. Johnson, Longmans, 1936.

The President was beset with pleas to remove Daugherty. Typical was the suggestion made by Raymond Robins, a follower of Theodore Roosevelt, who had nonetheless supported the Republican party in 1920. Robins has made a perfect and friendly picture of the President's position so far as it relates to Daugherty.[10] Let Robins tell the story of his evening with Coolidge and his attitude toward Daugherty.

When the Teapot Dome scandal was at apex, I grew weary of folks saying that Coolidge did not dare remove Daugherty because it would blow the roof off, and some leading Republicans would land in the penitentiary. I dined with him. We went upstairs to his study alone. I told him I understood that he had decided against removal of the Attorney General, that I plead for a rehearing and asked to make a statement. He said: "Make your statement." I told him I had known Daugherty ever since he was a lobbyist for eastern public service interests in the Ohio legislature, campaign manager in Henry Taft's campaign for the U. S. Senate, campaign manager for Harding for the State Senate, and each of his campaigns until elected President. I told him of my acquaintance with Jess Smith, and of certain transactions that involved both Daugherty and Smith. Then I said: "You have already discredited Daugherty." At this he flared and demanded: "When—in what way?" I recalled to his mind the afternoon I spent with him on the *Mayflower*, when he prepared the draft of a statement that appeared the next morning in every paper of the country in a little box saying in substance that he would engage competent counsel, not of the Department of Justice, to investigate the whole transaction of the Teapot Dome leases and recover any property of the United States wrongfully transferred to private ownership and prosecute any found guilty of wrong doing. I asked him: 'Why not of the Department of Justice unless that department was justly under suspicion?' I concluded pleading the good faith of the people in their government, the credit of the Republican party and his own fortunes as a candidate for the Presidency, and urged the summary dismissal of the Attorney General. The President replied: 'I will not so remove the Attorney General, for two reasons. First, it is a sound rule that when the President dies in office it is the duty of his successor for the remainder of that term to maintain the counsellors and policies of the deceased President. Second, I ask you if there is any man in the Cabinet for whom—were he still living—President Harding would more surely demand his day in court, would more surely not dismiss because of popular clamor than the man who was his closest personal and political friend? I am satisfied that you are right, the people would be pleased, the party would be helped, my campaign would be advanced, by the summary removal of Mr. Daugherty. We shall have to bear that burden. Regarding my being afraid to dismiss Mr. Daugherty, I can assure you, that if the Attorney General does any act I regard as wrong while I am President, he

[10] Letter of Robins to author, October, 1936.

will be removed.' (You will recall that later when Daugherty refused to
take some papers over to a Senate Committee, I think it was, Coolidge removed him overnight.) It was this experience with Coolidge that won my
allegiance and admiration, and that now makes me revere his memory.

So the holidays of 1923 passed. When Congress returned to Washington
in early January, it was evident that the members were alarmed at the
attitude of the country toward the sale of the naval oil reserves. The newspapers were adding incriminating details to the revelations of the official
investigators. It was rough sailing for the White House. In late January
Senator Burton Wheeler, of Montana, introduced a resolution calling
upon the President to ask Daugherty to resign for failure to prosecute certain forms of corruption. President Coolidge remembered Borah's reply to
the Presidential request for advice, the reply which urged the dismissal of
Daugherty. The President sent for the Idaho Senator one night and asked
him to come to the White House upon urgent business. When he had
been there a few moments without learning the nature of the urgent business, Attorney General Daugherty appeared. Miss Mary Randolph, the
White House secretary, was in the upper Hall of the executive residence
when Borah came in. A few moments later she saw "Daugherty come up
the grand stairway and go into the President's study, his jaw set and his
eyes like flint." [11] When Daugherty entered, he saw Borah, and Borah then
knew what the urgent business was. He was angry for a moment. The
President had sent for Borah to make Borah "say it to Daugherty's face."
In a few words the President started the inevitable row. Miss Randolph
declares that she heard Daugherty's "vociferous denials, his loud and angry
protests." [12] Borah remembers that the Attorney General started with
sarcasm. He said as the President indicated the subject of the meeting:
"Well, don't let my presence embarrass you!" [13] Apparently Daugherty
was daring Borah to tell him to resign. Like men of his type, Daugherty
always reckoned with ponderables. He said: "I know some people want me
to resign. Pepper wants me to resign because I wouldn't recommend one
of his men for Judge." He turned to Borah crying: "I don't know why you
should want me to resign. I have never turned you down. You have never
asked me for anything." Borah replied that it wasn't for him to tell the
Attorney General to resign; it was the President's duty. Never a word said
the President. And Borah and Daugherty had it out at the top of their
lungs for an hour, a duet, with the President sitting smoking in his chair

[11] "Presidents and First Ladies," by Mary Randolph, D. Appleton-Century Company,
1936, page 65.
[12] "Presidents and First Ladies," by Mary Randolph, D. Appleton-Century Company,
1936, p. 65.
[13] "Borah of Idaho," by Claudius O. Johnson, Longmans, 1936, p. 289.

on the small of his back, slumped as he was in Boston before the Storrow committee. When they had talked it out and Daugherty was at the door, the little man, standing to speed his parting guests, quacked: "Senator, I reckon you're right!" [14] Miss Randolph remembers that when Daugherty came out of the President's study, he was "white with rage." She watched him "stamp angrily down the stairs and out of the White House."

That short journey from the door of the President's study along the corridor down the great stairway was a Via Dolorosa for Harry Daugherty. For two years he had been passing across that wide hospitable hall. So many times he had been a welcome guest, the King's favorite in Harding's day such a little time past. But in those two years the disreputable gang of retainers who had followed Daugherty from Ohio to Washington—hale fellows with the President—all those familiar spirits who used to cross that hall and mount the regal stairway to the second floor to sit in the White House poker game, were dead, disgraced, or in disfavor.

It was the end of a dream for Daugherty, a vision not wholly unselfish which he had conceived, of raising a dear friend to great power. Being what he was, politics was personal to him. He never reckoned with the imponderable. Borah and his kind, and even Coolidge with his Garmanian philosophy, were not in the Daugherty world. But as he walked "white with rage" out of the White House, he was awakened to realities he had never known.

In the meantime, he did not resign in January. Instead he rounded up the two Ohio Senators, Fess and Willis, sent them to the White House vociferating that the removal of Daugherty was unthinkable, that it would split Ohio politics and give it to the Democrats. And the President, being canny "reckoned they were right" too, and did nothing.[15] But the White House was buzzing with the gossip of the row in the President's study. Miss Randolph remembers, and records in her book,[16] that she "felt his resignation soon must follow." But "soon" was two months distant. Daugherty's resignation had not become clearly a moral issue. The President seemed to be waiting for it to boil down to its essentials. It was the Boston police strike over again!

4

President Coolidge had high esteem for Borah's courage. Borah believed as a constitutional lawyer that the Senate had no right to demand the resignation of a member of the President's Cabinet. But he felt that the President should make the demand himself upon Daugherty and Denby.

[14] "Borah of Idaho," by Claudius O. Johnson, Longmans, 1936.
[15] "Borah of Idaho," p. 290.
[16] "Presidents and First Ladies," by Mary Randolph, D. Appleton-Century Company, 1936, p. 65.

A few weeks after the quarrel in the White House between the Senator and Attorney General Daugherty, Borah endeared himself again to President Coolidge when the Senators were clamoring for the removal of Secretary Denby of the Navy. Secretary Denby's connection with the oil scandals was that of a too innocent bystander who kept his back turned rather conspicuously and allowed Secretary Fall, of the Interior, to transfer the Naval oil reserves from the Navy Department to the Interior Department. A resolution was pending in the Senate calling for Secretary Denby's resignation. Before the vote was taken, Senator Borah was a White House guest with Silas Strawn, President of the national Chamber of Commerce, former Senator Pomerene, of Ohio, and Senator Brandegee, of Connecticut. In the conversation the Senate resolution came up. Borah proclaimed his belief that the Senate could not dictate to the President what the President should do in Denby's case. The President asked Borah to write a message which the President would send to the Senate, embodying his views. The next day when the President's secretary, Mr. Slemp, called Borah over the telephone, Borah prepared the message, read it to Senator Brandegee, and it was sent to the White House and then from the White House came back to the Senate. It was a strong message outlining the President's constitutional powers. It was a part of Coolidge's sentimental make-up never to forget a favor.

So the two men, Coolidge and Borah, in spite of much opposition to the White House from the Idaho Senator, remained personal friends. When Borah was asked, as he sometimes was, to support a Presidential measure, and when he refused, the bluntness of his refusal apparently pleased the President, who as all Presidents, was used to men who hemmed and hawed and wiggled off the hook.[17]

The oil scandals were overshadowing all legislative issues. This was due partly to the Progressives, partly to the Democrats who were playing good politics. For it was a Republican scandal, and the President by his characteristic silence cast himself as a Republican leader, rather than a national hero. In mid-February the Senate resolution passed, though it was patently unconstitutional, demanding the resignation of Secretary of the Navy Denby.[18] The revelations of the Senate investigating committee were prov-

[17] Letter to author from Senator Borah, September, 1935.
[18] An interesting story on the ramifications of the oil scandal went in a letter from the Chief Justice to Mrs. Taft at Barcelona, in May, 1924. The sons of Theodore Roosevelt, Archie and Theodore, had been in the oil business, more or less connected with Harry Sinclair. Young Theodore was Assistant Secretary of the Navy, a place his father had held. It was an uncomfortable berth. The Chief Justice declared that Young Theodore went to the White House to get the President to advise him "what to do in respect to Archy's story before the Senate Committee" and the Teapot Dome. He failed and "then," writes the Chief Justice, "Alice tried but Calvin flatly declined to

ing all that its Progressive members had charged and more. Harry Sinclair, one of the principals in the oil investigations, on the stand, had declared that he had probably contributed to the Democratic campaign fund in 1920, and that he did not remember how much he had contributed to the Republican campaign fund that year. But enough, probably, to give him the right to ask favors of the Secretary of the Interior, even before the iniquitous Continental Trading Company rose and faded. Then Albert Fall, appearing before the committee, shame-faced and grim, declined to answer questions which Senator Walsh asked on the ground that the answers might incriminate him. Two weeks later, being summoned again, four physicians swore he was unable to appear.

It was in February, two months after Congress convened, that President Coolidge finally made a major move in the oil scandals. First he called his Amherst friend, Harlan F. Stone, a New York lawyer, to the White House and went over the whole matter.[19] The President's embarrassment with Daugherty in office was obvious. The President had sent Chief Justice Taft to the Attorney General asking him to resign.[20] These things the President laid before his friend Stone, and Stone advised the President to appoint his own counsel and recommended Owen J. Roberts, of Philadelphia, as special counsel to prosecute the oil cases. This the President agreed to do after Stone had made it clear that he would not take the place. The President then appointed former Senator Atlee Pomerene, of Ohio, an honest, capable Democratic lawyer, and Owen J. Roberts, of Philadelphia, as special counsel to investigate and prosecute the malefactors in the oil scandal. This Presidential announcement created a tremendous sensation. The next day after Roberts' confirmation Secretary Denby resigned. In March, Harry Sinclair appeared before the Senate Committee and refused to answer questions about Teapot Dome.

Attorneys Pomerene and Roberts quickly moved to declare the leases on the Oil Reserves Numbers 1, 2, and 3 invalid. Receiverships were established to operate the Pan-American Oil Company's wells and plant on Reserve Number 1. And three days later when the country, lashed by the Progressive clamor in the Senate, had turned upon Daugherty, the President finally decided to act. Daugherty had refused to give the Senate committee certain papers in the Attorney General's office. That was the opening the President needed. He abruptly asked [21] for Daugherty's resignation, declaring that he could not be an impartial adviser of the government while

have anything to do with the Senate's investigation. Calvin," adds the Buddha of Wyoming Avenue, "was wise!"

[19] Letter to the author, 1935.
[20] The Taft letters in the Congressional Library.
[21] See letter of Chief Justice Taft to his son Robert, March 30, 1924.

under criticism himself. Politely the President explained in his letter that he did not question the "fairness or integrity" of Daugherty, but nevertheless demanded that he resign. Faced with the necessity of removing Daugherty, the President acted characteristically. In one of those moments of silent abstraction, he had thought it all out.

Perhaps no kindlier epitaph may be written at the close of Harry Daugherty's political life than the words penned by Chief Justice Taft in a letter to Elihu Root, in September, 1926. After Daugherty had resigned, after he had faced his accusers, after he had revealed his character and stood amid the ashes of his career, in this letter to Elihu Root, the Chief Justice discloses the fact that he talked with Daugherty at the suggestion of both Borah and Coolidge, trying to persuade him to retire. Taft felt that he should have retired when Harding died. "His fate," writes the Chief Justice, "is the story of ambition for too high a place, of lack of real ability and capacity, of careless courage and loyalty to his friend . . . with a lack of appreciation of his weakness before the public and of his love of office. . . . He appealed to me and he appealed to the President on the grounds that he could show himself as free as possible from the slightest participation in anything that was dishonorable. Then he would not face the probability that he would be entirely smeared before they got to him and gave him a chance to testify. He relied on Burns, the detective, to show up Wheeler, of Montana . . . as a defense against Wheeler's attacks. Burns, whom against the advice of many people, he insisted on taking into his department, entirely failed him. Altogether it is a tragedy!"

Later, after the testimony in the trial of Daugherty was in, the Chief Justice, writing to his brother Horace, expressed a belief that after all Daugherty was personally honest. "His destruction of evidence however is damning.[22] It was probably done for the purpose of concealing scandals connected with the Harding administration for which he will have to suffer. I am very sorry."

5

Statesmen came to the White House recommending a dozen possible candidates to succeed Daugherty in the attorney general's office. They received no encouragement, scarcely a pleasant look upon their endorsement. Evidently the President in his heart's secret council chamber had made his choice. When the President had first called Harlan F. Stone to Washington to discuss the oil cases, his Amherst friend remembers[23] that the President, for him, was unusually voluble. They talked of retiring both

---

[22] Daugherty burned all his files before his own trial.
[23] Letter to the author, July, 1935.

Denby and Daugherty, who were then in office. All day they devoted to the oil scandal. But just as Mr. Stone was leaving, while the two men were talking about the office of attorney general, suddenly the President said:

"How about your taking it?"

Mr. Stone replied he did not care for public office, explained that he was interested in his profession, had recently formed a new and satisfactory partnership from which he was receiving a large income. Then he turned to the President and declared that if he really desired Stone for the place he would regard it as a call to public service which he could not refuse to honor. But, standing by the door, he expressed the hope that someone else could be found. He heard nothing more of the matter until Daugherty's resignation had been forced. Then a night call by phone from the White House invited the President's Amherst friend over to Washington to a Presidential breakfast. On April 2 a group of senators, to wit, Senators Lodge, Wadsworth, Curtis, Cummings, Watson, Shortridge, Brandegee, Borah,[24] Willis and Moses, were assembled around the breakfast table. No one knew the guest who was introduced simply:

"Gentlemen, this is Mr. Stone."

No reference was made at the breakfast to the vacancy in the attorney general's office. The President had said nothing to Stone. The faithful Frank Stearns was there. Being from Amherst he knew Stone, and just before the breakfast assembled he asked Stone what the breakfast signified. Stone did not know.[25]

When the Senators had departed, the President led Mr. Stone into the northeast bedroom. They sat down before the fire. The President lighted a cigar. They chatted a while and finally he said abruptly:

"Well, I think I will send in your name!"

That was all. He finished his cigar, got up, left Mr. Stone in the White House, chaperoned by Mrs. Coolidge and Mr. and Mrs. Stearns, went over to his executive office, sent the name to the Senate. While the name was

[24] Excepting Borah, all right-wing Republicans.

[25] Parenthetically we should record here that the President was given to such quiet, puerile pranks. Later in that same year he invited from the War Department and the Navy Department a group of high placed officials, army and navy men, generals and admirals, most of them men who apparently were included in his list because of their knowledge of the pending army and navy appropriation bills. They came to one of these breakfasts like the one he gave for Stone. They came crammed to the gullet with information about the pending bills expecting to be quizzed. They brought portfolios under their arms to back up casual statements. The breakfast passed with no question, no significant remark remotely touching the army and navy bills. Indeed, the President sat looking down his nose into his grapefruit, his sausage, his fried potatoes, his buckwheat cakes and coffee and spoke hardly at all, let his guests break the ghastly silence that heralded the gloom of the occasion, before they began to cheer up and realize the joke. On the White House steps they burst into gales of laughter. (Story by General Briant H. Wells, U.S.A. retired.)

going to the Capitol, Congressman Snell, of New York, who had endorsed Mr. Stone as Daugherty's successor, appeared in the executive office on other business. Snell's endorsement of Stone, a few days before, had been greeted with foreboding silence. Snell did not mention Stone that April morning. Coolidge did not mention Stone. They transacted routine business. Snell departed for his office. When he arrived a telephone call came from Bascom Slemp, the President's secretary, saying that the newspaper reporters were clamoring for Snell to furnish them some information about the new Attorney General Stone, whom Snell had recommended. Slemp said:

"You have just made an Attorney General and the newspaper boys want you to tell them something about it."

In the meantime the Senate had confirmed Mr. Stone's nomination. When Snell came back to the White House a few days later and indicated surprise and possibly a little good-natured pique at the Presidential pro‹ cedure, the President commented:

"I thought you would probably find out about it," and let it go at that.

Republican National Committeeman Hilles, of New York, had an official right to be consulted about New York appointments. The Taft correspondence [26] indicates that Hilles knew "absolutely nothing about the new Attorney General." And he was also surprised at the appointment of Stone's first assistant attorney general, a New York lawyer from Buffalo. It was the Coolidge way.

When Harlan F. Stone took charge of the Department of Justice, he immediately ordered the spying, telephone snooping, and undercover work of the Department of Justice to cease.[27] A new regime began in that office that day.

In the early spring of 1924, President Coolidge did a most characteristic thing. He asked Harry Slattery to come to the executive offices and tell him all he knew about the oil scandals. Consider Harry Slattery: a left-wing, fighting Progressive friend of LaFollette, Wheeler, Kenyon, and Norris, associated with numerous left-wing liberal activities such as the National Voters League, the National Conservation Association, the National Popular Government League, and other groups trying to promote a larger use of government in projects for human welfare. Slattery was a Washington lawyer with a keen mind.[28] He rather than anyone else first uncovered the iniquity of the oil leases. He wrote to La Follette in the early winter of 1922 telling him the whole story of the plot, telling him how to get copies of the

[26] In the Library of Congress.
[27] Official documents, Department of Justice.
[28] A dozen years later he was assistant to Secretary of the Interior Harold Ickes, under President Franklin Roosevelt.

executive orders and leases. He penned the first rough draft for La Follette's resolution, directing the Secretary of Interior Fall to send to the Senate a list of all the oil leases made by the Secretary covering Reserves Number 1, 2, and 3 hereinbefore mentioned and "all correspondence, papers and files" showing the applications for those leases and the departmental action. Senator La Follette later read into the Senate record a tribute to Mr. Slattery, which declared that Mr. Slattery had been "the veritable watchdog of the nation's resources" and had been "in every contest over these resources, on the people's side ready to give his time without compensation." Shadows of men like Slattery had rarely fallen upon the White House desk since Wilson left. Why then did Coolidge send for Slattery? Was Coolidge at last coming to the boiling point, bubbling a little Puritan indignation? Remember that Fall had gone, Denby had gone, Daugherty had gone from the Cabinet. The newspapers were flashing headlines from the Senate investigation. The White House had appointed Senator Pomerene and Owen J. Roberts—later to go into the United States Supreme Court—as special counsel to prosecute the oil cases. Will Hays had told the story of Sinclair's donation to the Republican National Committee through the bonds of the Continental Trading Company, and the Democratic National Committee chairman had told of receiving thirty-nine thousand dollars from Doheny. On the face of it, the whole story was out. But who knew what else was waiting in the wings to pounce upon Calvin Coolidge in his campaign for re-election—who of all men but Harry Slattery, the prime agitator of the issue? The President also knew that Mr. Slattery was honest. In which case the dumber the President seemed the more he would learn. The President began with Mr. Slattery abruptly: "They tell me you are familiar with the details and background of the whole oil incident."

Then he almost curtly asked Slattery to tell him the whole story. Before beginning Slattery felt it fair to tell the President that he had told both Senator Walsh and Senator La Follette that he was called by the President to ask about the oil case. The President snapped his approval and waited. That news fitted into the Presidential strategy. Then Slattery told him the story in detail beginning in the early part of the Harding administration. The President pried in questions about Fall's motive.[29]

"It was evident to me that the President did not know much about the whole case. Nor was he familiar with the background, nor terminology of it although he had evidently gotten distinctly in his mind that the Navy oil leases played a most important part in the scandal. He asked particularly about the origin of the Naval Reserves and their legality." All his life we have seen that Calvin Coolidge would rather appear foolish than to be a fool.

[29] Letter to the author by Harry Slattery.

After Slattery had told his story and answered all his questions, the talk came to an abrupt, embarrassing end. Slattery had nothing more to offer. The President knew the worst. He had nothing more to ask. After a few seconds of silence, Slattery went away. Evidently the President had acquired faith in him for he sent for Slattery several times later to discuss important matters—for instance, hydro-electric power and conservation.[30] It may be significant to note that in the headlines of the *Washington Times* and similarly of other papers a day or two later, one finds this banner:

"COOLIDGE HITS OIL DEALS!
"Has quietly assumed control of situation—
"Surprise awaits his foes."

At last Calvin Coolidge, running true to his temperamental form, had sized up the traffic and was ready to cross the street of the oil scandals. From Senator Borah,[31] the President had learned all the politics of the situation. With Attorney General Stone [32] he mastered roughly the legal aspects of the story. After he talked with Harry Slattery,[33] an honest critic of Coolidge's slothful attitude, the President realized that no plot was abroad to ambush the administration, and he moved characteristically with circumspection but with neither fear nor joy in his gait! Thus he walked into his campaign for lieutenant governor, thus he entered the final phase of the police strike. In his great moments he was always prim, precise, prudent.

Bascom Slemp, writing of that period,[34] declares:

"No man is harder to get an expression of opinion from than President Coolidge on a subject to which he has not applied his mental and moral processes. No man's opinion is easier to know on any subject to which he has applied it."

And by way of a little friendly incense burning, Mr. Slemp writes: "He concentrates more intensely and more continuously than any man I have ever known."

Mr. Slemp, a fluent and delightful conversationalist himself, must have had moments like that of Frank Stearns when he tried to talk in those days to the President about the oil scandals for he continues: "Morning,

[30] He came to refer to Slattery when he answered a White House summons as "the oil and water [power] man"; and when Coolidge joked with a man it was a sign of confidence.
[31] Letters from Borah, Stone and Slattery to author earlier in this chapter.
[32] *Ibid.*
[33] *Ibid.*
[34] "The Mind of the President," by Bascom Slemp, Doubleday-Page, p. 6.

noon, and night he keeps thinking, thinking. He indulges in no distracting pleasures."

It is odd that amid all the turmoil of that time when the scandals of the late Harding administration were crying to Heaven, President Coolidge made no protest to indicate his unhappy plight as Harding's executor. He wrote no letter easing his inner wrath. When he snapped out "let the guilty be punished" in his press conference, he had spent his rage. Only once did he indicate to his associates, and then to one man, in the privacy of his study, his inner bitterness. Sitting alone with Herbert Hoover in his study one day, he broke out with this:

"Some people think they can escape purgatory. There are three purgatories [35] to which people can be assigned: to be damned by one's fellows; to be damned by the courts; to be damned in the next world. I want these men to get all three without probation." [36]

And then having eased the bowels of his wrath, he shut his mouth tightly in silence.

[35] In the end Denby went into obscurity, Fall and Sinclair went to jail, and Doheny to the heart break of a lonely old age! They did their penance.
[36] From *Good Housekeeping*, May, 1935.

CHAPTER XXV

But Recites Some Dull Lines

IT WAS safe to assume that when administration newspapers began to flash the story of the President's genuine interest in the oil scandals, the Presidential campaign of 1924 had begun. It had been eight months forming, but it was organized. After that the fight did not falter. Again we get a glimpse of the President from his secretary about this "outstanding quality of consistency." [1]

"He has never changed his mind on a fundamental public issue. This," quoth the nimble and happy Mr. Slemp, "is an extraordinary fact."

During the spring of 1924, President O'Neil, of the Prairie Oil and Gas Company, arrived from Canada and returned to his Company $800,000 taken out of the country by him in the Continental Oil deal. Fall, Doheny and Sinclair were re-indicted for conspiracy to defraud the government after the first indictment had failed upon a technicality. The government finally won its suits. All the oil leases were invalidated. But Doheny was acquitted. Fall went to jail and Sinclair was imprisoned for tampering with a jury. But the battle of the oil scandal left the White House early in 1924 and was conducted in the courts.

However, the moving spotlight of public interest left the courts and began playing upon the White House. The chief interest in the White House was the President's spring campaign in the primaries for delegates to nominate him at the Republican National Convention in June. Congress was in session. Small grain was grinding through the mill. The Johnson restrictive immigration bill excluding Japanese had passed. The soldiers' cash bonus bill was defeated in the Senate which then adopted the Insurance Plan known as the soldiers' bonus bill which passed both

[1] "The Mind of the President" by Bascom Slemp.

Houses, was quickly vetoed by the President, and repassed over his veto. Here enters a vote that made history and signed the political death warrant of Henry Cabot Lodge. Lodge voted for the soldiers' bonus bill upon its original passage. He voted for the motion to pass the bonus bill over President Coolidge's veto. Lodge claimed justification. He had promised the Republican State Convention in Massachusetts the year before that he would vote for the bonus. Lodge at the opening of Congress had been re-elected Republican leader. President Coolidge felt that Senator Lodge had another status from that as Senator from Massachusetts. He was Republican leader in duty bound to support the President's position. Moreover, Lodge, Republican leader, became factional leader of the bitter-enders who opposed President Coolidge's recommendation to bring America into the World Court. Senator Lodge, the Republican leader of the Senate, felt that this recommendation was an affront to his leadership. He defeated it by indirection but did not hide his tracks from Coolidge. Also Lodge opposed the President's effort to stop Japanese exclusion. Lodge was an ancient of days in Congress. He had been there forty years. Coolidge was new to national politics and to the White House. But he had his pride, and with his pride he had the power of a mighty office.

The Bursum Bill, increasing pensions for soldiers of all wars and war widows, was vetoed and the veto stuck. The week before the Republican National Convention met at Cleveland the first session of the 68th Congress adjourned.

The protective tariff, certainly one of the major bulwarks of American prosperity for half a century following the Civil War, also most certainly had become the nursery of special privilege in America. Early in Coolidge's Presidential term it became evident that the United States Tariff Commission was being packed by the tariff interests. The "packing" began and was well advanced under Harding. When Coolidge came to the Presidency, three high tariff members were already on the commission and Marvin of Massachusetts had been made chairman. Coolidge inherited the condition and it conformed with his philosophy. Men entirely honest but also deeply convinced of the wisdom of high protection and definitely representing certain high tariff interests, were being put into the tariff commission. Commissioner Marvin, of Boston, chairman of the commission, represented the Home Market Club of New England, the organization which looked after New England tariff interests. It was charged that Commissioner Burgess, an honest and zealous man, came to the commission with the endorsement of the pottery interests of Ohio and New Jersey. Similarly it was alleged that Commissioner Glassie came to the commission representing the sugar interests after certain Democratic Senators from the far South, particularly the Louisiana Senators, had supported the Republican tariff acts. Glassie through his friends and relatives was interested in stocks of sugar companies. He was related by marriage to the late Senator Donelson Caffery, of Louisiana, who was a conspicuous defender of the sugar interests of his home state. Two westerners, William S. Culbertson, of Emporia, Kansas, and Edward P. Costigan, of Denver, were members of the commission representing the liberal and western Republicans' ideas of a moderate tariff on the protective principle. Commissioners Culbertson and Costigan began to object when other members of the commission sat upon cases in which they or their relatives were known to have a direct financial interest. The matter was taken to the White House. It threatened a public scandal. The Progressives in the Senate, under La Follette's leadership, were preparing a resolution to prevent members interested in tariff items pending in the commission from sitting in a case in which the member had direct or indirect family interest.

Curiously the President could not see any moral obliquity in a man sitting in a case in which he was interested; Burgess on pottery, Marvin on textiles, Glassie on sugar. But the two westerners objected publicly. In a White House conference with the members of the commission, including Costigan and Culbertson, the President took the position that the eastern members of the tariff commission were particularly qualified to sit in matters wherein they were especially informed. He held that the commission was not a court nor were the commissioners supposed to be impartial judges. He began to receive letters of protest from western Republicans. Replying to these, he set down with entire frankness his views; that the commission was not a court, that the commissioners were not supposed to be judges, and that they were particularly valuable because of their intimacy and hence wide acquaintance with special matters in which they or their families had financial interests.[2] Finally a congressional resolution sponsored by Homer Hoch, of Kansas, stopped the practice, but before it was checked Commissioner Culbertson, who had been particularly, and probably from the high protection standpoint, offensively active, became the subject of attack from the right wing of his party in Congress. The attack was skillfully staged to appear that the President was anxious to get rid of Culbertson, because he was a low tariff protectionist. So specious charges were made that Culbertson accepted pay as a lecturer in a Washington school violating a law which forbade a commissioner to receive any salary or fee outside his official salary.[3] It was announced by right wing, high tariff senators that Culbertson was to go. Mr. Slemp, the President's secretary had asked the Attorney General to rule upon the Culbertson

[2] January 14, he wrote such a letter to the author.
[3] The law reads, "No member shall engage actively in any other business, function, or employment."

case. The Attorney General supposed that the suggestion for the ruling came regularly from the President. As the controversy grew warmer in the press, Attorney General Stone began to receive telephone calls from the White House urgently asking for the opinion and intimating that the President was anxious to act.[4] The Attorney General reached the conclusion that the statute did not authorize the President to remove Culbertson. A few days later the Attorney General pocketed his opinion, went over to the White House, and found that the President seemed to know very little about the charges against Culbertson. When the Attorney General handed the President his opinion, Mr. Coolidge said bluntly:

"I don't wish to remove Mr. Culbertson."

Then seeing the puzzled look on Attorney General Stone's countenance, the President declared that several complaints had been made about Culbertson but that he presumed they were based upon his tariff decisions and paid no attention to them. But the President told the Attorney General to intimate to Mr. Culbertson to get rid of the lecturing job to remove the cause of criticism. He refused to look at the Attorney General's opinion and that was the end of it. The episode is trivial, yet it indicates rather clearly President Coolidge's attitude in the conduct of certain public affairs. He was willing to stand by the principle of a protective tariff. He was even color blind to the moral turpitude involved in a man on the commission passing on matters in which he had an interest. In spite of his advisors, in spite of the White House coterie around him, he was not willing to tamper with the jury himself by removing an honest juror with whom he disagreed. Time and again in his public career this attitude of honesty according to his lights was revealed. In December, 1937, a Senate committee investigating the American railroads, turned up the curious information that President Coolidge, or someone in his office doubtless with his knowledge and consent, furnished to General W. W. Atterbury, of the Pennsylvania railroad system, advanced copies of the President's message to Congress, providing for the consolidation of all eastern railroads into four systems, which would be the Pennsylvania Railroad, the Van Sweringen group, the New York Central, and the Baltimore and Ohio. Calvin Coolidge could see no reason why the railroads affected by his consolidation proposal should not know well in advance of its announcement what his plans were. If the presidents of these roads benefited on the stock market, it was no affair or fault of his. He believed profoundly that the invested dollar was sacrosanct. It was a part of his philosophy. Yet he would not profit himself nor allow a friend to profit by information in his office. His integrity according to his light was impregnable. But men who did not agree with him could not see the nice distinctions that he made. These distinctions meant

---

[4] Letter from Justice Stone to author, May 23, 1934.

much to him, and the story of his life would be incomplete without these episodes which so exactly illuminated his viewpoint.

In his first year he was confronted with the Mexican situation. A new government more or less communist, after the Latin fashion of communists, was making things lively for American investors in Mexico, and the White House desk was often littered with strident calls for help from American residents in Mexico: superintendents of mines, owners of various business enterprises, holders of concessions who felt that the President should take a strong hand with Mexico. Returning tourists or observers for banking interests or journalists would call at the White House pleading for action, which would embroil America in the Mexican civil upheaval. Henry L. Stoddard, of New York, editor of the New York Evening Mail, was among those who, visiting at the White House, clamored for some sort of intervention. Stoddard had a right to talk plainly to the President. He had been one of Coolidge's early supporters. He tells the story himself, illustrative of the Coolidge attitude to a vital international question.[5] The President lighted a cigar, tilted back his chair, put his feet on the desk in the Lincoln room of the White House and said: "Fire away!" Stoddard let off his batteries. Finally the President cut in when his guest was going well:

"Now," he said, "look on the other side of the picture. Here we are the most powerful nation in the world. At this moment we have special representatives in Europe as well as all our diplomats urging reduction of armaments and preaching peace. The world of today would harshly condemn us if, despite our attitude, we should go to war with a neighbor nation not nearly our equal. What do you suppose people in years to come would say? Powerful United States crushing powerless Mexico! Don't you think it is better for us to find another way to handle the situation?"

He appointed Dwight Morrow, of the house of Morgan, Ambassador to Mexico. Of course Dwight Morrow, the conciliator, was the "other way." Morrow was an intelligent, diplomatic American with a courageous man's tact. It was a personal solution. Probably the Coolidgean foreign policy was dramatized rather simply around various persons who staged his American ideal: "The business of America is business;" in South America and Europe. His enemies called it "dollar diplomacy"! President Coolidge was not stopped by a phrase. He continued to apply the Garmanian philosophy to our foreign policy as he had applied it to all matters, great and small, through his career, believing on the whole in the mystical power of kindness and common sense. Common sense in his lexicon meant conserving the main chance. Curiously, with all his clowning capacity which sometimes made him see himself in grotesque perspective, he never seemed

[5] "The Mind of the President," by Bascom Slemp.

to dramatize himself to the nation as a St. George facing any dragon; either in his foreign policy with its obvious dollar-minded mixture of pawnbroker and samaritan, or in his role as defender of the rich. So his attitude toward malefactors in high places—in oil, in the tariff or in the various trusts—was difficult to comprehend at the time and later seemed reprehensible to many of his countrymen.

As the Cleveland Republican National Convention of 1924 in May began to take form, it was apparent that the Republican party was behind the President. It seemed likely that the country would accept Republican leadership again. The country did not burn with wrath against the Republican party because of the oil scandals. The country was slow to anger because a new President had come who was not to blame, and after a manner the old President by his death had paid the price. But the country's attitude of respect for Calvin Coolidge was deeper even than the acceptance of the death atonement of Harding. The country had come to respect the new leader. The people had given him their faith. The country had not refreshed its capacity for emotion since the war. It could not rise in its wrath. It had been disillusioned in the war. Deflation had broken the spirit of the farmer. The Allies were beginning to show their ancient animosities, their jealousies and worst of all their greeds. The rainbow of hope for a better world had faded by 1923. The rocket of popular faith in government as a savior of human destiny had burned out. No one was more keenly aware of this dead spent state of the public emotion than Calvin Coolidge. His twenty years as a vote-getter and a successful office seeker had made him clairvoyant in reading public moods and movements.

Yet so completely had the President changed the White House satellites —those who came to amuse, advise, inform and unconsciously guide the President—that a change of party would not have more sweepingly cleared the rubbish of the Harding regime out of the White House than Coolidge's casual order changed it when he talked quietly to Head Usher Irwin Hoover, that first five minutes in the White House in August, 1923. The grafters, the grabbers, the light-hearted, poker-playing, get-rich-quick denizens of the political underworld were completely banished. Respectability appeared—"the rich, the wise and the good!" [6] Because Calvin Coolidge always was regular, he consulted regular party men. There was no mug-

---

[6] John Adams suggested that "the rich, the well-born, and the able" should be segregated in a separate house of Congress—the Senate (as less likely to do harm thus), and the words "well-born" aroused much resentment. McMaster, "A History of the People of the U. S.," Vol. I, p. 469, says that the Federalists came to be called "well-born" by their opponents. The phrase was discussed in some of the debates on the ratification of the Constitution. See a discussion of it, with many references to sources on p. 47 and footnote, of C. M. Walsh's "The Political Science of John Adams" (Putnams, 1915).

wump nonsense, no Progressive favoritism in his administration. Politicians who had official business at the White House, Senators, committeemen, elder statesmen, members of the Republican samurai, were cordially received.

2

Soon after President Harding's death, newspaper reporters began to note a new face at the White House door—the smiling blondness of former President William Howard Taft, Chief Justice of the Supreme Court. Warren Harding's taste for trash seems to have offended the Chief Justice so deeply that his gratitude to President Harding for appointing him to the highest office within his power to give could not bring the Chief Justice to become a White House familiar where he would meet the President's cronies of the political demimonde. Taft knew Harry Daugherty in Ohio but chiefly as a politician, as his letters to his family indicate. The changed social atmosphere in Coolidge's regime attracted the Chief Justice.

As the power of Senator Lodge waned, the influence of Chief Justice Taft waxed strong. He was a frequent visitor but hardly an habitue of the White House. His good taste and his keen social instinct kept him from boring the President. Sometimes the Chief Justice dropped in for a chat, but as his letters reveal, he always had something on his mind. Often the President sent for the former President. For nearly a century no other former President had been a White House intimate.[7] Early death, partisan differences, factional alignment, and occasionally a hearty desire to be rid of politics had kept men who lived in the White House away from it when their official duties were done. But Chief Justice Taft could come to the White House with dignity. His office was almost as important as the Presidency. Since he had left the White House, ten years had trimmed down his paunch, but had lightened his step, had chiseled off his jowls and brightened his eyes, giving a gay resonance to his voice. Only death could muffle the joy in his laughter even in the decade that is "labor and sorrow." The Chief Justice personified in hearty yet simple felicity a Beethoven allegro. Sitting often before the broad flat desk which Calvin Coolidge always kept clear in the executive offices, facing the glum, sharp visaged, lean little Yankee in his fifties with parsimony printed in his sinewy hands, in his passionless mouth, in his rasping Yankee patois, what a comic contrast to him the Chief Justice made—the mountainous loose-jointed frame of the westerner with his mellifluous voice, his great soft hands, his twinkling eyes

[7] In President Madison's day former President Thomas Jefferson who frequented the White House occasionally was known by Madison's enemies as "the Grand Lama of Monticello."

and a laugh that was forever vaulting out of a diaphragm which once had supported a Buddha belly. The staid old executive mansion never had seen such an odd pair before.

But no idle caller was Taft. The President gave the Chief Justice the impression [8] that in matters of judicial appointment he would be consulted. Justice Taft was in communication through Charles Hilles with the New York Republican organization and thus with political Wall Street. But the relation between the Chief Justice and the President was not political.[9]

On the whole it was mutually respectful and close. The Chief Justice would drop in at lunch time, almost informally, or at dinner, and on one of those occasions wrote to his wife that it was a happy party with no one present who shouldn't be there. At another time at an informal Sunday dinner, he wrote smacking rhetorical lips over crab meat cocktail, broiled chicken, a green salad and ice-cream—his idea of a proper Sunday noon dinner, after which the Coolidges went to the Mayflower yacht on the Potomac. When Mrs. Coolidge wished to entertain the wives of the Justices of the Supreme Court, soon after the Coolidges came to the White House, the ladies of the court had not called. Mrs. Taft was an ex-President's wife, the wife of the Chief Justice, and more than both, a Herron, of Cincinnati—a royal highness in her own right. Mrs. Coolidge was a former teacher in a deaf and dumb school. William Howard Taft was a social Solon—something wiser than a former President, and a Chief Justice. He was a diplomat with great gifts. So Mrs. Taft called and left a card. The tea came off, a felicitous gathering. And Taft wrote chuckling to his children about it, indicating that Mrs. Coolidge was a Nice Person!

Politically William Howard Taft had no submerged political past. He was the same yesterday, today and tomorrow, the honest, profoundly convinced intellectual conservative, a statesman of charm who could meet his political foes at a punch bowl with graceful naivety and eviscerate them in the forum wearing the same dimples the while. Coolidge was safe in Taft's hands.

Early in his administration, the new President came to grips with the Senate over the matter of federal judicial appointments. Under a practice dating back a hundred years, judicial appointments in the federal district and circuit courts had been considered patronage of the Senators who were members of the President's party. This rule of patronage had been responsible for many weak judicial appointments, at least since Van Buren's day. It was obvious that certain members of the Supreme Court, notably Chief Justice Taft, desired to bring the President to a realizing sense of the weakness and the wickedness of this liaison between the legislative branch of

[8] See letters of Taft to his son and family in the Congressional Library.
[9] Ibid.

the government and the administrative branch in naming federal judges. In his correspondence with his brothers, with Charles D. Hilles, with his daughter and his sons,[10] it is obvious that Chief Justice Taft felt it a bounden duty to labor with the new President to break him loose where it was possible from Senatorial control in naming federal district judges. The Chief Justice even suggested that the President appoint a number of Democratic federal judges and went to Harry Daugherty, Attorney General, with the proposition in the autumn of 1923.[11] Daugherty told him that the President was not quite up to appointing Democrats yet, but he thought he could do so after making a few Republican appointments. When the new President had been in the White House less than six weeks, the Chief Justice took time by the forelock to speak to Coolidge about a vacancy in Missouri. The Chief Justice had a low opinion of the Missouri Senator's candidate. But having a love for intrigue, the Chief Justice set out some backfires and wrote to Gus Karger,[12] about Senator Spencer's candidate, that "Coolidge will send for the record and examine it and if he does, I think he will find some facts there that will shake a man who has Massachusetts or Vermont ideals of what a judge should be."

Six months later, the Chief Justice wrote to his son Robert,[13] about this matter of judges. "Think of estimating the importance of a federal judge by the amount of receiverships or other things he may have to fill, but that's a good idea of this Spencer, of what he can do if he has the control of appointment of a judge." And there the naive Chief Justice put his hand upon one powerful lever which swayed the Senatorial interest in judicial appointments. The Senator who appointed the judge had control over the appointment of the receiverships in the judge's court.[14]

But as the administration grew older, the Chief Justice wagged his head

[10] This correspondence is on file in the Congressional Library.
[11] Letter to Charles D. Hilles, December 5, 1923.
[12] September 14.
[13] March 15, 1924.
[14] Letter to author, March 9, 1937, from Everett Sanders, former secretary to President Coolidge: "It may give you a little insight to the Coolidge character when I tell you that he became convinced that it was useless to send to the Senate a nomination for an appointment to the Bench in any state where the Senators would make real objection. He told me once that the Senate with its power of confirmation had by a course of construction tried to make it an appointing power rather than a confirming power and added that he had to deal with the situation as he found it. He pointed out, however, that he had remaining a considerable power in the matter because he did not have to appoint someone selected by a United States Senator. Any appointments recommended by a United States Senator were submitted to the Attorney General, Stone, and—after his appointment to the Bench—Sargent. They would make careful investigation and report to him and nearly always the matter would be a subject of a personal conference between the President and the Attorney General. Recommendations

a good deal and his letters indicate rising doubts about the President's capacity, and finally about his desire to break the bond between the Senate and the White House in naming federal judges. An apocryphal story of the day declares that in a certain California appointment, the President was convinced that a Senator had named a man unfit for the place. The President called in two progressive Senators, Borah and Norris, told them of the situation, asked them to prevent the confirmation. He was using their nobility of purpose and high patriotism to get them to start a fight on a fellow Senator's prerogatives. And they told him frankly that they did not propose to wash his dirty linen, in so many words. The Chief Justice was worried about that appointment. In a letter to his son, Robert,[15] the Chief Justice cleansed his hands of the whole matter.

"I have tried in the past to influence the President with reference to judicial appointments when I thought an opportunity existed to secure

were not infrequently rejected and the Senators asked to recommend someone else.

"The selection of Judge Wallace McCamant of Oregon, who had nominated Coolidge for the Vice-Presidency in 1920, was a personal selection of the President's although he secured the assent of the Senators from that state. You will recall that the appointment was to fill a vacancy which occurred during recess and that McCamant had gone on the bench. Hiram Johnson who was on the judiciary committee, I think, asked him some questions which proved that McCamant had once criticized Theodore Roosevelt and used this as a lever to prevent McCamant's confirmation. About all Coolidge said concerning this was 'I told McCamant not to go down to the Senate hearings and talk too much.' . . .

"A curious quirk on this particular Coolidge psychology occurred when he appointed to the Federal Bench in Arkansas a man who was a personal friend of Joe Robinson's now Democratic leader and as I remember it, then Democratic leader. The President, who liked Robinson, had made the nomination as a personal tribute to Robinson, and the nominee was a Democrat. We sent the nomination down at noon and without referring it to a committee, the Senate brought it upon the floor and confirmed him within an hour after the appointment reached the Senate. Instead of being pleased with the quick action on his nomination, he was greatly displeased and in almost a rage told me that they held hearings on usual nominations and rejected his nominees almost without any excuse; that they held up his appointments for months and months and now that he had sent one down for Joe Robinson, a Democrat, they confirmed it without any investigation. He added 'The man will be a good Judge, but so were many of the nominations I have made.'

"As a matter of fact few of his nominations were rejected, but this was due to the fact that after he had served a while as President he refused to send down a name unless he knew that the Senators from the particular state were agreeable. . . .

"Just how fully the President reconciled himself to the rule was sharply shown to me when I presented a petition to the President urging the appointment of some judge. He said to me 'You tell them that their Senators will not agree to this appointment. You may also tell them that they know that the Senators will not agree to the appointment and you may further tell them that I had nothing to do with the election of their United States Senators. If they want this man on the bench they must elect some Senators who entertain their views.' Of course, I censored this message and when it finally reached the petitioners it was in more diplomatic but less accurate terms."

[15] March 15, 1924.

good men, but I think the President thinks I am too insistent on having good men and am not sufficiently sympathetic with his trials with the Senators. I am going to keep out of judicial selections hereafter."

So the federal judiciary, under the administration of the man who knew the backstairs of Massachusetts politics so well, remained a part of an intricate sinister political machine.

But alas for high resolves, when it came to naming an Attorney General to succeed Stone, the Chief Justice toddled over to the White House and tried to give the young President good advice as we shall see later in this narrative. And a few weeks after that, his resolve apparently weakened again, for we find him mourning in a letter to Charles D. Hilles [16] that the Senatorial "support is helping to lower the standard of selection for district judges." He writes sadly: "I am not called in now in respect to judicial appointments. I think my constant interest and my attitude of opposition to Senators have tired the appointing power." A week later he writes to his son, Robert: [17] "The President has not consulted me so much about judges as he did. When Stone was Attorney General, he was anxious to know what I knew. I warned the President about Charley Warren but since that time he has not been anxious to have my advice so I do not tender it. He made a very poor appointment in the district of southern California."

But these matters were purely political, and matters political seemed of first importance to President Coolidge whose name had been nineteen times on the political ballot. He gave them politically his closest attention, handled them with his best judgment. The really vital affairs that were engaging the American people in those days were not political. They were financial—stepchildren of the President's heart. He gave them a lick and a promise and let it go at that. Because he did not believe that politics should be in business, he let business go its way without hindrance from politics.

[16] April 24, 1925.
[17] May 3, 1925.

CHAPTER XXVI

Cosmic Scene Shifters Sweat at Their Work

IF THE President seemed to be oblivious to business, he certainly did not close the White House doors to business men. Bankers, financiers, industrial entrepreneurs, the wizards of Wall Street, so long as their wizardry was not advertised as unethical, came and were cordially, if circumspectly, received at the White House. They brought their women who were charmed with Mrs. Coolidge. The men were invited into the Lincoln study after lunch or dinner. There they met a new man—the loquacious Coolidge. When the President had decided to release speculative credit, he talked freely of his plans. His friends in Wall Street and in Boston knew rather definitely that the lid was off; that the Federal Reserve Bank in New York would permit speculative loans of a certain rather conservative pattern. The purr of the gold making machine grinding out credits, by creating debt, pleased him. Apparently his advisors from Wall Street had persuaded him that depressions, like the two minor depressions that had come with the credit deflation during the first three or four years after the war could and should be avoided.

Of course, what the Reserve Bank did (one needs no inside information to verify this) was to buy several hundred million dollars' worth of government securities. The effect of this, combined with the incoming gold, was to make money rates drop. Then, as merchants had no use for additional funds, the funds began to flow into (a) collateral loans against stocks and bonds, (b) bank investments in bonds, (c) installment finance paper, (d) real estate mortgage loans in banks. Simultaneously, the Dawes Plan restored confidence in Europe.[1] The year 1924 showed a jump in foreign loans made by the United States of from around three hundred millions in

[1] See Dawes letter.

289

1923 to a billion dollars in 1924, the greater part of which came in the latter half of 1924—after Coolidge's nomination, after the Democratic row in Madison Square Garden, which we shall consider later, had made Coolidge's election certain. Times were good. Business was booming. Simultaneously, also, America had a large wheat crop while there was a partial failure of the Canadian wheat crop, and the position of American agriculture was radically elevated. The balance between agriculture and manufacturing was temporarily restored, and precariously maintained from the middle of 1924 on through 1929 by an ever increasing volume of credit to Europe to take out exports. Meanwhile, however, our high protective tariffs, which had been raised in 1921 and 1922, prevented the backflow of European goods which was needed to make these credits safe and which was also needed to supply a real basis for our export trade.

The fundamental criticism of economic experts upon the policies of the Coolidge administration [2] from 1923 to 1924 seems to be this: The administration allowed a combination of cheap money and high tariffs to dislocate the balance between agriculture and manufacturing. The high tariffs prevented a normal, dependable export market. And the cheap money temporarily offset the balance and created a far worse problem when pay day came. Coolidge was the idol of the New England protectionists. He was happy in the holy temple of a high tariff. He could not see that credit expansion was making cheap money. Cheap money was going out in foreign loans and flowing into collateral loans against stocks and bonds, into bank investments in bonds and into installment finance paper, and into real estate mortgage loans in banks. This credit expansion was growing a bubble innocent and iridescent which finally was to blow up the temple of protection.[3]

---

[2] And, indeed, in the policies of the Harding and Hoover administrations.

[3] The incoming gold and the excess credit created by the Federal Reserve banks created excess reserves in the hands of the member banks of the Federal Reserve System. These excess reserves became then the basis for an expansion in the superstructure of commercial bank credit in a multiple ratio. The reserve requirements of the member banks were, for demand deposits, 13 per cent in New York and Chicago, 10 per cent in other reserve cities, and 7 per cent in country banks; and for time deposits they were 3 per cent in all banks. These requirements were far lower than they had been in pre-Federal Reserve days. They were lowered at the time the Federal Reserve System came in, and they were lowered again in 1917 as a war measure. They were much too low, and they made it possible for bank credit to expand appallingly on the basis of excess reserves when confidence was high and the demand for credit strong. The ideal of the Federal Reserve System was that there would never be any excess reserves and that there would never be any reserve deficiencies, that elastic Federal Reserve credit would control the volume of reserves of the member banks, letting them expand when they were needed and causing them to contract when the need was over. This would have been the case, had rediscount rates been held above the market and had the Federal Reserve banks used their open market policy to take up slack when slack appeared. But instead, the whole drift from the middle of 1922—the close of

To those who wish to be on with the narrative, perhaps this paragraph is unnecessary. But it seems wise to include here some explanation of the banking setup which was in the basement of the Coolidge boom. In the period from 1922 to early 1928, almost identically the Coolidge years in the White House, the commercial banks of the country expanded their deposits thirteen and one-half billion dollars, and their loans and investments fourteen and one-half billion dollars. Here was the magic that made the great boom of the "new era"—the Coolidge bull market. With this magic we made over a billion dollars a year of foreign loans; this credit inflation made real estate mortgages and installment finance paper easy to float. Under this magic we offset our unsound high tariff policy and kept the farm exports going, without receiving manufactured goods from Europe in exchange. But alas this magic left the country overburdened with debt when the great smash came—for boomers always forget that the family name of credit is debt.

American dollars, not the product of savings but, rather, of bank expansion, flowed to Europe, to South America, to Russia, to Central America, to the Near East. Here was the heart of the Coolidge boom—and much he really knew about it, or anyone else for that matter.

Under the dollar diplomacy of the Coolidge administration, we tried to make those dollars seem safe, by credit devices like the Dawes plan and the Young plan. In Europe, Charles G. Dawes, President of the Reparations Commission, was Calvin Coolidge's vicegerent, appointed in December, 1923, to play providence with the credit of Europe.[4]

Here was the beginning of the end of Europe's dream of battening on Germany.[5] But a few years later the Young Plan superseded the Dawes

Harding's administration—down to early 1928 toward the close of the Coolidge administration (with an exception between March of 1923 and March of 1924)—was to make reserves increase. Coolidge, blinded by the Mellon influence in the treasury department, could not see this trouble.

[4] The Dawes plan was, in many ways, a brilliantly conceived scheme, its greatest drawback being that later experience proved that Dawes set the schedule of reparations payments too high, just as our settlements of interallied debts with Europe set the schedule of payments too high. The world was illusioned by the cheap money policies of the Federal Reserve banks and the Bank of England. But the Dawes plan contained safeguards which might have enabled it to survive, permitting reduction or even cessation both of payments and of transfers of reparations if they proved to be impossible. Unfortunately, as the crisis came on, the Young plan was substituted. The Young plan reduced the schedules for payments but also removed the safeguards. The adoption of the Young plan in 1929, which seemed to be a feather in the American cap was in reality a striking illustration of the ignorance in high places among distinguished men of the world's economic picture. Dr. Schacht, head of the Reichsbank, fought the Young Plan bitterly and accepted it only under protest. He knew that the safeguards were needed.

[5] Stresemann, the German statesman, declared that Americans regarded the Dawes plan experts' dictum as "a political Bible." Probably the Young plan was regarded as merely a revised version. And when a few years later the State Department issued a

plan, and informed Americans knew definitely that the Germans could not pay fantastic reparations to the Allies and the Allies could not pay their debts to America. So burst the bubble of Versailles. We had financial observers at Lausanne and Locarno. Our finger was in every foreign pot—not only our financial finger but our naval arm! The Coolidge administration backed our dollars in Central America with the marines. Potentially we were the protector of "the rich, the wise and the good" [6] in Europe. Our admirals passed more than the time of day at the International Arms Conference with lobbyists of the munition makers who were bent on stopping limitation of armaments. In the old world empire we were swaggering imperialists—clanking dollars as well as sabers. "The business of America is business."

It was a great day for Boston, Northampton, Ludlow and Plymouth! So our savings accumulating for a generation, in the little towns in the hills, on distant prairies from the farms, the shops, the stores and the great surpluses of industry stored in trusts, in insurance companies and in banks in America, moved at first slowly but after Coolidge came to the White House, gathered in a flood of credit that swiftly flowed in a mill race.

It headed through Wall Street to the ends of the earth. Riding on that racing flood, Uncle Sam became the world's creditor. We Americans were proud—even vain—of our position as creditor, and Calvin Coolidge, typifying by some mystic coincidence or curious prescience the spirit of American democracy, as its leader and master gave his blessing to the new order. Thus sitting in the Lincoln study of the White House at odd times, he sat chatting away his evenings with his country's financial captains.

Coolidge talked economy and believed in it. He and Mellon reduced the debt of the United States government by vast sums, because taxes kept rolling in out of the proceeds of gigantic stock market profits and other profits made in a great boom. Coolidge did not see that while he was getting the federal government out of debt, the cheap money policy

---

statement that loans to Germany must not be of a character to interfere with Germany's reparations plans, Europe was greatly upset. The German democracy began to crack, and before President Coolidge left the White House it was obvious that Germany would soon cease paying reparations and that England and France would sooner or later repudiate their debts with some polite diplomatic formula.

[6] The Federalist (old number LX, modern number LXI) contains this passage of Hamilton's: "great injury results from an unstable government. The want of confidence in the public damps every useful undertaking, the success and profit of which may depend on a continuance of existing arrangements. What prudent merchant will hazard his fortunes in any new branch of commerce, when he knows not but that his plans may be rendered unlawful before they can be executed? What farmer or manufacturer will lay himself out for the encouragement given to any particular cultivation or establishment, when he can have no assurance that his preparatory labors and advances will not render him a victim to an inconstant government?"

(which, by the way, he specifically and publicly endorsed in a campaign speech in 1924) was plunging the country as a whole into debt—including the states and municipalities, which overborrowed greatly because money was so easy.

In the majority of honest bankers directing the mill race of credit were many greedy men. Never on earth before had an acquisitive civilization offered so much possessive advantage, such crass power to the few men who had no immediate talents save the grabber's virtue. So while we Americans were sending out our economic lifeline to the world, American capital, the savings of millions of small investors, simple people, not rich, not exceptionally wise, was furnishing the funds to make lovely suburbs rise around the German cities, to build great hydraulic projects in Russia, to save the French franc and put French millions into industry. And all this time in America, in Washington, the little tin Croesuses, wise and foolish, honest and rapacious, were beating a path to the White House door. There they were received with what pomp and distinction the crusty, taciturn, undramatic, frigid, but inwardly proud and sentimental little New Englander could accord to them with the scant ceremony which delighted him. To him they seemed to epitomize business virtues, and "the business of America is business."

American billions [7] set up great plants for making capital goods to reduce America's world market for machinery and to fertilize the roots of industrialization in far places. Bond issues to strange lands, promoting strange enterprises, began to appear in this country. Passing through New York brokerage and banking houses these curious bonds and stocks went to the banks of Boston, Philadelphia, Chicago, Atlanta, Pittsburgh, Kansas City, St. Paul, Portland, San Francisco, Los Angeles, New Orleans, Houston, and they in turn sent the bonds to the county seat banks, established investment bankers who in another day had been careful and critical of credits.

A new crew of reckless plungers appeared in Wall Street who did an increasing percentage of the business that was done. As always in time of inflation the "financial wizard" wrested control of financial institutions from more conservative and cautious men or built a mushroom growth of financial power. Partly through skullduggery and abuse, but much more from loss of perspective and from trusting the insecure foundation of excess reserves and easy money and apparently inexhaustible investors' money, the situation grew more and more unsound and a greater and greater flood of securities, unsound securities,[8] were placed in the hands of the American public. The Coolidge bull market was well on the way.

[7] See "The Hoover Administration," by Wm. Starr Myers and Walter H. Newton, Scribners, page 8.

[8] Incidentally, many of the foreign securities which did in fact turn out badly would not have turned out badly if we had not raised our tariffs still higher in 1930, choking

The Garmanian philosophy was getting out of its depths, tiptoeing and teetering in the tide at its armpits into deeper water all unbeknown to the philosopher who was smugly telling Bruce Barton in those days:

"When I am doing the right thing a great many unforeseen elements come in and turn to my advantage."

President Coolidge pitchforked the muck of oil and petty graft from the Augean stables of the White House, but he let in, all smartly frock-coated, plug-hatted, high-collared, bespatted and smugly proud, another crew which was to devastate his country more terribly than Harding's greasy playfellows. All the day the friends and emissaries of Kreuger, the match king, Insull, the utilities king, Wiggin, the wizard, Mitchell, the manipulator, Doherty, the monarch of gas, the Morgans, the Rockefellers, and Mellon in person, the bad and the good, unchecked and unidentified sat at his council table. Harding's rascals faced death and indictment in the first ten months of Coolidge's regime. And also those unctuous and complacent pillars of the temple of Croesus, all profoundly ignorant of the havoc they were making—the business leaders whom Coolidge welcomed, had their day of doom.

still more the one avenue through which foreigners could make payments to us, namely, in goods. The following table from the *Commercial and Financial Chronicle* shows the record of new securities issued in the United States from 1921 through 1929:

Publicly Placed Security Issues in the United States
(Refunding Excluded)

1921	$ 3,581,958,000
1922	4,304,363,000
1923	4,304,426,000
1924	5,593,180,000
1925	6,223,865,000
1926	6,344,134,000
1927	7,791,130,000
1928	8,114,395,000
1929	10,182,767,000
1930	7,038,615,000

CHAPTER XXVII

The Big Scene in Act III

Now let us turn from business to politics. The spring of 1924 saw a major miracle in American politics. It was Coolidge's own miracle. No old stager of national politics ever went more directly to his end and aim, for a Presidential nomination, than Calvin Coolidge. Two days after Harding's death, Richard Oulahan, of the *New York Times*, dean of the historians of the daily press in Washington, wrote that Harding's death had left the Republican party in chaos, divided into many factions, with many candidates for the Presidential nomination. He listed Senators Hiram Johnson, La Follette, Borah and Watson. He included Herbert Hoover, Lowden and former Justice Hughes in his list.·Coolidge evidently realized what Oulahan sensed. Characteristically when only his own fortunes were at stake, he acted with decision. The President managed to have his boom for nomination launched from the front steps of the White House early, even before frost, when Senator George Moses, World Court irreconcilable, after chatting for an hour with the President, came out and declared:

"I shall call on all New England to stand behind President Coolidge for a second term."

Senator Borah declared a week later: "Give him a chance to make good. I think he is an able man. Give the man at the helm a chance."

And the next day a Coolidge for President club was organized in Kansas City. Before Thanksgiving, tried and true party men like Senator Smoot, Senator Keyes, of New Hampshire, Senator Weller, of Maryland, Wm. S. Vare, the Philadelphia boss, James S. McNary, a Republican field agent in Texas and New Mexico, and Representative Martin B. Madden, of Chicago, had managed to proclaim their support of Coolidge. Late in September, the President planned to restore to Senator La Follette, of Wisconsin, federal patronage taken by Harding from LaFollette and given to Lenroot. The olive branch was waving. It was announced that LaFollette would be a welcome caller at the White House.

In ten months, between August, 1923, and June, 1924, the miracle was achieved. Calvin Coolidge had remade the Republican party. He definitely took it away from national committee control, from the old guardsmen who had nominated Taft and had given an empty nomination to Hughes and who finally had rewarded Harding. But Coolidge gave his leadership frankly, openly, proudly to American business by direct rather than by indirect control. Heretofore for fifty years his party had served business through the leadership of politicians. They assumed to arbitrate between capital and government, between the people and organized plutocracy— government invisible and never quite brazen. Coolidge in ten months had wrecked that political liaison. He destroyed the arbitration myth by putting much of the control of the Republican party directly into business without mediation of the political machine. The National Republican Convention that met at Cleveland, in June, 1924, was a Coolidge convention. New England was there—not the old New England of Lodge of Massachusetts, Proctor of Vermont, Aldrich of Rhode Island—nor its residuary legatee, but a new New England led by William M. Butler, a Boston industrialist, friend and fellow worker with the late Murray Crane (also an industrial captain). The young legionnaires were in evidence; snappy bond salesmen from the cities in saw-toothed, blue-banded straw hats and in clothes of light weight and gay colors.

It was evident early in May that the revised New England delegation was at outs with the old guard which had nominated Harding. The oldsters had to take orders from the new leaders. The Sunday night before the Cleveland Convention opened, and just before the crowd left Washington, the President invited William Butler and Senator Lodge to dinner at the White House. Coolidge sat through the whole evening without referring to the convention even by indirection. Butler felt [1] that Coolidge was trying to soften the blow for Lodge—the blow that was to remove him from leadership, by avoiding any discussion of convention affairs and putting the hospitality purely upon a social basis. Butler felt that Lodge would realize that he was out of the convention program. Butler also felt that this was a merciful way to break the news. But alas, after nearly forty years as a White House guest,[2] it was no treat to Henry Cabot Lodge to dine

---

[1] Conversation with Senator Wm. M. Butler, Boston, June, 1935.—Recorded immediately.

[2] About that time, indeed that very week, the Chief Justice was writing to Mrs. Taft at Paris, that he saw Lodge at the White House looking "very well pleased with himself unconscious apparently how much he is disliked." So possibly Lodge did not take the Coolidgean hint!

socially at the White House. And the realization which doubtless came to him surely was painful even though Coolidge's intention was merciful. Whatever it was, some sentimental quirk in Coolidge's mind persuaded him to send for Lodge; and then perhaps the President changed his mind. His sentiment cooled and he sat eating in silent competence while Lodge and Butler made the best of an appalling evening! Senator Moses and a group of Senate irreconcilables made it their business to call on Lodge and comfort him before his fall. Said Moses [3] to Lodge when his comforters called:

"They intend to stick an elbow into your ribs at every turning in Cleveland. They will permit you to have no position in the convention except that as a delegate from Massachusetts. I want to counsel you to hold this in mind—on your way to Cleveland and after you get there—to remember that you are Henry Cabot Lodge, the Senior Senator from Massachusetts, the senior of all the Senators in the United States, and the leader of the majority in the Senate. Remember also that National Conventions are sucked oranges to you.

"The first one I ever attended in Philadelphia in 1900 you were permanent chairman. In 1916 you were the chairman of the committee on resolutions. In 1920 you were both temporary and permanent chairman of the convention. This convention has nothing to give to you; and if you will hold these things in mind and remember that you are Henry Cabot Lodge, you will find yourself surrounded at Cleveland by a group of devoted friends who will be resentful of the reprisals put upon you and who will sustain you against any buffeting which may be designed for you."

Going from Boston to Cleveland on the train, Lodge's friends learned definitely the whole truth! He would not be temporary nor permanent chairman, nor chairman of the committee on resolutions. He would not be given any place or voice in the convention. He would not be consulted about the platform. He would know nothing and be nothing but one lone, lorn delegate. A story only fairly well authenticated declares that when he learned the truth he cried out futilely: "But they can't do that to me!" Yet they did.

When the Massachusetts delegation assembled in their hotel in Cleveland, Lodge, a member of the delegation, sat on the front row of seats. Senator Butler was in the chair. Lodge heard the motions made, seconded and carried, to select the Massachusetts representatives for the various convention committees—resolutions, credentials, rules, and order of business. His name was not spoken! No one proposed him for any place in the convention organization. Without blinking an eye, or moistening his lips, he

[3] From a letter to the author by Senator George Moses.

stared politely before him and gave his foes no sign of his pain at the per-
formances.[4] He drained his cup alone and without a grimace, a gentleman
in defeat.

The new Massachusetts leader, William M. Butler, would give orders,
would pick the key men of the convention, would dominate its proceedings. Industry was to take direct charge of the Republican party. Lodge
and his kind were to be super-cargo and, being deeply, instinctively regulars, they could not summon either the courage or the technique of mu-
tiny. Through the three days of the convention at Cleveland they went
where they were sent. They said what they were told to say. They suffered
and were strong.

So now read the story of the eighteenth Republican National Convention held in Cleveland from June 10 to June 12, 1924, which nominated
Calvin Coolidge for President and Charles G. Dawes for Vice President.
It met in a new, modern convention hall seating fifteen thousand people.
For the first time a great national gathering was heard upon the radio. For
the first time at a convention of a major political party nominating a President, the bowers of flags and bunting that had given gala atmosphere to
the hall, outside and in, were missing.

One great red, white and blue wall hanging, too large for a flag, perhaps
40 by 25, partly covered a silk curtain back of the stage. On this red, white
and blue background were suspended portraits of Lincoln, McKinley and
—mirabile dictu—Theodore Roosevelt. Enter Roosevelt the bolter into
a Republican convention for the first time since the split of 1912. Eight
years before in the National Republican Convention of 1916, Warren G.
Harding all panoplied in his frock coat, inside a white vest, bedizened
with a red carnation and with heavy braid draping from his pince-nez, as
befitted the chairman of his party convention, had stood at its highest altar
and had poured the burning oil of his sludgy rhetoric upon Theodore
Roosevelt, heaving over his head the black curse of the party regulars as
a renegade and bolter. In those eight years, Harding had joined the ranks of
the forgotten and Roosevelt had entered the trinity: Lincoln, McKinley,
Roosevelt, the new gods of a new, but alas a short and stormy day for the
party; the party that mocked Lodge, the ancient mediator to mammon,
and exalted Butler, his high priest, instead.

A soft, arched skylit dome lighted the center of the great hall. In the
arched dome a gray light filtered through clouded glass. A gray sand-colored,
rough-finished plastered ceiling surrounded the skylight, producing a most
beautiful effect, new and striking. The new priesthood was doing itself
well! The whole effect was Coolidgean, the effect of decent repression of color to serve by suggestion rather than by direct appeal. The gray

room, solid in cement, gave the sense of permanency, of reality, and took away definitely the spirit of carnival that had pervaded convention halls for so many decades. It was as though Coolidge had ordered it, his own mild, soft spoken spirit architecturalized. Of course it was nothing of the sort. The day and the time were leaving the rococo; so was politics. Exit Harding, Penrose, Barnes,[5] Hanna, "yielding place to new." Enter Coolidge, the man of the hour, by some sort of natural spiritual selection.

Of all his predecessors for nearly a century, only Wilson, Grover Cleveland and U. S. Grant, like Coolidge, had never attended a national convention. Yet Coolidge controlled this convention of 1924 at Cleveland with Coolidgean technique. His domination was absolute. No important candidate appeared against him. The national committee, which organized the convention saw to it that a Methodist Bishop, William F. Anderson, of Boston, made the opening prayer and that Senator Theodore Burton, Coolidge's fellow Congregationalist of Ohio, was made temporary chairman.

Burton's speech was scholarly. He once ran for mayor of Cleveland, and the opening sentence of his opening speech was a Latin quotation. At the convention that June day, he spoke for an hour before he drew any applause and that was for prohibition. Whereupon he warmed up. A decent but not fervid response came when he first mentioned the name of Coolidge. But the band which was under orders did not play. The applause was measured in terms of seconds. Great applause, however, greeted Senator Burton when he criticized partisan investigation at Washington which had uncovered the machinations of Fall and Daugherty and the connection between the Republican organization and the oil scandals. Generically a Republican of that era was shocked to tears at anything that tore apart the identity of wealth with brains. Probably if Fall and Daugherty had been in the convention at that hour, they too would have received accolades. For the old spirit of the party was not dead—only dormant. As it was, the ovation of the hour came when a frail, gray-haired wisp of a man rose in the Pennsylvania delegation waving some sort of a paper. Instantly the convention recognized Andrew W. Mellon, Secretary of the Treasury, known in the calendar of Republican saints as "the greatest Secretary

---

[5] William Barnes, the New York boss who had been so potent a figure a dozen years before when he and Penrose overcame Theodore Roosevelt in the Chicago convention which renominated Taft, sat in the gallery at Cleveland with his new wife looking on with a sad, perhaps sardonic expression, cutting no figure in the proceedings. Henry Taft saw him then and wrote about the lonely figure to his brother, the Chief Justice. He adds that even Reed Smoot did not have much to do with the convention. Then he declares that he noted the curious change by which Lodge and his kind were retired to the sidelines. The Chief Justice passed the news on to Mrs. Taft, in Paris: "The old Senatorial gang, Lodge, Brandegee, Watson and Curtis, seems not to have much to do with the platform."

since Alexander Hamilton." Here as Squeers would say, was richness, the richest man in the world. For the first time the owners of America were appearing in Republican conventions. And Andrew Mellon was an owner. The chairman recognized him and the official proceedings of the convention [6] record: "Applause, long and loud, many delegates rising and waving flags." The actual presence in the flesh of a major party saint overcame the convention. His presence restored the Republican *amour propre* shaken by the conduct of Fall, and Daugherty. The proceedings continued:

"The Temporary Chairman: 'It is evidently the desire of the delegates that the gentleman come upon the platform. Will he please come forward?' (Enthusiastic applause while Mr. Mellon was mounting the steps of the platform.)"

The slight, well-tailored figure of the man who weighed scarcely a hundred pounds, edged modestly, hesitantly down the aisle. The delegates cheered; fifteen thousand people in the gallery roared their delight. He mounted the steps and stood bashfully revealed as some subterranean creature unused to the light.

"The Temporary Chairman: 'Ladies and Gentlemen of the Convention, I now have the honor to present to you Mr. Mellon, a delegate from Pennsylvania,' (Renewed applause)." [7] A paroxysm of joy at seeing in the flesh so rich and powerful a being.

Then in a frail voice that could not be heard fifty feet from where he stood when he moved from the microphone, "Mr. Mellon, a delegate from Pennsylvania," said:

"I desire to offer the following resolution which I ask will be read from the platform and then move its adoption:" [8]

The temporary chairman made proper parliamentary genuflections. The reading clerk appeared and bellowed forth his entirely innocuous resolution providing for the appointment of the usual committee on resolutions. It was staged. As Andrew Mellon shrank away from the desk cheers rolled across the auditorium. The new Republican party, constructed by the waving of Coolidge's wand in ten months, the party in which the owners had appeared to take direct charge of the party machinery, was in full swing.

The man who represented Coolidge was William M. Butler, of Boston, an authentic capitalist and industrialist in the flesh, an honest forthright man of consequence. To him the reporters went for news. To him delegates went for orders.

[6] Official Proceedings, page 40.
[7] Official Proceedings.
[8] Ibid.

2

Under orders from William M. Butler, the convention made Frank W. Mondell, of Wyoming, permanent chairman of the convention. He made a desiccated old guardsman's speech. Also under orders, Charles B. Warren, of Michigan, whom we shall meet later in this narrative, was made chairman of the committee on resolutions—not Mellon who offered the resolution providing for the committee. He presented the Coolidge platform which was adopted, a decent platform. Dr. Marion L. Burton, president of the University of Michigan, who had served as president of Smith College, Northampton, Massachusetts, while Calvin Coolidge was rising from mayor to governor, was selected to make the President's nominating speech, a speech which did not mention the Republican party but was devoted entirely to Coolidge's life, character and career. It was a good, but not a political speech and after four orderly minutes [9] of most respectful cheering at its climax, the convention resumed its regular order. That was the shortest acclaim that any Republican nominee for President ever received in the history of his party. In his program the frugal President made no time-allowance for applause. The Presidential nomination aroused no interest. The only dramatics in this situation calling for free play and climax, however small, came in the nomination for Vice President.

Wednesday night when it was known that the program called for a Thursday Presidential nomination, the delegates were confronted with the Vice Presidential candidacy. Before the convention the apocrypha of politics relates that President Coolidge, in Washington, sent for Borah and asked him to run on the ticket, and Borah is said to have asked: "Well, at which end?" Which closed the Borah episode. The Progressives desired Kenyon, feeling that he would make a good running mate for the President. The old guard had lined up behind Senator Charles Curtis,[10] of

[9] Against hours for other names in other days—McKinley's, Roosevelt's, Blaine's for instance!

[10] Let Nicholas Murray Butler in his memoirs "Across the Busy Years," Scribner's, 1936, recite this footnote to the story of the convention:

"On Saturday morning before the convention met, James R. Sheffield of New York and I arrived in Cleveland and called upon our friend, Secretary John W. Weeks, of Massachusetts, for a friendly chat. Quite casually, we happened to say to Weeks, 'What are you doing about the Vice-Presidency?' He responded, 'Nothing. What are you two doing?' We replied, 'Nothing.' Thereupon we three decided to do something, and during the next few hours we brought together under the informal chairmanship of Weeks a small group, the purpose of which was to select a candidate for the Vice-Presidency whom we might all support. By Monday morning the group included Secretary Mellon of Pennsylvania, Secretary New of Indiana, Senator Curtis of Kansas, Speaker Gillett of Massachusetts, Charles D. Hilles of New York, Congressman Longworth of Ohio, Senator Watson of Indiana and three or four more. We discussed every

Kansas, one of its members. But the White House thumbs had turned down on Curtis. As Senatorial Republican whip he did not save the Administration program. He allowed the President's veto of the soldiers' bonus bill to be overridden, and he was marked off the slate. Until midnight the corridors of the political hotels were abuzz with Vice Presidential rumors. Senator Butler's room was crowded with delegations, emissaries, deliverers of ultimata and various heralds of peace and good will. After midnight came an hour when Manager Butler's control seemed to be slipping. Senator Wadsworth and young Teddy Roosevelt, with a group of middle of the road politicians and friends of the Old Guard, met to put up a candidate for Vice President. On the phone, the word came back from the White House that Borah had consented to run. The names of the conferees were given out in the early morning press report and Borah's nomination was formally ordered. But before daylight Borah, from Wash-

sort and kind of name, hour by hour, by night and by day. Some one had objections to make to every name that was proposed. Finally, at about four o'clock one morning, Harry New rose and said solemnly: 'I am going off to bed. The kind of man you are looking for as Vice-President was crucified nineteen hundred years ago.' Among others we discussed former Governor Herbert S. Hadley of Missouri, who had played so important a part in the Convention of 1912, and who was now Chancellor of Washington University, St. Louis. I telephoned to him to ask whether we might consider his name, but he said, 'No.' Finally, we came to the conclusion that probably the best thing to do would be to offer the nomination to Senator Curtis of Kansas. When that conclusion was reached, it was proposed that we should then go down to the room of Mr. William M. Butler of Massachusetts, who was the Coolidge manager at the convention, and tell him what was proposed. Objection was made to this on the ground that if we did so, the several score of newspaper men who were waiting in the halls outside our rooms would say that we had gone to Butler to get our orders, and that, therefore, we must bring him up to see us instead of our going down to see him. This was done and between one and two o'clock in the morning Chairman Butler arrived, escorted by Charles D. Hilles. John Weeks told him of our discussions and deliberations and of our conclusions. Butler looked at Weeks with a perfectly impassive face and said: 'I have just been talking by telephone to the White House. We must nominate Borah for Vice-President.' This statement was received with a silence which could have been cut with a knife. Finally, Chairman Butler said to Weeks: 'Mr. Secretary, what do you think of Borah?' Weeks, coming forward to the front of his chair, replied with a stream of adjectives and expletives which would not look well in print. There was more silence. Then Chairman Butler turned to Secretary Mellon and said to him: 'Mr. Secretary, what do you think of Borah?' The Secretary, looking dreamily off in the distance, took his characteristic little cigar from his mouth and said placidly: 'I never think of him unless somebody mentions his name.' "

After this, history, with prophetic eyes, considering how she would summarize the two careers in her big book, smiled a twisted smile at Andrew Mellon's vain persiflage. The footnote is set down here to prove how far out of the situation President Coolidge had elbowed the old guardsmen who had ignored his candidacy in 1920. "He smiled with satisfaction," wrote Taft to his wife in Paris, after the Cleveland convention when the Chief Justice explained to the victorious candidate "that the Senatorial gang with Lodge at the head was relegated to the back benches." In another letter to Mrs. Taft, we read: "Coolidge's secret thoughts of Lodge are not fit to print!" And it mattered much to Lodge who only eight years before had stirred the Massachusetts delegation in Chicago to raucous derision when he cried· "Calvin Coolidge—Oh my God!"

ington, being told what Butler was saying and doing, cried denial in appropriate wrath. Borah asked former Senator Beveridge to speak on the floor the next day withdrawing his name. He called the Washington press associations and broadcast his absolute and final denial. When Butler heard this he was puzzled. Shortly before dawn Thursday morning a group of westerners from Idaho, Kansas, Michigan, Iowa, and Missouri led by Hanford MacNider, national commander of the Legionnaires, marched into Butler's room to demand the nomination of Judge Wm. S. Kenyon, of Iowa. They tried to show the New England leader that Kenyon was a liberal and not a radical. The New Englander swallowed a grin. Reporters edged into the conference. They heard Butler still palavering about Borah. The Borah talk from Butler at that hour puzzled the reporters. They knew of Borah's rage, of his telephone message to Beveridge. They slipped out to reverify the Borah denial. In the corridor of the hotel they met Senator James Watson in his shirt sleeves, suspenders down, necktie gone, parading with Senator Lodge who was also in that chill hour before the dawn sartorially at half mast. A third Senator came alongside and asked his two wise colleagues:

"Well, who is it?"

Sardonically Watson answered: "We don't know. But Butler is in here now telling the delegation that it is Curtis."

Whereupon a reporter cut in: "Well, Senator, I don't want to contradict you. I have just come out of Butler's room there and he is telling the crowd that it is still Borah."

Senator Watson's face lighted up more in sorrow than in anger and he said to Senator Lodge:

"Senator, these are strange days and stranger nights when three United States Senators, the night before the nomination, get the news from headquarters about the candidate for Vice President out of the mouths of reporters." [11]

And so saying, the Senatorial trio went to bed. It was low tide, with many crawling things on the fringes of Lodge's memory—low tide and the end.

It was the President's strategy to close the convention Thursday. The idea of dragging out a pale and uninteresting pageant for four days, offended the Coolidge sense of economy—economy of time and economy of the money of a thousand delegates and of ten thousand spectators. So he hurried the proceedings and came to the ballot for President at the morning session Thursday. In the rollcall which followed swiftly, the President received 1065 votes, Senator La Follette 34 votes and Hiram Johnson re-

11 Story from the *Detroit News*, the *New York Evening Post*.

ceived the ten votes from North Carolina. The permanent chairman announcing the result declared:

"By your votes you have nominated Calvin Coolidge, of Massachusetts, as your candidate for President of the United States."

So short was the demonstration which lasted less than five minutes, that the recording angel of the convention noted the parsimonious cheers merely as "(applause)." [12] And the convention upon the motion of Charles B. Warren, of Michigan, adjourned until 3:30 that afternoon.

Immediately after luncheon, the nominations for Vice President proceeded. Hanford MacNider, making his first appearance in a Republican National Convention, opened the ball by presenting the name of Judge William S. Kenyon, of Iowa in a speech as brief as a day letter. Arizona unexpectedly presented the name of Former Governor Lowden, of Illinois. Nebraska, his former home, offered the name of General Charles G. Dawes. Kansas' gift to the convention was the name of Senator Charles Curtis. Seconding speeches resounded in the hall for half an hour or so, and the first ballot for Vice President was taken, Lowden leading with 222 votes, Kenyon with 172, Dawes with 149, with a long line of minor candidates. Quickly the next ballot was taken where Lowden led, Senator Burton, of Ohio, followed, and Kenyon had third place. A stampede began before the vote was announced which gave Lowden a majority. Then out in the country millions of listeners-in upon the radio heard the cheering that followed the nomination of Lowden, and under the cheering got this:

A Voice: "But I tell you, here is his letter!"

Another Voice: "Well, what can we do? They have nominated him."

First Voice: "Yes, but here is his letter. He says he won't take it."

And listeners-in identified the First Voice as that of Chairman Mondell, of the convention.

The First Voice continued: "What are we going to do?"

The Second Voice: "I tell you I won't let this convention adjourn with the nomination refused by the Vice Presidential candidate."

First Voice: "Why not get him on the phone?"

Second Voice: "I got him on the phone. You can talk to him three minutes. Adjourn the convention for an hour."

First Voice: "But how are you—"

And then some blat of a band drowned out the sentence. The next thing the radio auditors heard was the motion to adjourn until eight o'clock. They knew that something had happened. Then came the night session, the reading of Lowden's letter of refusal, and the nomination of Dawes, practically by acclamation, but without drama. The administration apparently had tried in the recess rather feebly to name Hoover. It had

12 Official Report.

abandoned Kenyon when his name had failed to rally a majority after the first ballot.

Charles G. Dawes was nominated by Congressional leaders, the men who nominated Harding. It is interesting to note that when a majority in a Republican convention, or at least when amalgamated minorities, have nominated a Presidential candidate, almost invariably they have turned the nomination of Vice President over to some unexpected and often unorganized minority. It is the way of politics.

But in spite of the meager drama which landed Dawes at the end of the Coolidge ticket, the convention adjourned without a sense of climax. It backed out shyly and faded from the cold gray hall whence the spectators had disappeared during the early afternoon. It was a typical Coolidge exit. Was it some accidental combination of a thousand circumstances which always kept Coolidge and his works slightly out of focus, always drab and cheerless, or was it some touch of his own aura, some curious emanation from his own slight public personality, which slowly uncorked whatever vessel he used for his political purposes and let the flare and fizz and culminating zip dribble out in a luke warm serenity?

CHAPTER XXVIII

Upon Which Our Hero Plays Second Fiddle

WHAT was Coolidge luck? The theory of his enemies during his lifetime that he was the product of the misfortunes of others is supported by certain facts. Nevertheless, these facts do not convey the truth. But they are interesting, these facts; convincing unless one puts into the equation another fact. Time and again in his career, when Coolidge was elected mayor, when he went to the State Senate, when he was made president of the State Senate, when he was elected lieutenant governor, an examination of the circumstances surrounding the promotion shows that Coolidge's rise came when someone else stumbled, though never did Coolidge trip the stumbling man. But the other important fact to know about Coolidge which changes the luck theory is that when the other man fell, Coolidge had within his own mind and heart certain important survival qualities, qualities which gave him powerful friends, which set him down in a powerful party and made him an active, useful and sometimes a necessary member of the dominant faction of that party. So that when the other man fell, Coolidge's friends, moved somewhat by his obvious loyalty and by his capacity to make his loyalty useful to them, gave him a hand and helped him to rise where another had fallen. Coolidge himself was Coolidge luck.

This run of luck favored him after his nomination for the Presidency in June, 1924. The Democratic National Convention of that year, which met two weeks after Coolidge's nomination, staged a fall, such a fall as Coolidge's luck was always able to capitalize. A row started inside the Democratic party in the national convention in Madison Square Garden, New York, which made Coolidge's election certain, but only because Coolidge had built himself up as a trustworthy figure before the country. A united

Democratic party with a vigorous leader—a man like Governor Smith, of New York, if he could have been nominated quickly and with some unanimity, might have made real headway against President Coolidge by making the oil scandals the only issue of the campaign. But for two weeks the Democrats wrangled in their national convention, and so bitter was the fight that when it was over and a compromise candidate, John W. Davis, a man of great ability and high integrity, was nominated he could not lay the evil spirits of race and religion that had been conjured up by the animosities of the Democratic convention. For in the long drawn out contest primary and convention between Governor Smith and former Secretary of the Treasury McAdoo, the great cities of the nation with their urban populations scarcely more than a generation from Europe supported Smith, and the South and the West somewhat under the influence of the Ku Klux Klan backed McAdoo. Each candidate had other support than the Tammanies and the Klansmen, but to the Klansmen it seemed that Smith, the Roman Catholic, was the candidate of the big cities and to Smith's supporters McAdoo represented the bigotry of the anti-Catholic, anti-Jewish and prohibition forces. Ten days' balloting in Madison Square Garden with the galleries packed by a provincial mob from New York City which Tammany assembled, split the Democratic party so wide and so deeply that eight years were required to heal the breach. And no matter how logically the Democratic candidate talked, no matter with what justice he assailed the Republican position, no matter how hard he tried to make the corruption of Republican leaders in the oil scandals a vital issue, his cause was foredoomed. People were feeling, not thinking. In August, Senator La Follette entered the fight as an independent candidate for President. Burton Wheeler, Democratic Senator from Montana, was La Follette's independent candidate for Vice President. La Follette and Wheeler drew from Davis the support of the liberals and radicals whom a Democratic candidate normally might have called to his banner. But no one could call anyone but a wool-dyed Democrat to the Democratic banner. Wool-dyed Democrats were in a sad minority. Probably fear that La Follette would carry a few western states and deadlock the electoral college also helped Coolidge. The Coolidge luck, whatever it was, which had followed him from Northampton to Boston and on to Washington, protected him and gave him an easy victory in the campaign of 1924. In the November result he received 382 electoral votes. John W. Davis received 136 electoral votes. The victory was more significant than the score would indicate; for the Republicans carried again the border states along the South and by this overwhelming vote the Republican party had absolution for its sins, and the oil scandal as an issue was wiped out of American politics.

The victory was not unalloyed. While the Democratic convention was

in the midst of its turmoil, it paused to send a kindly message to the
President and Mrs. Coolidge sympathizing with them over the death of their
fourteen year old son and youngest child, Calvin, on July 7. Calvin was particularly his father's boy. The President moved a dozen times a day back
and forth from his desk to the boy's sickbed. One day, remembering little
Calvin's love of animals, he coaxed and caught a small brown rabbit among
the plants in the White House garden, picked it up gently and came
trotting through the White House to the sickroom. In return, across the
pained young face came a smile and they took the bunny away and the
President went back to his work. It was all that he could do. Four days
later the lad in his delirium imagined he was fighting a battle against great
odds. Finally with all his waning force he could command, he cried out:

"I surrender!" And turning to the nurse beside him said: "Now you say
it too, say you surrender!"

The nurse replied: "All right Calvin, I surrender."

And a moment more he had slipped into his last long sleep.[1]

After Calvin's death his father liked to talk about the boy: "He was
working in a tobacco field the day I was made President. Some of the boys
said: 'If my father were President I wouldn't be working in a tobacco field
in the Connecticut Valley!' Calvin said: 'If my father were your father
you would!' He looked like my mother's people, particularly my mother."
A pause, then: "Playing tennis on the soft south grounds of the White
House he got a blister on his toe. Blood poisoning resulted." Another
pause: "When he was suffering he begged me to help him. I could not."
Another long pause, he had told his story.[2]

Dr. Edward Brown, the Coolidge family physician, turns the flashlight
on Calvin Coolidge in another hour of agony: "It had come time for him
and John to leave the hospital. I happened to glance out of the window
and there below in the street, standing quite still with John's hand in his,
was Mr. Coolidge looking up fixedly at the boy's room—a forlorn and
touching picture." [3]

Five years later, the wound was still in the father's heart and he wrote:
"When he went, the power and glory of the Presidency went with him.
. . . I don't know why such a price was exacted for occupying the White
House." [4]

[1] "Presidents and First Ladies," Mary Randolph, p. 82.
[2] Conversation with the author in December, 1924.
[3] Good Housekeeping, May 1935.
[4] "Autobiography," p. 190. Also Chief Justice Taft was a White House caller in
mid-July and he writes to a Cincinnati friend: "I saw President Coolidge today and
talked with him for a few minutes. Slemp had asked me to remain behind in order that
I might ask him a political question. He bears up with wonderful courage. 'I do not
mind the loss of the boy so much,' he said. 'My sorrow is occasioned by the belief that
he possessed great power for good that would have made itself felt, had he lived.' "

It was under the terrible shadow of her son's death that Grace Goodhue Coolidge burned like a glowing altar flame to duty. She became a mistress of the White House whose charm, whose tact, whose outward gaiety, whose amiable diligence in what is a trying position at best, made her a White House hostess who will rank with the most gracious and most important of wives of the Presidents. Her task and its achievements were heroic. The President often showed the world his sorrow. Mrs. Coolidge—never! It was her pride that kept her face up and smiling.

It was in November, 1924, that Mrs. Coolidge began to crochet a bedspread, eight squares long and six squares wide, one square for each of the months which the Coolidges would have to live in the White House after the election. She and the President alone in their sorrow—talked it over many times, month by month, as she finished each square, and they counted the months until they should be home in Northampton and away from the grinding pressure of the Presidential job.[5]

2

No Republican ever came to the White House, except possibly Theodore Roosevelt, who was elected in his own right with fewer strings on him. Calvin Coolidge had remade the Republican party in his own image. He was an organization man but the organization of the party owed more to him than he owed to it. He was a natural ally of organized capital, those vast amalgamations of wealth which controlled the banks and so had suzerainty over major commodity industries of the land. But there again Calvin Coolidge was free. He had befriended the bankers and their industrial lieges. They had done little for him. No scandal surrounded the campaign fund of 1924. It was an inexpensive Republican campaign. The President for the most part stayed in the White House and paid his own way when he went out to make the few speeches which graced the campaign. His old friend and fellow worker in the forgotten Crane machine in Massachusetts, Wm. M. Butler, was chairman of the National Committee and made no commitments binding the new President. On the day after the election, Senator Lodge died and was spared the chagrin of knowing how overwhelming was the Coolidge victory. For the Coolidge victory probably was greater than Lodge feared. Later the Coolidge influence in Massachusetts sent Butler to the Senate as Lodge's successor. He became the vicegerent of Coolidge in the Senate, his leader in the Republican party national organization where Butler owed nothing to anyone save Calvin Coolidge.

After the election was a busy time for the President-elect. Congress met

[5] Conversation with Frank Buxton, editor of the *Boston Herald*, to whom Mrs. Coolidge told the story.

for the short session in early December and the President's message outlined the issues of his new administration. Briefly they were these:

Reduction of public expenditures. "The government," he declared, "can
do more to remedy the economic ills of the people," by "rigid economy
in public expenditures" than by anything else.

Taxation "scientifically revised downward," referring to high income
taxes. He said: "The present method of this taxation is to increase the cost
of interest on productive enterprises and increase the burden of rent."

Legislation for waterways—the Great Lakes to Gulf project, the St.
Lawrence project, the Mississippi basin project, and the Cape Cod Canal;
also for reclamation of arid land.

As to agriculture, chiefly he asked for an appropriation for further fact
finding. He declared against agricultural price-fixing, but Congress devoted
much time to discussing it.

Muscle Shoals [6]—the sale of the property "under rigid guarantee for
nitrogen production" for fertilizer.

Railways—consolidation of lines and further examination into new and
better methods for rate-making. He also recommended a labor conciliation
board "retaining the practice of collective bargaining," with the public's
right of "uninterrupted service."

In the matter of departmental reorganization, as a "companion piece to
the budget law," he asked for "authority for thorough reorganization of the
federal structure with some latitude to the executive" to shift government
activities which "follow changes in a developing country."

In the passages of the message referring to the army and navy, he stressed
disarmament, and referred to the disarmament agreements. He added: "I
want the armed forces of America to be considered by all people not as
enemies but as friends."

In foreign relations he was in favor of keeping out of the League of
Nations, but he recommended American membership in the World Court.
Also he favored further disarmament conferences and he declared definitely
for some kind of an international agreement "outlawing war"—the first
time the phrase had appeared in a Presidential message. He said: "I am
opposed to the cancellation of the foreign debts to America." He declared
they should be "paid as fast as possible."

In the midst of these busy times the President's father died. He was
unable to go to his father during his last illness. It is absurd to infer that
he was heedless of his father's need. All his life the father and son had
cherished an intimate and affectionate relation. The son was proud to lay
his achievements at his father's feet from the time he won his first medal
back in his law clerk days until he turned to him in pride after the Presi-

[6] Later the soul of TVA.

dential election. He was an old-fashioned son who kissed his father when he was in his fifties. Once visiting Plymouth the son kissed the father the second time for camera men; so that there are two filial kissing pictures, one representing the President with his hat off, one with his hat on. Certainly his father's death moved him deeply following so closely upon Calvin's going. But Congress was bedevilling him and he turned to his job, submerging his sorrow.

With Spartan brevity he writes of his father's death: "During his last months I had to resort to the poor substitute of a telephone. When I reached home, he was gone. It costs a great deal to be President." [7]

3

Inauguration Day, 1925, dawned clear and fair. "Coolidge weather," remarked Frank Stearns, who was really the belle of the ball and seemed to be, more than anyone else in the White House, radiant and happy. The President rose at seven o'clock, took his usual stroll about the White House grounds with the secret service men, slipped around to review the vacant pineboard stands in front of the White House, sent Frank Stearns to the station to meet young John Coolidge arriving from Amherst on a belated train, finally greeted his house guests, kin and old friends: his mother-in-law, Mrs. Goodhue; Mrs. Laura Skinner, Mrs. Coolidge's distant cousin; Ralph Hemenway, his law partner; and President Olds of Amherst College, Amherst. Certainly a sentimental galaxy! No politics, no business, no "pride of heraldry or pomp of power" in that list, the countryman's idea of a cozy family party.[8]

He went to his office as usual, ground through his mail and at ten o'clock, when Washington was alive with the stir and bustle of the great quadrennial event, while crowds were moving along the Avenue and stray bands from far places went swinging down the pavements looking for their stations in the parade, while soldiers were trimming their kit and accouterments, and statesmen in the Capital were blowing dust from their silk hats, and diplomats were feathering their plumes for the great show, the President was slipping into his official morning suit. At half past ten he was talking to Frank Stearns, waiting for the word to start to the Capitol. Guests, Cabinet members and their wives began to assemble on the main floor of the White House. Vice President-elect and Mrs. Dawes, with Senator Curtis, leader of the Senate, came punctually a few minutes before eleven

---

[7] "Autobiography," page 191.

[8] Hoover, the White House usher questions this newspaper report. Hoover declares that the Coolidges had their breakfast together in the privacy of the President's bedroom, probably correct, while the Stearnses and the house guests ate in the larger family dining room.

o'clock, the zero hour. The President and Mrs. Coolidge greeted the Vice President-elect and Mrs. Dawes in the blue room and so led the way to the automobiles lined up on the driveway of the White House, behind the white columns. Under the guard of Governor and Mrs. Billings, of Vermont, with a military escort for which the Vermont legislature had appropriated prodigally twenty thousand dollars, the President and his retinue swept their way to the Capitol amid the acclaim of the multitude.

Calvin Coolidge was a popular President. He had made good and had been elected in his own right. He was no longer an accident. He had grown in the esteem of the American people and in the respect of the world. He had earned his day of glory, as those days go in a democracy.

While the crowds are assembling in the plaza before the Capitol building, here we may record something of his achievements as President. Not the least of these achievements were his Presidential messages, which gave color to the superficial political thought of his day. His was partisan thinking, more or less—thinking in newspaper editorials. In these messages he had turned America toward thoughts of economy in government. His leadership incarnated economy. In March, 1924, he had vetoed the soldiers' bonus bill, which later was adopted over his veto. He had vetoed the Bursum bill increasing pensions. He had vetoed a bill raising postal employees salaries sixty-eight million dollars yearly. He had signed the naval appropriations bill with its implied call for another arms conference to cut down armament. A bill, providing for increase of salaries to Representatives and Senators, the Vice President, the Speaker, the Cabinet, had been passed by the House without a record vote, and later passed overwhelmingly in the Senate; and a modified postal rate and pay bill containing many of his economies had been passed and signed. He had submitted a rigid economy budget for the following year.

In addition to economy measures, President Coolidge had settled the United States-Mexican claims agreement. He had pardoned the last of the prisoners convicted under the espionage act, had secured a successful ratification of arbitration treaties with Great Britain, France, Japan, Norway and Portugal. He had suggested recognition of Russia, which offended his more conservative supporters (and his suggestion was later interpreted into desuetude by Secretary Hughes). He had secured the prohibition of the sale of arms to the Mexican rebels; had seen Dawes open the Paris conference on reparations; had severed relations with Honduras because of revolt there; had signed with misgivings the restrictive immigration bills with provisions excluding the Japanese; had signed the relinquishment of more than six million dollars of Boxer indemnity due from China; had seen the Dawes plan go into effect September, 1924; had seen the ratification of a commercial treaty with Germany; and had signed the federal labor arbitration act. He had appointed, without serious question of Senate confirmation, four members of his Cabinet, Harlan F. Stone, as Attorney General, Curtis Dwight Wilbur, as Secretary of the Navy; and upon the death of Secretary Wallace he had appointed Governor Howard M. Gore, Secretary of Agriculture (to be succeeded that day by William M. Jardine, of Kansas). In 1924 he had named Frank Kellogg (now about to succeed Secretary Hughes in office) as Ambassador to Great Britain. He had also appointed former Senator Miles Poindexter, as Minister to Peru, and reappointed J. C. Davis, as director general of the railroads, and former Senator Harry New, as Postmaster General.

To the Interstate Commerce Commission he had named McManamy, of Washington, D. C., and Mark Potter, of New York (a reappointment); to the Federal Reserve Board, George R. James, of Tennessee, and Edward Cunningham, of Iowa (reappointments); to the World War Funding Commission, Edward Hurley, of Illinois; to the Shipping Board, Frederick Thompson, of Alabama, and Bert Haney, of Oregon; as Ambassadors to Mexico and Belgium, Charles B. Warren, of Detroit, and William Phillips; and as Minister to Switzerland, Hugh Gibson.

So much for the work of his nineteen months in the White House—a busy but not an important time.

Now we may turn to the waiting inaugural party which went from the White House to the executive chamber in the Senate. There the President left his Cabinet members and his official retinue and went directly to the President's room in the Senate to sign or to veto the sheaf of seventy-seven bills which had been passed during the last hours of the retiring Congress. He sat in the little room under the queer ceiling scroll which the newsmen have nicknamed the "Eye of God." There he signed, in that delicate hairline signature which reveals so much of his own thin, repressed, self-reliant, passionless but highly sentimental character, the words "Calvin Coolidge," seventy-six times. At the last bill he hesitated. It was the bill increasing the salaries of Congressmen, Cabinet members, and the Vice President. He knew well that the increased pay was earned, that these public officers, particularly Congressmen, needed for decent maintenance, the salaries provided in the bill. Yet his whole philosophy, his unique place as an apostle of economy, was mocked by this bill. It would increase the individual's federal tax bill less than a hundredth part of a penny, yet it was not economy. As he sat under the "Eye of God," characteristically he spoke no word. He gazed vacantly around at the walls, took off his glasses, wiped them with his new, clean handkerchief, put them on again, began to twirl them nervously. He beckoned Budget Director Lord, traced his finger over three lines of the bill and asked in a whispered voice a short question. Still the President hesitated. Minutes passed. He drank the glass of water which

McKenna, the doorman, had brought him, looked up at the wall clock. He saw that he had been debating with himself ten minutes. He fidgeted a moment, went to the toilet, came back again, looked at the clock. It was 11:56. He settled down, put on his glasses, sighed a little short sigh and signed the bill, the last official act of the accidental President.

By one of those curious ironies of fate, the new President-elect could not function in all his power and glory even on his inauguration day. Vice President Dawes burst into the Coolidge spotlight. As soon as he was sworn in as Vice President, Dawes proceeded to deliver an inaugural speech. Instead of the perfunctory salutatory inanities which Vice Presidents deliver to the Senate, Dawes, being Dawes—the irrepressible, proceeded to harangue the Senate, shouting at the Senators, shaking his finger at them and banging the desk for emphasis as he declared they were wasting their time. He assumed to condemn the Senate rules which permitted one Senator to stop the machinery of government. Dawes probably was justified by the facts. It was questionable taste—only that. But it was a spectacle, this Vice Presidential outbreak. It became a sensation. The Senators glared back at Dawes. Coolidge's face hardened, expressing disapproval. Of course the Dawes performance was the big news in the American papers.[9] The Presidential inaugural became an incident. So strangely enough Coolidge began his first elected term—a secondary figure, slightly Cinderellaish, a bit reminiscent of "the singed cat" of his first public appearance. The curve of his path did not swerve!

In his "Autobiography," he does not mention the glory of the day or the pomp of the hour. If Dawes irked the new President, nowhere does he record his irritation. If the spectacle of the acclaiming multitude pleased him, he gave no hint of his pleasure.

As to the ceremony itself, former President Taft, who as Chief Justice was administering the oath, gave it his blessing in a letter to his son Robert[10] in which he declared that it went "off very well indeed. The arrangements were good and Coolidge made an admirable speech." The Chief Justice tells his son that he delivered the oath slowly, "fearful lest I might forget it." But he got it right and "the slowness of the delivery made it so distinct that I have heard from many places in the country that they heard every word and recognized my voice.[11] . . . Coolidge's response was so low that I didn't hear it. But when he began to give his address his voice sounded quite clear."

[9] The Chief Justice wrote to his son, Charles P. Taft, March 8: "Dawes made an excellent speech but he disgusted a good many people. But the matter was true in every respect." He hoped that the Senate would use it as a justification for introducing cloture in debate.

[10] March 8, 1925.

[11] This was the first Presidential inauguration ever to go on the air.

If the exultation of power came to him in the high moment when before a hundred thousand people in the great plaza in front of the Capitol he took his place for four years as the most powerful ruler of the earth, he has left not a scratch of the pen to recall his satisfaction, much less a man's natural vanity. His voice, as he stood there before the microphone, was broadcast over the earth.

After Chief Justice Taft had administered the oath, which the Chief Justice had taken as President sixteen years before, the new President kissed the Coolidge family Bible, the one owned by his grandfather that had been given to him by his grandmother. He had kept it in the White House. He touched his lips to the verse opening the first chapter of St. John:

"In the beginning was the word, and the word was with God, and the word was God."

Long before the forty-seven minutes had passed which were required to deliver his inaugural address,[12] the crowd began milling about, scattering to their places in the parade, drifting to their mid-afternoon trains. If he saw the crowd melting before him, he was unmoved. But probably the long row of microphones around him shut off his vision. At one o'clock, he started down Pennsylvania Avenue at the head of the parade and for an hour rode amid the greatest acclaim he had ever received, the greatest he would ever receive. He smiled now and then, and doffed his silk hat, holding it out impulsively, awkwardly at arm's length, denoting some unusual emotional response for Calvin Coolidge. Mrs. Coolidge, in her moonstone gray, later to be called the Coolidge gray, a wistful color recalling the sad funeral of Young Calvin a few months before, was able to smile, even to seem gay, the one unwavering note of joy in that exalted hour.

[12] The last paragraph of this inaugural is Coolidge at his zenith. It reads:

"Here stands our country, an example of tranquillity at home, a patron of tranquillity abroad. Here stands its government, aware of its might but obedient to its conscience. Here it will continue to stand, seeking peace and prosperity, solicitous for the welfare of the wage-earner, promoting enterprise, developing waterways and natural resources, attentive to the intuitive counsel of womanhood, encouraging education, desiring the advancement of religion, supporting the cause of justice and honor among the nations. America seeks no earthly empire built on blood and force. No ambition, no temptation, lures her to thought of foreign dominions. The legions which she sends forth are armed, not with the sword, but with the cross. The higher state to which she seeks the allegiance of all mankind is not of human, but of divine origin. She cherishes no purpose save to merit the favor of Almighty God."

Probably nothing he said or wrote elsewhere represents so perfectly the Coolidge ideal, the Coolidge literary style, which in itself deeply reveals the man; a sentimentally aspiring man, full of good will, a man not without an eye to the political main chance, a man always considering the vote-giving group, shrewdly eloquent about accepted beliefs, never raising debatable issues, a good man honestly proclaiming his faith in a moral government of the universe.

The inaugural party arrived at the White House at two o'clock with only time for a quick luncheon. In the interests of Coolidgean simplicity, the elaborate semi-public luncheon customarily given by the new President to his Cabinet and to the official guests was omitted. The new Vice President and his wife joined the President and Mrs. Coolidge in the quick luncheon of coffee and sandwiches on the second floor, while the Cabinet members and their wives waited down stairs, without food. However, Colonel Sherrill, the President's military aide, apparently without the President's knowledge, had prepared a luncheon across the street in the State, War and Navy Building, for the aides. He invited the hungry Cabinet members to his party and when the President and the Vice President came down with their wives to go to the reviewing stand, they found the rooms empty. His guests came straggling to the reviewing stand during the next hour after the President had arrived. They made no apologies. He made no comment.[13]

For some reason the parade was delayed. The President waited patiently. He looked silently down his nose with his own peculiar petrified grimace on his face. The reporters noted a certain lack of enthusiasm in the parade as it passed the reviewing stand. It takes two to wake up the hurrahs of a crowd, the hurrahers and the hurrahees. That fine, fair Coolidge day the hurrahee's emotions—never tenacious—were spent by four o'clock. He was tired. An hour later he returned to the White House, ate a bite of lunch while his guests made their way to the state dining-room to partake of an elaborate buffet luncheon. As soon as he could he withdrew from the public part of the White House and threw himself across the bed, worn and weary.[14] With the innate sentimentality which so often inspired him, he went to a party that evening which he had called together at one of the hotels in Washington, a delegation from Massachusetts, men with whom he had been associated for twenty years in official life in his home state. And a curious gathering it was; Massachusetts congressmen and senators, Massachusetts federal officials living in the Capital, a few Boston politicians, one or two of a shady sort, and one or two loyal honest friends like Tom White. But he came away from the Massachusetts party at 9:30 and went to bed—a spent rocket. And at ten o'clock he was asleep. And so ended his day of glory. After that for four years it was all work. Here was the meridian procession of Calvin Coolidge's life. He had no taste, no talent for and little patience with raw splash, the strut and splendor, the solemn mockery of a triumphal pageant. So while the drums were thumping, the cymbals clashing, the trumpets blaring at the inaugural ball, the Yankee slept.

[13] "Forty-two Years in the White House" by Irwin H. Hoover.
[14] Hoover's, "Forty-two Years In the White House."

CHAPTER XXIX

And Publicly Stubs His Toe

WHEN Calvin Coolidge began his first elective Presidential term the tide of the American industrial and commercial boom of the twenties was almost at its full. Prices and employment continued to rise slightly during the year that followed his inauguration.[1] But as he sat at his desk punctiliously prompt at 8 o'clock the morning of March 5, 1925, the lusty breakers of prosperity roaring in, sounded sweetly in his ears. Here were the savings of multitudes—vast Puritan self-denials—functioning in an economic structure that was producing a rising standard of living; slowly and with what seemed approximate justice, distributing the gross income of the nation even though great fortunes were being prodigally enlarged, even though nearly a fifth of his countrymen in the South, in the mining districts, in the ghettos of the great cities, and in shabby smoke-stained industrial hovels were living on a shamefully low standard. But the net of it, the full-throated chorus of the breakers on the strand washing in the steady swelling tide of middle-class abundance, must have throbbed like a cosmic lyric in Calvin Coolidge's heart.

If he had cared to listen he might have heard the moan of an undertow. In the agricultural West the farmer had become a free hand soil despoiler and the lumberman a freebooter. Western grain lands were showing a constantly decreasing yield per acre. Pastures were shrivelling; forests disappearing. The water level in the Mississippi Valley was steadily receding in the wells, in the creeks, in the rivers.[2] Great floods, uncontrolled, were tearing their disastrous way through the valleys. Mortgages were increasing. Tenantry was on the rise. People were moving restlessly about the land

[1] "Audit of America," Edward E. Hunt, McGraw-Hill Company, New York, 1929.
[2] U. S. Geological reports.

seeking to better unhappy conditions. Migratory labor had begun to work in the wheat harvest, in the citrus country, in the fruit orchards, in the lumber camps, on the docks; a thing scarcely known in the America of Coolidge's childhood when the home was the very present anchor of the laborer whether in field or factory.

Swifter and more powerfully ran the vortex current of bank expansion through all the Coolidge years. Money in terms of hundreds of millions, even of billions, was coming from all over the continent into the speculative orgy of stock selling.[3] The great banks of the Atlantic seaboard and as far to the west as Chicago were swollen with brokers' loans. In those years the turgid flood of liquid bank resources, cheap money and credit expansion, covered America. No wonder swindlers reached hand over fist for the gold to be had for the asking.

Yet in that day when Calvin Coolidge came to his desk in his own right in March, 1925, the watchman on the tower from hour to hour was calling, "All is well."

The Republican party was triumphant. Its leaders, impeccable regulars, were business men of parts and consequence, highly esteemed. Yet because this was a representative democracy, the President had to face a minority in the Senate and the House—men who represented in their ideals and with their votes, some honestly, some with obvious cant, the under-privileged, those living on low standards, the exploited workers, the discouraged farmers. This irritating Progressive minority which had formed itself into the farm bloc, still held a balance of power between the conservative Democrats and the regular Republicans. After Coolidge's election, the Progressive group was no more bound by party tradition than before. It was anchored in a homogeneous majority in several states, chiefly midwestern and far western states. This homogeneous non-partisan majority— operating largely in the Republican party—could and did express itself with reasonable intelligence through the direct primary, and for the most part this Progressive congressional minority was a more secure group with the electorate than the conservative leadership in either party. These Progressives at first irked, then annoyed and finally angered President Coolidge. During the early days when he followed Harding, President Coolidge tried to placate them. Conspicuously he consulted with Senator Borah. In his elective term he remained on good terms socially with Senator Capper. He did not quarrel personally with any of these Progressive leaders, reasoning with good Puritan logic that sugar catches more flies than vinegar. The Progressives nibbled at the sugar, avoiding the trap, and the President was puzzled. They were a new kind of bird in his sky.

When Coolidge faced Congress after his election of 1924, a new Repub-

[3] "Recent Economic Changes in the United States," McGraw-Hill Book Co.

lican Senate leader greeted him. With the death of Senator Lodge, Senator
Charles Curtis, of Kansas, the party whip, the legs of Lodge and most of his
organization brains, became the titular leader of the Senate Republicans.
The dynasty was falling: Hanna to Aldrich, to Penrose, down to Lodge,
again down to Curtis. Hanna, Aldrich and Penrose were political masters
of the plutocracy. Wall Street went to these Senatorial leaders sometimes
for orders, often for advice, but hardly to command them. Lodge was a
patrician first, only incidentally a plutocrat. If he served Wall Street, he
also despised it. The Progressives in the Senate got out of hand under Lodge.
So Coolidge, seeing the disintegration of the plutocratic vicegerency in
the Senate, sensing the threat of the Progressive domination, not realizing
at first how much cannier than Lodge, Curtis was, how much more effective a good errand boy was than the late ossified ideal in a gray Vandyke,
called in Butler, the captain of industry, and gave him command of the
Republican party. Calvin Coolidge made Butler the President's spokesman
in the Senate. Curtis, titular Republican leader, wielded only a tin whistle
scepter under Coolidge and wore a messenger's cap for a crown.

The Progressive, or liberal sentiment in America was dammed up back
of the reactionary leadership of the Republican party. President Calvin
Coolidge in his White House bower saw only that the wall of the conservative dam was strong. When the Senate under the nimble leadership of
Curtis surrendered to the Progressives, providing a spillway for the dam by
passing the McNary-Haugen bill, establishing farm loan banks, and challenging the President's program with various forms of state socialism—the President got release by quacking to the Chief Justice: [4]

"The Senate is a lot of damn cowards!"

2

Coolidge suffered his first definite defeat from a congressional Progressive-
Democratic fusion in March, 1925. When Attorney General Harlan F. Stone
was promoted into the Supreme Court, the President suggested the name
of Charles B. Warren, of Detroit, as an attorney general to succeed Stone.
To understand the bitter fight which followed, complicated by the President's stubbornness and by the militancy of the Progressives, one must remember and identify this Charles Warren. From the President's standpoint,
he was an ideal candidate for attorney general: Michigan born and bred, a
graduate of the University of Michigan with honorary degrees in plenty from
other important colleges, a lawyer of distinction in Detroit, counsel for the
National Bank of Commerce and the Union Trust Company, counsel for the
United States before the joint high commission to determine the Bering
Sea claims, counsel for the United States in the North Atlantic coast

[4] Taft's letters in the Congressional library.

fisheries arbitration with Great Britain before the Hague Tribunal, delegate at large from his state to the Republican National Convention that nominated Taft, national committeeman under the interregnum between Taft and Harding, Ambassador to Japan, head of the High Commission to Mexico that negotiated terms for the resumption of diplomatic relations, the first Ambassador following the resumption of relations, and chairman of the committee on platform of the Republican National Convention which nominated Calvin Coolidge. He presented the Coolidge credo to the Republican Convention at Cleveland, guided it through the sub-committee, thence through the general committee and read it with great gusto on the floor of the convention. He was the wearer of a distinguished service medal after an honorable service in the world war, member of the Detroit and Michigan state bar association, a Phi Beta Kappa, and an Episcopalian. But alas! the Progressives rejected him because of his connection with the sugar scandal of 1910. In an investigation before the Hardwick sugar committee of the House, the testimony pointed to a transaction which in the minds of the Progressives disqualified Mr. Warren. They contended on the basis of that testimony that as a representative of the Havemeyer interests in New York, Mr. Warren held in trust a large share of stock of the Michigan Sugar Company, that as an undercover agent of these interests he had sought to control the eastern sugar beet industry. Also he had been president and general counsel for the Michigan Sugar Company which at that time was alleged to be affiliated with the sugar trust. The sugar transaction carried no turpitude in the President's mind. Here was only sharp trading. Perhaps—indeed probably—in the rarefied air of high finance where larger sums are sacrosanct, it followed the customary practice. The pernickety attitude of the Progressives who tried to stop the President in a laudable endeavor to reward an old friend and respectabilize a Cabinet place that had been clouded by Harry Daugherty, seemed to Calvin Coolidge a personal insult. When the Democrats, always ready to make trouble for the Republican majority, saw the Progressives led by Borah and La Follette and the farm bloc lining up against Warren, they quickly and gleefully, joined the Progressives en masse and made a majority. The President issued a White House statement in mid-March defending Mr. Warren's Michigan sugar transactions and denying the truth of a Federal Trade Commission charge that certain of those sugar agreements had violated the anti-trust laws. Nevertheless the Senate majority prevailed. Warren's nomination was rejected. Following his rejection, five leading Republicans, including Curtis, the Republican leader, called on the President and officially informed him that a renomination of Warren would be fatal.

Chief Justice Taft called on him while the Warren matter was pending and declared [5] that he felt President Coolidge did not consult enough people . . . and "often consults the wrong people." "I think," writes the Chief Justice, "that he is very much afraid of seeming to be under the control of somebody . . . and Butler is finding that to be the case. It isn't important except that it leads him into mistakes, and the Warren incident is a glaring one."

Taft's instinct may have sensed the reason why he rejected the advice of these five Republicans including the canny Curtis who knew the Senate better than the President. But Coolidge told James Derieux [6] that the reason a man has "so much trouble with the Senate is that there isn't a man in the Senate who doesn't think he is better suited to be President than the President, and thinks he might have been President except for luck."

In this whole Warren episode, the President was stubborn to his own hurt and without much principle involved. After Republican leader Curtis had come to the White House to warn the President against the Warren appointment, Jud Welliver, the literary executive secretary, made bold to tell the President about Warren's part in helping the American Sugar Refinery get virtual control of the beet sugar industry, not a pleasant story. In 1910, Welliver, who was working with La Follette, had written a series of articles for Hampton's Magazine about this questionable sugar episode and told the President that it would turn up. Sure enough Democratic Senator Joseph Robinson read from the files of Hampton's Magazine, the story Welliver had written, "and for two or three days," writes Welliver, "I was fearful that Coolidge might resent the fact that my name was used in the Senate debate. But he was the most unpredictable person I have ever known. I don't think he ever mentioned the episode to me afterward." He wouldn't. The President knew that he was wrong. But he was the kind of man who would never admit it. Not only did Curtis, as congressional leader, indicate that Warren could not be confirmed, but Chief Justice Taft, as the fidus Achates of the administration, went to the White House several times, warning the President against presenting Warren's name or trying to push him against the conviction of the Senate. He wrote to his son, Robert, in February, 1925, that Warren was "a very agreeable man . . . a man of acute mind and executive ability, but he hasn't the sturdy character of Stone. . . . Warren resorts to evasion and concealed methods." Also when in March, Warren's appointment was kicking up a row, the Chief Justice, writing to his son, Robert,[7] complains of Warren's "subterranean methods," but adds that the "President for some reason has

---

[5] Letter to Charles D. Hilles, March, 1925.
[6] Letter from James Derieux to the author, September, 1936.
[7] March 15.

conceived a very high opinion of him and his abilities and is determined
to stand by him. . . . The President evidently has a feeling that the country is with him against the Senate and that the union of the Democrats
with the radical Republicans, La Follette and his crowd, is apt to arouse
continued support for him (Coolidge) in the controversy with the Senate
which he has now begun and which he will probably continue for four
years. In that fight," concludes the Chief Justice, who was blind to public
sentiment, "he will have the people from the beginning. He will be able
to put the Senate in the wrong in many cases."

But alas, in putting the Senate in the wrong, while the country did apparently support the White House, the President was deceived. For popular faith of America in the Republican party was dying at the roots.

The President obstinately again sent Warren's name to the Senate amid
the plaudits of the delighted conservatives. The *New York Times* thundered its approval and declared that "at least the middle west was going
to find out that the President was not lacking in the elemental force necessary to maintain party leadership." The *Manchester Union*, in New Hampshire, declared that Coolidge's leadership was of the true New England
type. "It does not bluster, but it is tenacious. It makes few gestures but
it never slackens its grip."

After all this fine fluttering of battle flags, the President went down to
defeat on a fluke. The final vote for confirmation of Warren was divided
forty to forty with Vice President Dawes taking his noon nap at the Willard Hotel. Senator David A. Reed, conservative of Pennsylvania, before
the vote was recorded changed his vote to "aye" that he might move for
reconsideration. Word was sent to Dawes. He charged down Pennsylvania
Avenue, rushed into the Chamber just as the vote on reconsideration had
again resulted in forty to forty and Senator Overman, conservative North
Carolina Democrat who had voted for confirmation, arose and changed
his vote from "nay" to "aye," making the vote stand 39 to 41 against Warren. It was charged that while Dawes was sleeping at the Willard, he
might have been dreaming of that hour in the Republican national convention at Cleveland in 1924, when the one vote on the Michigan delegation against Dawes for Vice President was cast by Charles B. Warren, who
voted for Herbert Hoover. This charge was never substantiated. It was a
curious coincidence—probably nothing more—but justifying a giggle! The
President in the White House blustered, sputtered, threatened, but finally,
facing the inevitable, took his first defeat from the Progressive bloc. Norris
was avenged. That it was dramatic, even ridiculously dramatic, by reason
of the Dawes episode, did not dignify the President's position. But it did
dramatize the fact that the President was not in control of the Congress,
that the representatives and advocates of the underdog minority even in

the big boom of the twenties held the balance of power. It was the beginning of a long drouth cycle for the Republican party. When Coolidge in the late twenties sat glumly looking at the messenger who brought the news of Warren's defeat, he could not realize that the news was only the call of the Bull Moose echoing in his ears. The President's liberal yearnings too often remained futile. For he could not or would not, being a regular party man, call those recalcitrant Republican Progressives to his aid and finally would not consult with them en bloc. They scattered their votes generally on all issues except economic issues. Sometimes they supported the President's foreign policy but generally were divided.

The President early warned the Congress against further reduction of the army and navy [8] but he remained cautious of military expansion. Probably the President was cautious about military expansion because of its cost. He was intent upon reducing the national debt, determined to scale down taxes. Under Mellon's advice he would scale down the taxes of the rich rather than of the middle class. And as he cut down the debt, the House appropriations committee tried to edge up the army budget with the President's disapproval. He was tight with the navy, but eight new cruisers were authorized in 1924 and a few years later the big navy interests tried to wheedle him out of nearly half a billion for the three remaining cruisers called for in the quota of an arms agreement made by President Harding and Secretary Hughes with Japan and England. Aviation began to cut into the Coolidge economy program. And in mid-year, 1926, an extensive program of aviation unit expansion was launched. But unquestionably, in spite of the strong military machine which was growing up in Washington, President Coolidge believed in the ascendancy of the civil over the military authorities. He was the first President since Jefferson to stress this point in a public address. This address was delivered with characteristic courage to the graduating class of the naval academy. Yet that year found him sending details of the army and navy officers to the governments of South America to cooperate in their military affairs. They were not so welcome as the President imagined. For at the beginning of the Hoover administration these military missions had to be withdrawn, several of them upon request of the beneficiary nations.

President Coolidge on the whole promulgated a dollar diplomacy in South and Central America. He assisted the de facto governments of Nicaragua and Mexico by selling them arms and munitions and in Nicaragua recognized a constitutional government, one not recognized by the rest of Central America and Mexico. He withdrew marine occupation from Santo Domingo in 1924, but continued the occupation of Haiti. The marines landed in Honduras in 1924, and American troops intervened in

[8] Annual message 1923.

the Republic of Panama to stop rent riots in '25. Marines also landed in Nicaragua in '26, and an expeditionary force followed the next year. The South American press, particularly that of Colombia, Peru and Chile, denounced this intervention as American imperialism. Historians must record the fact that Calvin Coolidge was never well beloved in South America.

As to his general foreign policy, President Coolidge kept on the whole a middle course between the irreconcilable isolationist Senators and the mild internationalists. He recommended joining the World Court and in the four year controversy which raged mildly in the Senate, the President was more frequently with the advocates of the Court than against them. On cancellation of debts he was adamant. There, New England spoke. Said he: "They hired the money didn't they? Let them pay it!" He was generous however in the matter of terms, interest payments, extension of time and irrelevant details.

In the meantime as President Coolidge went further into his elective term, Europe became more and more suspicious that he was trying to dominate the world. The feverish resentment which was evident in South America began to be manifest in Europe. The legend of America's desire to dominate Europe seems to date from the days of the Dawes plan and the settlement of foreign debts. Coolidge's obvious desire to bend European statesmen to his own construction of certain clauses limiting American responsibility if America entered the World Court, strengthened that position. He wrote of our proposed Russian relations that "Our country ought to be the first to go to the economic and moral rescue of Europe," but declared that "Russia must pay the debts of the Czarist and Kerensky government to America" before we do any considerable moral and economic rescuing. He illuminated this statement by writing: "I don't propose to make merchandise of American principles," then added, "but while the favor of America is not for sale, I am willing to make large concessions" for a chance to do a little profitable economic and moral rescuing.

Dennis in "Is Capitalism Doomed?" calls attention to the fact that the Dawes and Young plans had been really "the technique for financing German reparation payments since 1919." And when we made it clear that we would no longer hold the European bag, when the President wrote of these post war loans: "The money we furnished we had to borrow; someone must pay it. It cannot be cancelled," European diplomats were baffled. Jonathan Mitchell [9] declared that the diplomats of Europe believed President Coolidge to be truly dangerous and watched him "the way mice watch the cat." It was a period of financial turmoil, slow economic decay: decay camouflaged in Europe by various opportunist schemes and devices to hide the truth.

[9] "Goose Steps to Peace," Little, Brown & Company.

At home we kept our feet somewhat because we were exhilarated by speculation and had the drunken man's luck. The President's foreign policy in those years from '23 to '29 was not intelligent enough to be called heroic. It was only successful in postponing the inevitable crash that came out of Europe in 1931. How much the President may be praised or blamed for the result of his foreign policy no one now can honestly say. He had no diplomatic training. He believed in the holy dollar. His State Department was not a tower of wisdom. He took advice from American money lenders; but they led him only where his instinct would have guided him and his instinct was probably reflected in the feeling of the American people in those days. "Get the money" was the chief end of man in the third decade of the new century. It dominated our foreign policy. It was the first clause in the catechism of prosperity.

Calvin Coolidge as President will not be remembered by his foreign policy. He will be remembered as the high priest of prosperity and the prophet of economy. "Work and save,"—upon that foundation he built the structure of his kingdom.

Alas the country did work, and Coolidge did save—but as President he saved only part of what the country was squandering. American business, under President Coolidge, went into debt to create the stock market profits which Coolidge taxed to cut down the government debts.

At home we kept our Republican debate as near equilibrium by sympathising and the fruit, turned his first hand a foreign policy to those points so to educate our students, even at the cost level. It was only a second in separating the two holic much that hatred of Fascism at bay. How many the President may be called or blamed for the ruin of our foreign policy no one can honestly say. He had to deplore treating the old ore in the last decade, His State Department was no force of wisdom. He was able to and from Americans among leaders, but they lost first only in the majority would have an end through his regime was probably relaxed in the feeling at the American people in these days. And the money was the road ended now in the good times of the new century. It does not like one temptations, if this the one to lose to the extent of its presence.

Capital Coolidge is sometimes still not to be acquitted, by his long or policy. He will be tormented by this high priced prosperity and the period of economy. "Waste and save"—more that temptation he built the measure of his ruin now.

Mrs. the widow's oil, and Coolidge still sat sat as President he lived only part of winter's journeys most smugly American business since President Coolidge came into public service. His stock market profit wish Coolidge taxes kept down the government claims.

BOOK FOUR

PREFACE

Coolidge escaped in time. He was out of office before the final mad orgy of the 1929 stock market in the United States, and before the heaviest pressure began upon the debtor countries of Europe. But he lived, none the less, long enough to see the collapse of a mad world.

Financial pressure began for Germany, and for most of the debtor countries of Europe, with the tightening of the money market in the United States and the great reduction of foreign loans by the United States in *1929*. Germany turned from borrowing to repaying in the middle of *1929*. and in the last half of that year developed an export surplus in so doing. The Young plan was substituted for the Dawes plan in *1925*, over Schacht's protest against the abandonment of the protective clauses of the Dawes plan. All debtor countries of Europe felt the tightening of the money markets.

The British stock market crashed at the time of the Hatry scandal, preceding our own stock market crash. Following our stock market crash, an ominous price collapse came in almost all of the great international staples, including especially farm products. But international confidence remained, and in '30 another great German loan was placed under the provisions of the Young plan. The depression of 1930 was relatively mild. It was more severe in England than in the United States, because British costs and prices were very inflexible and would not yield fast enough to prevent diversion of trade from England to other countries. Debtor countries were paying with goods. It seemed possible that the great mushroom of international credit might be sufficiently liquidated in this way. Then we raised our tariffs in 1930, and the whole world began to raise tariffs and other trade barriers to heights never dreamed of before, and the best intentioned debtor struggled with difficulties.

In the spring of '31, a new factor entered. Austria and Germany, for what seemed to them sound economic reasons, entered into a customs union which involved a large measure of free trade between them. France interpreted this as the annexation of Austria by Germany, and took violent exception to it. Then and there international financial policies came to be dominated by political considerations. France had become powerful financially following Poincaré's reforms in 1926. In the past, in all great crises, New York, London and Paris had counted on cooperation from one another. In 1931, this failed. Paris would not join London and New York in saving a threatened bank in Austria. New York then failed to join London. London alone could not save the Austrian bank. It went down, and with it the credit of the Austrian government itself.

Then came great fears regarding Germany and a run on Germany which pulled her down despite the one year "Hoover moratorium" on reparations payments and on interallied debts. This froze great sums of money which the British and other countries had lent to Germany. Then finally came a run on England to which England yielded despite international help, with the abandonment of the gold standard in England.

Finally came the worst of all the troubles. The shock to world credit of England's action was incredibly great. If sterling was not safe, what was there that was safe? Was the dollar safe? A great international run on the dollar began in the autumn of '31, which we met successfully. No fear can rise so demoralizing in economic life as fear regarding the future of the currency. We, and the world at large, plunged to new depths. President Hoover's proposal of a one year moratorium on German reparations and on debts due from the European governments to the American government at first brought a hope. But the tardy and reluctant acceptance by France came too late to avert the German breakdown. In the autumn of '31 the French Premier Laval visited President Hoover and the two issued a joint statement at Washington, looking forward to a revision of German reparations to be followed by a new discussion of intergovernmental debts due the United States "within the Hoover year of postponement." European governments did their part at Lausanne in the early summer of 1932, virtually abandoning German reparations. President Hoover, however, had a political campaign on his hands by that time, and despite his understanding with Laval in the preceding autumn he did not, indeed, could not, reopen the question of interallied debts. When, two days after the November election, England and France brought up again the question in connection with the December 15, 1932, payments, Mr. Hoover's view seemed to be that the question had been raised too late.

A second run on the dollar came in the spring of '32, which again we met successfully.

Three things came together in the early summer of 1932; (1) we had beaten off the second great run on our gold—had paid foreigners every cent in gold they could demand, and could announce that we still had a billion dollars more in our gold reserves than the legal requirements; (2) at Lausanne, the Germans, French, British, and others made an agreement virtually abandoning German reparations; (3) the Senate of the United States by a decisive vote defeated the proposal to pay the soldiers' bonus with paper money. There came a new hope, a rally in all markets, and a strong rally in business and industrial activity in the United States. This was the turning point for the world as a whole, though for the United States, another great setback came culminating in the closing of the banks in March, 1933.

Nineteen thirty and the two years following were gloomy years for Europe. Mussolini, following a visit from Secretary Stimson in 1931, turned statesman, and for the next two years was one of the best citizens of Europe, trying to pacify, trying in particular to effect a conciliation between France and Germany. An Austrian house painter, who served Germany as a world war soldier, was rousing the people against communism, Jews, and France. He demonstrated that he had dangerous political power. Germany's hope lay in Hindenburg, an aged and rapidly aging soldier.

Ramsay MacDonald's socialist experiment in England went by the board. French statesmen were parading double quick into power and out of favor. Lines tightened everywhere. Much of Continental Europe slumped into a hopeless sort of apathy, which made it easy to fall into the arms of the tyrants. The dominions of the British Empire huddled together for safety.

Japan, with the military party increasingly in the saddle, seeing Europe preoccupied, defied Secretary Stimson's protests, and, in defiance of treaties, seized Manchuria. Respect for international understandings and engagements, for promises to pay debts, for promises to pay gold, for solemn formal treaties, stood at an appalling and an increasing discount—though the worst of this was yet to come. War and post-war hatreds, inflation boom, and violent depression, shaking the world from 1914 into the 1930's, have left a dreadful mark upon morality, personal, national, and international. The world was terribly shocked in 1914 when a solemn treaty was called "a scrap of paper." But the world is littered with scraps of paper today. The rebuilding of good faith, personal, national and international, and the restoration of confidence that men and nations will keep their word to their own hurt, are the most vitally necessary tasks of the next decade.

Calvin Coolidge back in Northampton from his march to glory saw with a breaking heart in an old man's dreams, the visions of his youth dissolve.

CHAPTER XXX

The Big Bull Pageant Swings Across the Stage

THE middle of 1927 saw a reaction in American business. On the domestic side the explanation is in large measure to be found in the fact that the Ford automobile establishment was going through a drastic reorganization. Model T was being discontinued and to produce new models and gear up the establishment for production of changing models meant Ford's withdrawal from production and sale from June, 1927 to December.[1] On the foreign side increasing evidence was manifest of strain on the part of the Bank of England. This strain grew out of England's effort continued from 1921 on, to avert internal readjustment of prices and costs by an expansion of bank credit which England's gold reserves did not at all justify. Britain had gone through the prolonged coal strike of 1926 without readjustment through a forced expansion of bank credit, buying with sterling—that is to say with checks drawn on British banks—quantities of commodities, even of coal, from the Continent. This meant that the Continent had large deposits in London. The depositors were of course free to take these deposits in gold whenever they chose.

Uneasiness appeared in world finance because of this over-extended position of England and a complication grew out of the fact that the Bank of France, in order to hold the price of the franc down, had to buy a heavy volume of sterling in Paris—sterling which it did not want.

Growth in American unemployment in the early part of 1927 showed itself in the heavy industries and appeared in the building trades, and some weakening was evident in agricultural prices. The red signals from agriculture had been winking trouble during all of Coolidge's years in the White House. Mr. Coolidge looked at the farm facts, and helped in the develop-

[1] The passing of the old "T" Model, the appearance of Model "A".

ment of banks for farm credit, but what the farmers really needed was not more opportunities to get into debt but rather better markets.

The cheap money policy of 1927 of the Federal Reserve banks was a further makeshift in maintaining the foreign market for farm products and easing the position of the Bank of England, only by letting the burden of debt grow more and by setting a wild stock market going. Mr. Coolidge studied this situation. He learned many facts. His tough-fibered brain could assimilate facts and daily he was in conference with men who knew facts, men from Wall Street and men from the Government where these facts were assembled. But Mr. Coolidge probably failed to see the underlying principles which these facts embodied.

A record was made by Irwin H. Hoover, head usher of the White House, of the White House guest-list during the year of 1927, a sample year in the days of the big bull market. The records show that the secretary of agriculture or someone from his office was a frequent White House guest. It was his problem to keep the White House advised about agriculture. Certainly he saw clearly the menacing signs of the times. Statisticians from the labor department were in and out during that year. And socially we find from the record of the head usher that Clarence W. Barron, the grand old gossip of Wall Street, was three times luncheon guest in 1927—in January and twice in mid-April. He came again in January 1928. W. H. Grimes, Editor of the *Wall Street Journal*, a compendium of economic information, was a guest, and early in February, J. P. Morgan sat with the President at luncheon. Those were days and times when the Wall Street market was booming. Those were the times when the Federal Reserve Bank in New York was pouring millions into Wall Street, not directly into brokers' loans but indirectly into an already glutted money market, which could only use the money in brokers' loans to stimulate speculation and maintain prices. Curiously the Federal Reserve Board in Washington seems to have been politely trying to check the speculative trend in the Wall Street market. Yet early in the year 1927, President Coolidge issued a statement predicting that the year would be "one of continued healthy business activity and prosperity." This statement exhilarated stock exchange trading. But a counterstatement from S. W. Straus [2] expressed the view that the country had reached "the saturation point" in office buildings, apartment houses, and hotels. The statement depressed stock prices. A few days later Secretary Mellon, always on the side of the market, expressed an opinion, through the *New York Times*, that: "While the saturation point may have been reached in some cities, that condition was not reached in others." He

---

[2] Mr. Straus was at the head of one of the most important mortgage loan companie. in America, and intimate with the building trades. These statements of the President and others herein quoted will be found in the files of the *New York Times*.

cited Pittsburgh. Within a week after Mr. Mellon had put his shoulder to the
wheel, the stock market began a definite upward movement, stock prices rising rather steadily from an average of 171 on January 25, to 189 on March 1.
But in early March, prices began again to decline. The New York Times financial editor spoke of "an anxious and a nervous market." But it predicted
that "the tone and tenure of the news will to a larger extent than has been
usual of late, guide its fluctuation." That editor probably had a tip. For
blithely, a few days later, Secretary Mellon came out with the announcement
of his intention to replace the second liberty four and one-half per cent bonds
with three and one-half per cent notes. This announcement was a definite
indication that in the opinion of the Secretary, money rates would remain
low, which opinion stiffened market prices. And on March 9, the New York
Times observed that Wall Street was "disposed to draw highly optimistic
inferences from the Treasury plan . . . and this may have revived buying
interests in the market." In the meantime, J. P. Morgan lunched at the
White House in February. Morgan was in speculation as deep as any other
big investment house, but no deeper. The Morgans financed the Van Sweringen railroad combine, United Corporation, and other long shots including Standard Brands. They took heavy losses, grinned and bore it. Of course
they were in on the foreign government loans in a big way. So Morgan did
not do much in the White House to quench the speculative fire.

The New York Times noted, March 10, that "the main advance has
been resumed in Wall Street." Stock prices jumped ten points, then sagged
a little, and in late March 1927, Secretary Mellon, departing for Europe,
left amid a fountain of iridescent rockets as he sailed down the Bay on the
Olympic, indicating that "the stock market seems to be going along in an
orderly fashion, and I see no evidence of over-speculation."

Under the canopy of the Secretary's popping rockets we may recall that
the New York Times, a conservative and reliable newspaper if one was ever
printed, had been recording the heavy buying of the Mellon interests in
railroad, power and light stocks. The clever device which Mellon's attorneys had rigged up, by which Mellon, entering the Cabinet, had turned
over his investments to a family corporation which kept right on buying
and selling stocks at that time made a mild sub rosa scandal in financial
circles. The New York Times, in March, 1926, had reported that the Mellon family had made three hundred million in the great bull market on
aluminum and Gulf Oil alone. Mellon's market optimism in 1927 under
those circumstances is not so difficult to comprehend. Not that he was
dishonest. He knew "that his redeemer liveth"—and where!

By mid-summer, 1927, the leading industrials had reached an average of
217. The pace slowed a bit. Then the President got out again and pushed,
and in June predicted a future of sound prosperity and sound business con-

ditions. Again the market jumped twenty-six points. Indeed a careful search
of the market during those halcyon years between Coolidge's inauguration
and Hoover's reign, discloses this fact: Whenever the stock market showed
signs of weakness, the President or the Secretary of the Treasury or some
important dignitary of the administration, but significantly rarely if ever
the Federal Reserve Board at Washington, issued a statement. The statement invariably declared that business was "fundamentally sound," that
continued prosperity had arrived and that the slump of the moment was
"seasonal." We read from William E. Humphrey, of the Federal Trade
Commission, in April 1927, this pronouncement:

"The President instead of scoffing at big business does not hesitate to
say that he proposes to protect the American investor wherever he may
rightfully be," which was, of course, the slogan of the dollar diplomacy of
our foreign policy. Naïvely, Mr. Humphrey added this gem to the casket of
administration jewels which were pawned to sustain the market. We read:
"Little slips of paper—stock certificates and bonds have been the talisman
of this revolution, the wholesale fusion of political and economic life."
Here is an attitude which certainly, just as Coolidge's attitude more than
any official act, became public policy.

Commissioner Humphrey may well be regarded as a minor prophet of
the Coolidge administration. But here speaks a major prophet—Chief
Justice Taft after a pleasant visit in the White House sat down and wrote
to George Burton, of LaCrosse, Wisconsin, that he thought that "general
business seemed to be settling itself into wise progress without feverish
speculation." He added that the country had concluded that it is well to
be "conservative and avoid the isms" and expressed the opinion and the
hope that there would be "a demand for another term" for the President.
The White House atmosphere was keenly partisan in the midst of the
Coolidge regime, partisan and inclined to be class conscious. The Chief
Justice wrote to his son, Robert, in February, 1926, that the Attorney General was trying to cooperate with Senator Thomas Walsh of Montana, a
Democratic Progressive, in bringing certain unpleasant matters to the surface. The Chief Justice railed against the Attorney General, declaring it was
painful to see how "awkward and clumsy" he was and adding that he felt
like "going to the President and telling him so." About the same time he
was writing to National Republican Committeeman Hilles, in Wall Street,
that he was glad to see that certain Republican Senators had finally summoned pluck enough to fight Walsh in his effort to drag out an aluminum
investigation,[8] adding that it was none of the Senator's business.

[8] Probably scandalizing Mr. Mellon!

2

In those days when the snail was on the thorn and God was in his Heaven business men crowded into the White House until the luncheon guest-list looked sometimes like a chart of interlocking directorates of high finance. At a press conference a reporter, anxious for a story, quizzed the President:

"Why is it that your White House guests are limited to men of business?"

He referred to such names as Morgan, Barron Collier, Lewis Pierson, President of the Chamber of Commerce, President Loree, of the Reading Railroad, Harvey Firestone, President W. W. Atterbury, of the Pennsylvania Railroad, E. T. Stotesbury, the Pennsylvania financial and political pundit, A. P. Giannini, the San Francisco banker, George F. Baker, Sr., the richest man in Wall Street, Paul Shoup, the Colorado capitalist, President Bardo, of the New York Shipbuilding Company, Julius Rosenwald, Owen D. Young, William Butterworth, President of the United States Chamber of Commerce, who were White House guests at that time. The reporter queried:

"Why don't you have artists, musicians, actors, poets around the White House as Wilson and Roosevelt did, and sometimes Taft and Harding?"

The President pulled his solemn clown face and looked down his nose, as he drawled:

"I knew a poet once when I was in Amherst; class poet, name of Smith." A cud-chewing pause, then: "Never have heard of him since!"

So much for poets. It was supposed to be a dumb answer. It was the Coolidge estimate of the arts, in his dry, shrivelled Yankee wit, in the days of his triumph when the "business of America was business," days to which Charles Schwab looked back wistfully when he told the Senate committee in 1933: "I was not engaged in making steel. I was making money!"

3

Once, however, in that booming period, the President did take the hint at which he sniffed in the press conference. He sent for Professor William Z. Ripley, of Cambridge, who had written an important widely quoted book [4] upon the dangers of the vast holding companies that were aggrandizing wealth. These holding companies were hiding away the power of wealth in the irresponsible hands of men who had control of wealth through the holding companies without ownership. Ownership was invested in common stock distributed among the masses. This device—the holding

---

[4] "Main Street and Wall Street."

company—Professor Ripley had called: "the anonymous industrial autocracy."

Much that was in Ripley's book had appeared in the *Atlantic Monthly*, the court bulletin of New England. The article made a furor all over the country. Among those who were close to the President was Judson Welliver, who had been White House literary clerk from March 4, 1921 in the Harding administration to November 1, 1925. He had helped two Presidents with most of their public utterances. He had written many paragraphs and many speeches for President Coolidge.[5] Welliver was worried about the bull market in Wall Street. He turned to Senators Curtis and Smoot. They sympathized with Welliver's misgivings. Both of them went to the President and urged him to take some action to check the stock market inflation. To each of them he listened patiently. He assured both of them crisply that his advices were that "everything was all right in the market." He quoted Secretary Mellon. And when Welliver went directly to the President with his anxieties, the President sent Welliver to Mellon who insisted that the stock market "was not quoting prices in excess of actual value, at least on the substantial and worthwhile stocks." When Ripley's article appeared, Welliver had left the White House, but he called one day when the press was just beginning to discuss the article, read parts of it to the President and suggested again that the President take some action which would mitigate the stock market menace. Mr. Welliver remembered that the President replied:

"Well, Mr. Welliver, even if you and Professor Ripley are right about it, what is there that I can do? The New York Stock Exchange is an affair of the State of New York, not of the Federal Government. I don't think I have any authority to interfere with its operations."

"But, Mr. President," replied Mr. Welliver, "whether through the Federal Trade Commission, the Federal Reserve Board, or any other agency, you have legal authority, you are the President and have a measure of moral authority that would be effective. President Roosevelt and President Wilson many times invoked this moral authority, and made it effective. Frankly, I have always felt that neither you nor Mr. Harding have appreciated the extent or effectiveness of the moral force you might wield in such situations as this. A few words from you at a press conference would be a warning to anybody who may be manipulating the market. Or, if you want to do it by indirection, I suggest that you invite Professor Ripley down to dine with you some evening soon, while his article is receiving so much attention. That gesture of interest in him, at this time, will be all the hint the stock market boys will need."

That evening, after dinner with the President, Mr. Welliver found in

[5] The facts in this episode are found in letters and in conversation with Mr. Welliver.

the executive office that an invitation had gone to Mr. Ripley. Whereupon, being a good newspaperman, he planted the Ripley article generously among his friends, the Washington newspaper correspondents, so that they would give the kind of interpretation to the Ripley visit that Welliver felt was needed.

Mr. Ripley had been professor of economics at Harvard for nearly a generation; a lecturer at Columbia on business corporations and trusts. He had been a member of the executive committee of the Chicago, Rock Island and Pacific Railroad. He had written three standard source books on railroad economies. In Boston, Governor Coolidge had voted for Mr. Ripley as a member of the minimum wage commission, at the suggestion of Charles Hayden of the Hayden-Stone banking house. So Coolidge knew that Professor Ripley represented New England liberalism, and with the natural Yankee curiosity in his blood, the President wanted to hear the other side of the story of the Coolidge bull market. The Ripley interview [6] carries a picture of the Coolidge of that day, the Coolidge whom Art Young, the cartoonist, once called "the slit-mouthed Puritan." He sat in the great barny executive office of the White House, tilted back in his chair, smoking a waning stub of a cigar, his feet cocked and embedded in the top drawer of his desk. He waved Ripley a welcome and steered the Professor into the heart of his story. He smoked his stub out, put down his feet, opened a lower drawer, while Ripley talked on, took out another cigar from the box, closed the box, shut the door, offered Ripley no convivial cigar, parked his feet in the deep drawer again, leaned back, looked out the window, smoked, listened, said no word. Finally as the second cigar was well burned down and as an attendant brought in a waiting Senator's card, Coolidge straightened up, put his feet on the floor and remarked:

"Whatcha doin' this noon? Can you come to lunch and talk this thing over some more?"

So that afternoon, when lunch was finished, the President and the Boston Professor adjourned to his study in the residential part of the White House. There, because his day's work was done or nearly so, and he was a social host and was not afraid of his guest overstaying his time, according to habit he opened a drawer in the mahogany desk, took out a box of cigars, passed it to the Professor who took one and the two smoked while Ripley talked. He told the President the truth about the crooked organization of some of the great public utility industries. He explained carefully the sources of the vast surpluses that were draining into Wall Street and were furnishing what might be called outside capital for the tremendous volume of broker's loans that was whipping up the bull market. He detailed the "prestidigitation, double-shuffling, honey-fugling,

[6] Conversation with Mr. Ripley, 1935.

hornswoggling and skulduggery," [these were Ripley's words] which characterized so much of the business in the Wall Street market which the Coolidge administration was supporting, and as a symbol for which Coolidge was to go into history. The President again leaned back in his chair, put his feet on the desk, closed and opened his eyes occasionally, clearly absorbing the Ripley story. His face was troubled. Finally he asked sadly:

"Well, Mr. Ripley, is there anything we can do down here?"

In 1927, Ripley did not consider the organization of public utilities a federal matter. He answered:

"No, it's a state matter."

Trouble rolled away like a cloud, and the sunshine of sweet complacence bathed the Yankee head like a sunset in the Green Mountains. It was the ideal situation, a chance for that masterly inactivity for which he was so splendidly equipped.

The day after Ripley left, the newspapermen, who had been primed with the article by Welliver, turned loose much of Professor Ripley's criticism of the stock market. A visit to the White House gave Ripley's story a tremendous multiplication. On the second day after the Ripley visit, the New York stock market had one of the greatest blow-ups it enjoyed during the entire Coolidge boom era. It was first page news for two or three days.[7]

4

In the late spring and early summer of 1927, in Boston, Senator William M. Butler, Chairman of the National Republican Committee, was

---

[7] When Welliver was in the White House several weeks afterward, visiting with the White House secretaries, Everett Sanders, the President's secretary, said:

"Aren't you going to go in and see the President?"

"No," said Welliver, "I haven't anything much to talk to him about and I know he must be awfully busy. I had rather not bother him this time."

The fact was that Mr. Welliver was literally afraid to go in and take the scolding he suspected the Ripley episode would bring upon him.

"Well," said Mr. Sanders, "I guess you had better drop around about 3 o'clock; the President wants to see you and very particularly told me to have you come in the first time you were in town."

So at 3 o'clock Mr. Welliver was ushered into the inner sanctuary. In a moment, to the relief of Mr. Welliver, the President, contrary to his custom, with Mr. Welliver at least, rose, came around his desk, shook hands, smiled his thin-lipped and enigmatic smile, and said:

"Well, Mr. Welliver, our experiment with Wall Street and the moral influence of the President was quite a success!"

Mr. Welliver writes: "The President was amused, and apparently no more than amused, over what had happened. We talked for some time, and again he reiterated that he didn't believe there was anything wrong with the market. But he was plainly pleased with the demonstration of how significant even a hint from the President might be."

beginning to prepare the ponderous machinery for producing the second nomination of President Coolidge in 1928.

The announcement of the candidacy of Dr. Nicholas Murray Butler, which enraged Chief Justice Taft, apparently did not throw consternation into the Coolidge camp. It was taken for granted that the Doctor was nobly dramatizing the fight on prohibition. But serious talk of Lowden did disturb the friends of President Coolidge, and a widespread agitation for Herbert Hoover brought to Senator Butler and Chief Justice Taft, and those near to the White House, a lively sense of danger in delay. Senator Butler [8] assumed, and the President definitely permitted Butler's assumption to go unchallenged, that the campaign for the second nomination should proceed vigorously. And when Butler approached the President upon the subject of the campaign of 1928, he received, as the Chief Justice had received, every encouragement except the direct word "go ahead." It was Frank Stearns who declared that when he needed a direct statement from Calvin Coolidge he wrote him a letter. But in the matter of the second nomination Stearns was as positive as Butler and Taft were that the President was anxious for the plans to move. Similarly Everett Sanders, the Presidential secretary, who deserved and had Mr. Coolidge's confidence, had every convincing evidence that the plans which Senator Butler and his friends were busily making, met with the approval of their chief.

The Taft family correspondence of that date [9] shows that the Chief Justice who was a White House intimate, and all of the Republican temple high priests, were going ahead, and that the President knew they were proceeding to nominate him for the Presidency in 1928. The Taft correspondence shows that Mr. Hilles and the powerful Republican state bosses almost without exception were gradually convincing themselves and so were leading their party to believe, that the renomination of President Coolidge was logical and in a sense inevitable. Being familiar in the White House, Chief Justice Taft would have known if the President was at all displeased with this Republican ground swell moving in his direction. Moreover it was evident that the Progressives who would like to have opposed him were without a candidate.

They were considering Herbert Hoover, Secretary of Commerce. But they were by no means united upon him, and in the left wing of the Progressive group enthusiasm for Hoover was not boisterous. Yet he was the looming candidate of the Republicans who opposed the President and the most determined opposition to the President came from the Progressives. Official progressive support of Hoover had not been tendered. It was one

[8] Conversation with Senator Butler, June, 1935.
[9] Now in the Library of Congress.

of the first chores of Senator Butler, the President's campaign manager, to see to it that Hoover and the Progressives did not get together. Indeed that was the only serious problem that confronted the President's friends and supporters.

The spring flood of the Mississippi River in 1927, was unusually devastating and the President before leaving for his summer vacation sent the Secretary of Commerce into the flood area as a national relief agent. Herbert Hoover rose to his opportunity. He revived almost overnight the memory of Hoover of the Belgian Relief. Being human, the revival spurred a laudable ambition to rise. But Hoover had a decent sense of loyalty to his chief in the White House. The chief had made no sign to the Secretary of Commerce that his track was clear. Hoover saw the preparations for the Coolidge renomination going forward in Chairman Butler's offices in Washington and Boston, and withheld his own plans. No one knew better than Calvin Coolidge how delicate the situation was becoming. And how the old New England mouser loved to toy with his prey!

CHAPTER XXXI

Being an Intermezzo While Only the Stage Hands Work

THE four years from March 4, 1925, to March 4, 1929, marked the years of the big bull market—the Coolidge market—the era which preceded the crash of the speculative boom. Calvin Coolidge will be remembered as the last President of the old era. It is a vital part of the Coolidge story to know something of the causes that made for the vast credit expansion that produced the boom of those years. The boom had its roots in many causes hereinbefore set down. Expansion of credit seems a primal cause. But perhaps it also was a symptom. Foreign loans which were draining money and credit from American industry also were merely significant episodes, not first causes. Bank expansion made phenomenal increases in loanable funds. Naturally the outlet was Wall Street, and a glutted Wall Street. It sent much of these savings of America abroad in 1914 and '15 and '16, and thereby somewhat we were drawn into the War. But war finance and war profits were only symptoms. America's real troubles just then seemed to grow out of unsound money market policy and unsound tariff policy— rather than failure of production and distribution.

According to the tables in "Recent Economic Changes in the United States," [1] the output per wage earner in the ten years preceding 1929, grew rapidly. Frederick C. Mills, of the research staff of the National Bureau of Economic Research, definitely indicates the belief that the output per

[1] This book contains a report of the Committee on Recent Economic Changes of the President's Conference on Unemployment, Herbert Hoover, chairman, including the reports of a special staff of the National Bureau of Economic Research. Incorporated; is a two volume work carefully, conservatively planned and executed. It seems to be a standard source book for those who would study the period closing with the crash of the late Twenties—McGraw-Hill Company, 1929.

wage earner in the United States from 1919 to 1929, increased forty-three per cent or at the rate of three and three-eighths per cent a year. The railways handled a growing volume of business. Mining and agriculture showed enormous increases in their output. This increase in production for each wage earner was in degree, at least, peculiarly an American phenomenon.

America had developed a machine-minded man who instinctively knew how to handle screws and levers, belts and wheels, cams and cogs, almost as he knew how to handle his fingers, his eyes, his feet. He had extended in effect his instinctive coordination far beyond his body. His daily toil, whether as a wage earner, or a salaried man, or a profit sharer, or a merchant, is speeded up with machines. His social organization is machine conscious. Overnight a new device is adopted not with protest but with acclaim, quickly, eagerly. The farmer is also ductile when machinery brings change. His peasant prejudice against new things is gone.

What was happening to America's distributive machine? How was America handling the problem of getting this flood of new-made goods and chattels parcelled? The answer to this question must remain unproved for a decade—possibly for a generation. The problem only may be stated and that statement in its simplest terms, subject to many qualifications. But this statement may well engage the attention of those who would understand the realities of the Coolidge administration. They may profitably spend a few minutes with these pages.

We were not entirely static. The marvelous distribution of automobiles in those fateful ten years before 1931—the Coolidge decade in Washington—indicates that American genius could distribute where it would. At the end of the period we had an automobile for every five and a half persons in the land, almost one to a family. In consuming electrical goods America was a fairly competent distributor. We had multiplied the output of household electrical heating and cooking apparatus by two hundred. Radios had been hurried into at least a third of the homes of the United States, largely middle class homes, homes of wage earners, farmers, salaried people, and of course of the well-to-do. When Coolidge came to the White House, sixty-eight per cent of our urban homes were well plumbed, had stationary bathtubs. The distribution of telephones more than doubled in those ten years from 1919 to 1929. And there was a telephone for almost every other residence in the country. It was the farm homes—and only the underprivileged farm homes—that showed the lack. Town and city homes were almost universally wired. We were consuming two or three times the water per capita that the cities of Europe consumed. But the ghettos of the metropolitan areas and the homes of the sharecroppers and tenant farmers indicated a shamefully low consumption of water per capita.

Indeed the farmer's living standard, and farmers composed two-fifths of our population, was not abreast of the consumption of material things which his city brethren enjoyed.

The farmer's net income had begun to drop—actually and relatively. There our distributive system clogged. The farmer's income was not adequate to keep agriculture from declining relatively to manufacturing, mining, trade and finance. The bright and the alert young men of the farms were drawn off to the city for better wages and better opportunities. Our high tariff policy tended to prevent the export trade upon which agriculture depended largely for its income. After the middle of 1924, we tried to make our foreign loan policy offset the tariff. The outside world roughly borrowed from us a billion dollars in 1924 and much larger sums in the years that followed, the peak rising between the middle of 1927 and the middle of 1928, to an astounding total of $1,800,000,000. It is odd that Calvin Coolidge in the White House with all his hatred of debt and extravagance did not see the truth of this situation. Yet he promised this cheap money policy and this rapid expansion of bank credit which made it easy also for agriculture to get deeper into debt and promoted the wild stock market and the various other phenomena of the ill-fated Coolidge boom.

Moreover the restriction of immigration which the war brought, which the immigration restriction laws produced, created a labor shortage, and a rise in wages in the cities. So that industries which once drew labor needed for expansion from Europe began to draw labor in from the farms. Up went labor costs to the employing farmers. Railroad rates increased in 1920 and remained for the decade of the twenties well above the levels of pre-war days. Agriculture also was burdened with rising taxes as counties and road districts and States borrowed to build roads and for other purposes. All this prodigality handicapped the American farmer—Coolidge's own people. The horse and mule were displaced by the automobile and the automotive truck. In pre-war days cities bought from the farms hay, oats and corn to feed the mules and horses, and during the twenties the farms were buying automobiles and trucks from the cities and gasoline to feed them. And agriculture, unlike corporate industry, had built up a great fabric of mortgage debt as land speculation, on credit, ran to extremes in 1916–20. The only reaction Calvin Coolidge's conservatism displayed toward this was in his stubborn refusal to buy an automobile. Was it a subconscious revolt against something which his instinct told him was deeply wrong?

One gets a flashing illumination of Coolidge's real views about the agricultural problem in a side remark he made to R. A. Cooper, who was chairman of the Farm Loan Board. He and Chairman Cooper were considering the farmer's woes. Coolidge paused a moment, opened his eyes and looked at the chairman of the Federal Farm Loan Board as he twanged, "Well, farmers never have made money. I don't believe we can do much about it. But of course we will have to seem to be doing something; do the best we can and without much hope."

He paused a moment, then fell into a reminiscent mood about his father and Young Calvin's experiences on the farm at Plymouth Notch, then added dryly: "The life of the farmer has its compensations but it has always been one of hardship." [2]

Which was candid enough and probably represented the honest views of honest American politicians in that day. Here's another view of Coolidge's ideas about farming. One day, talking to the same Judge Robert A. Cooper, the President recounted:

"At every Cabinet meeting for a year or so back, Secretary Henry Wallace [3] used to be grumbling and complaining about the price of corn and was always wanting the government to do something about it. Then corn took a rise. The government didn't do it. I noticed," said Coolidge, "that Wallace had shut up on the price of corn. Naturally I thought he was satisfied. One day when the Cabinet meeting had adjourned, I said to Mr. Wallace: 'I suppose your corn farmers are very happy over the present price of corn?' 'No,' said Mr. Wallace, 'the price is not so material because we feed corn to hogs anyway.' " And this is one of the few Coolidge jokes which his friends remember that he liked to laugh at. He generally pulled a straight face on his own jokes. And this also was the measure of his sense of the reality of the American farm problem.

And that very day, more than thirty-eight per cent of our farmers lived in counties where the average value per acre of farm lands and buildings was less than forty dollars. It was the inability of that thirty-eight per cent (and crucially the backward status of the lower half of that thirty-eight per cent) to find money to pay for their share of the vast increase in production which our social and industrial order was making possible through our machine-minded workers that stopped the flow of things from the factory to the home and the return of money from the consumer to the producer.[4] There was Coolidge's real problem. He did not see it.

[2] Letter from Robert A. Cooper, September, 1936.

[3] Secretary of Agriculture under Harding and the father of the Secretary of Agriculture under Franklin Roosevelt.

[4] The figures quoted above may be found in "An Audit of America" by Edward Eyre Hunt, McGraw-Hill Company, New York, in Chapter II, "The Consumer and His Standard of Living." This book is a summary of a book previously mentioned, "Recent Economic Changes in the United States."

But the plight of the submerged farmers, the share-cropper, the hill-billy, still does not explain entirely why in the Coolidge era we were unable to absorb the increased output of the American wage earner. It may be well to inquire how the forty-three per cent increase in the productive capacity of the American wage earner was obtained. The obvious answer, of course, is by electric power. By the time Coolidge got to the White House, fifty per cent of manufacturing power was furnished by electric motors operating on purchased power—the centralization of power. The Diesel engine appeared. And curiously only a slight advance occurred in the number of wage earners per establishment, but a striking advance in horse-power, more than double the value of horse-power machinery per establishment. And here is a significant thing: The income tax returns for those middle years of the Coolidge administration show that these larger enterprises furnished about one-half of the income tax returns, while the enterprises connected with financing the country stand second in the list, and trade and merchandising third. Then note this fact in the midst of the Coolidge era: The number of income-paying concerns in manufacturing had been static for several years. The incomes from trading concerns showed a slight increase, but the incomes from the financing companies, banks, trust companies, insurance companies, brokers, speculators, had more than doubled in number while Coolidge was in the White House.[5] It is evident that those who were making the mergers, amalgamations, and combinations of industry were taking an increasing toll from the national income as a reward for their services.[6] City dwellers of the middle economic group were bettering their standard of living, while farmers on the whole, considering the tenant farmer, and the share-cropper, were scarcely rising in their living standards and certainly not increasing their reward. Thus while everyone else in America—consumers, laborers and bankers and the public service stockholders—were profiting, the farmer was going to seed. Immigration practically had ceased and the birthrate declined. So we had a static population increasing manufacturing productivity with marvelous speed; indeed the output per worker on American farms, in mines, factories and railways "increased twenty-seven per cent" in Coolidge's day of glory.[7]

[5] "An Audit of America," p. 31.

[6] Ibid., p. 78.

[7] Probably this inequitable distribution of income had little to do with the collapse of 1930–33. But when barricades rise in the street, when revolutions change the bases of governments and re-color the maps of continents, the psychological effect of the maladjustment of wealth is vastly more important politically than is the real cause of economic disorders. In our boom and collapse, though truth and statistics point to other causes than the maladjustment of income, it is after all the fact "that the rich are growing richer" that gets into politics and overthrows political dynasties and changes the course of history. The ancient and probably eradicable evils which make the unfair differences

2

This shift of national income in the 1920's from the farm to the factory, the bank and the public utility, flashed a danger signal which thousands of Americans were reading. The danger signal flashed in the colleges, from the statistical research departments of banks and of great industrial concerns. Many of these economic experts issued statements explaining the impending danger. Dr. Benjamin M. Anderson, economist of the Chase National Bank, in his economic bulletin, was prophetic in his warnings. Jeremiah, at the Court of Zedekiah, was not more disturbing than this seer of Wall Street. But the President should not be blamed too heavily for failing to heed the jeremiads from the colleges and the clearing houses. The day of experts had not yet dawned.

But what if the President, through some miracle of insight and by some onrush of will, had started to stem the flood? His action would have been an order for a minor panic! How quickly the Congress would have left him! And the Supreme Court, holy of holies, for the priests of offended minorities, how it would have clamped the hobbles and handcuffs of injunctions upon any agent of the White House who would dare to lay hands on the altar of prosperity in that day. And indeed back of congress, back of the courts, supporting both in their check of the executive, would have been a raging populace roaring at the White House to let the carnival go on. The Coolidge boom was ordained by the American democracy. It came out of the people. It was woven inextricably into the lives of the people. It was their conscious purpose, their highest vision. In the bull market of the Coolidge era, democracy was having its will and going mad in its appointed way.

While these great affairs were moving on the face of the waters in the hearts of the people, a national myth was arising, the Coolidge myth. The President was becoming an ideal. It is odd that his thin, dry, harsh personality should have been dramatized. But the people took his very tenuous qualities, his curt manner, his drawling Vermont *patois* by which he was able to make three syllables out of the word cow, his lank body, his parsimony of word and deed, his obvious emotional repression, the glints of his sentimentality which they saw fleetingly at times, and piecing all these together they made a man, a little man to be sure, but one of their kind, a hero in fact. He was a sort of midget Yankee Paul Bunyan, and stories began to rise to illustrate his ashy humor.

in the distribution of wealth and which do not probably make our booms or our economic catastrophes, nevertheless do make history. Maladjustment of wealth and incomes of course are symptoms of more fundamental disorders, but how the quacks who prescribe for mobs do love to doctor symptoms.

Old stories were revived and fitted into the Coolidge vernacular. But they were all clean stories. In the lot was never a word implied to Coolidge that had to be said behind one's hands. This essential hickory cleanness of the man became one of his mythical qualities. Ordinarily Americans do not create fictional characters out of their President. They idealized Lincoln. They created "The Teddy" out of Theodore Roosevelt, but Washington, Jackson, Lincoln and "The Teddy" are about all the White House gods that America has set up. But everything this President of the 1920's did or said or seemed, fitted in or was molded to fit the fictional sculptured character America was chiselling out of this bit of Vermont granite. He tried rejecting a private car when he traveled—"just a little stage play, I am afraid," wrote Taft.[8] He used the *Mayflower* to go down on the Potomac River every Sunday afternoon but he always went to church in the morning after the fashion of the God-fearing country townsman. That fitted into the picture that was growing in the popular heart even though a sophisticate like Chief Justice Taft sniffed [9] that "he has a deuce of a time having it made public that his trip on the River is after he goes to church." However, Coolidge knew how to deal with the sophisticates. The Chief Justice, having an appointment with the President, had to break it because he was laid up, and when the Chief Justice wrote apologizing, "the old man sent me some flowers in the absence of his wife which shows he knows how to do things when the ladies are away." [10] Horace Taft wrote to his brother after hearing the President talk at Andover, that his voice was "about as musical as the sound made by a buzz saw." Of course, that rasp voice also was a part of the Coolidge myth. It delighted the people who were on the whole rather tired of silver tongues and wanted just what Coolidge had to offer.

In the preliminary byplay a year before the campaign of 1928 for the Presidential nomination, whether Coolidge wanted it or not, he put on a charming little play of hide and seek. Chairman Hilles, the super-politician of Wall Street, according to the Chief Justice [11] was trying to draft Coolidge. William M. Butler, the President's political shadow, was, according to the Chief Justice [12] more or less in the plot. But Coolidge was as crustily coy as a New England school teacher. He wrote a message in the spring of 1927 blistering the McNary-Haugen Farm Relief bill, which the Chief Justice called a "sockdolager," [13] and even the farmers delighted in its blaze of wrath

[8] To Mrs. Taft, October, 1925.
[9] Letter to Mrs. Taft, April, 1924.
[10] Mrs. Coolidge was attending the Commencement of her boy, Taft wrote to Mrs. Taft in Europe.
[11] Letter to Moses Strauss, May, 1927.
[12] Letter to Henry Stimson, January 16, 1928.
[13] Letter to Robert Taft, May 27, 1928.

from a man who had been for five years swallowing his emotions. That also became part of the Coolidge myth. And here is the core of it. The reputation that he was building up in the hearts of the people was strangely the exact figure which men would create in their inner hearts who were living in another world, a world deeply in contrast with the narrow circumspect spiritually stingy figure that they still held as the American ideal in their hearts. Their real world was running madly extravagant, wanton in its waste not only of money but of every quality of mind and heart, a world gone gaudy, even morally bawdy in its Babylonian excesses. A sin-sick world in the nature of things would erect this pallid shrunken image of its lost ideals and bow down before it in subconscious repentance for its iniquities.

3

A striking consistency is revealed in the relation between the financial world and American politics as it was represented and dominated by President Coolidge. The regular Republican organization of Congress in those days was definitely conservative, at times distinctly reactionary, and from that conservative group came the names submitted to the White House for the federal judiciary. So we have here in the heart of the Coolidge era an unyielding conservative President, a definitely conservative congress and a federal judiciary springing out of congressional patronage which was not only conspicuously conservative, but which Chief Justice Taft vainly was trying to keep at least decently honest. His efforts were only fairly successful. He was ignored from time to time by the President who had to take names of inferior men for the judiciary to keep recalcitrant Representatives and Senators in line for the Presidential program. Occasionally, as we have seen, the Chief Justice tried to dilute the President's partisanship. As late as the spring of 1927, the Chief Justice wrote to Thomas Shelton, a West Virginia friend: "The President looks askance at my non-partisan feeling about judges." In Taft's letters of that period, one realizes the tremendous pressure that is brought upon a President when a federal judge is to be appointed. In the interest of Judge Augustus Hand, of New York, the Chief Justice writes that he himself wrote to the President, that he stirred up the Attorney General who went to see the President, that he wrote to George Wickersham, who wrote Coolidge a letter, and he wrote to former Chief Justice Hughes, who wrote a letter to the President, and to Charles Hilles. "The latter," explains the Chief Justice "was for the purpose of taking care of the political end." Thus the financial and commercial world and the political world in the days of the Coolidge bull market grew in beauty side by side in the creation of a federal judiciary.

In spite of the rebuffs which Chief Justice Taft felt he was taking from the President when he volunteered his aid in selecting federal judges in the first year of Coolidge's occupancy of the White House, we find in these summer days the Laughing Buddha of Wyoming Street visiting the White House again and again. Evidence in his letters to his family will be found that the Chief Justice was interested in the Wall Street boom. He indicated that he hoped, and probably suggested to the White House, that the bull market was one of God's good and perfect gifts and doubtless arrayed himself with the secretary of the treasury, who felt that the market should be supported. Chief Justice Taft was never among those who raised jeremiads when people of his class and caste were making easy money.

For some reason the Chief Justice fretted about a federal judgeship in Georgia and wrote to his friend, Charles Hilles, in Wall Street, that the President had not consulted him about the Georgia appointment. He adds in sadness, this "present administration makes its selections in a very curious way and I have very little to do with it." And to his son, Robert, the writer reflects his impatience that the President will not "agree in advance to appointing Democrats" to the federal judiciary, even if Congress passes a bill giving the country ten new federal judges. The dissolution of Taft's dreams of a bipartisan federal judiciary must have been one of the major irritations of a beautiful friendship—irritations which the President, a strict Massachusetts partisan Republican bore with Christian fortitude and patience. For the Chief Justice was no hand to bow and kotow before the mahogany desk in the Presidential office where he sat so often gossiping and gassing with his New England friend. Referring to another matter, not long after this letter, he revealed his relationship to the President in a sentence: "You can bet I shall tell Calvin just what I think about it, no matter what he thinks, because I know what I am talking about!" Taft had sat on the other side of that desk himself and his own desk in the Supreme Court was one of power and distinction.

In February, 1927, it was obvious that President Coolidge, as a potential candidate for the Presidential nomination in 1928, was gaining strength among the regular Republicans. On the other hand it was equally plain that Republicans of the progressive type were organizing against him. In Congress, one heard talk of a joint resolution against a third term. The proposed resolution had support in certain more conservative quarters, where Coolidge had made enemies with patronage. His veto of the farm measure known as the McNary-Haugen Bill strengthened him greatly as the Presidential nominee in regular Republican circles, though many strait-laced partisans voted for the measure. Chief Justice Taft in writing to his son Robert expressing his joy at the veto, registers a belief that many of the regular Republicans who voted for the bill felt confident that Coolidge would veto it, or failing such a happy event, that the Supreme Court would kill it. And in the letter he exclaims against the "exhibition of political cowardice" in both Houses.[14]

But in the matter of his supposed interest in a third term nomination, even his best friends could only guess what was in the President's mind. The Chief Justice who knew him in his daily walks declared: "if I were with him alone I would talk frankly on the subject and I think while he would not express himself, he would leave me to infer with certainty what his attitude is!"

From which remark it is evident that even the ingratiating Chief Justice was never able to break down the Chinese Wall of uncandor which was a part of the Coolidge birthright.[15]

[14] The Chief Justice notes the fact that the McNary-Haugen Bill was passed by "bribing the cotton people and the tobacco people" in the South, putting these crops in as beneficiaries of the measure. It started out as a wheat and corn measure. The President's veto was one of the few vituperative public documents which he ever penned.

[15] The Chief Justice could not separate himself from national politics. He represented typically the intelligent conservative view and his opinion about current politics was the White House opinion. His attitude therefore is important in the Coolidge story for probably no other mind reflected so exactly the Coolidge attitude, hence this footnote detour into the Taft correspondence. He wrote to his family that summer that with the growing strength of the Progressive group in Congress and the probability that the autumn elections would increase it, the President would be compelled to accept the Presidency for a second elective term. He felt that as the election approached it looked as if "business would be better . . . and that Coolidge's strength will reassert itself." Yet he also wrote to his wife, that summer, that when he viewed the Republican party, it seemed to be "all broken up"; and one can see him smile as he adds that when he considers the Democratic party "the fissures in it are just as great." But a few days later he writes to his friend, Associate Justice Sutherland, that this is probably "as good a situation as the Republicans can expect." And looking at the market from the White House viewpoint, the Chief Justice indicates that a deadlock in legislation "in these times with prosperous business is not a bad thing." We find him writing again to Charles Hilles, down in Wall Street, that summer that "continued good business will wipe out the slips" that Coolidge has made and will make him not only available but inevitable for the Republican nomination two years later.

These views certainly reflect something of the White House attitude which the President probably radiated in a thousand unconscious gleams from his inner heart. Yet Chief Justice Taft was no hero worshipper in those days. We find him writing to Hilles in mid-summer that the President's weakness is due to his "uncertainty as to appointments." He feels that this Presidential uncertainty rises from the fact that Coolidge knows he hasn't good judgment and is addicted to bad luck in appointments, and dillydallies, yielding too much "to Senators and Congressmen in their demands for patronage in respect to positions as to which he should feel independent."

As the summer closes and the autumn of 1926 appears, the political philosophy of the White House is reflected in Taft's letters to his sons indicating that although the Republicans are in danger of losing the Senate to the Progressives, a Republican majority entirely undependable when a vote on a partisan measure occurs is of no particular use to the President and also that despite the success of the Progressives in Congress Coolidge's nomination in 1928 seems certain. And the Chief Justice adds: "He is the strongest man we have and he is safe."

The regular Republican losses and the Progressive gains in the election that autumn

In America all the world, politics, business—big and little—wanted above everything the status quo. America was worshipping on padded knee the God of things as they are, and these hopes of Taft, high priest of prosperity in the temple of legal stability, reflected only the hopes and aims of ten thousand acolytes, altar boys, elders and deacons in the great acquisitive Hamiltonian cult—to honor the name of Coolidge and adore him forever. So early in 1927 a second elective term was looming on the horizon despite the Presidential veto of the bonus, and the McNary-Haugen Bill in the spring of that year—offending the farm bloc, and despite the personal dislike of him which animated many conservative leaders. But public sentiment that led the leaders, and guided American destiny, was pulling like an irresistible undertow "from out the boundless deep, too full for sound and foam."

apparently did not disturb the White House. For we find the Chief Justice, after a visit there, writing to his brother that "the President now will not be responsible for legislation that he cannot control" and he adds naively, certainly not without a knowledge of the aims in Coolidge's heart and the hopes of his supporters around Washington and in New York (meaning the Wall Street group) that if it is finally decided "to put him in the running for 1928, his record is already made." In his correspondence, Chief Justice Taft notes that various candidates are erecting lightning rods. He lists Lowden and Dawes among others, and Nick Longworth; and writing to his daughter chuckles: "Since Harding was nominated I have just lost confidence in my political judgment as to the nominations of either party."

CHAPTER XXXII

Our Hero Plays Sphinx

THIS was the world of the summer of 1927, when Calvin Coolidge, followed by heralds, news hawks, camera men proclaiming his divine vicegerency in the best possible world, with experts, soothsayers and wise men in his entourage, wended his way in a special train from Washington to the Black Hills of South Dakota. There he established his summer White House. The executive offices were set up in the school house at Rapid City, South Dakota. The executive residence was a ramshackle, pine board, paper-lined, hotel-like structure, eight or ten miles further up in the hills. It was surrounded by pine woods. The altitude was around five thousand feet. Being in the north, it was reasonably cool, save in the heat of the day. The natives of the western hills were happy about the royal invasion. For the most part they were not unlike the President's neighbors in Vermont, except that their fathers, grandfathers and great-grandfathers had moved westward with the generations. They were a kindly, hospitable people as the Vermonters are, and the President sincerely tried to neighbor with them. If they brought him a cowboy outfit, including sombrero, leather chaps, a tight jacket, and a red shirt, by way of being kind—and they did so—to be appreciative he wore it. He kept on wearing it when the photographers assembled, and he stood and let them photograph him, looking just as miching as he felt. He had no notion he was costuming himself as a bold, bad man in those western movie cowboy togs. He ambled lumberingly down the steps of his summer house, looking every bit as awkward and unhappy in his clanking spurs as he was. But he made this holy show of himself only by way of indicating his appreciation of a bit of neighborly kindness. When the pictures appeared in the newspapers and when he clickety-clacked down those steps in spurs and the creaking leather armor of the cowboy, in the sound pictures, his enemies chortled and his friends groaned, all *unbeknownst* to him!

The westerners' togs sported by this stern Puritan struck Gamaliel Bradford, who focussed his psychograph on Coolidge playing cowboy, as "ludicrously inappropriate decorations." He saw "the garish cowboy rig and in the midst of it the chilly Vermont countenance wondering painfully, wearily what it was all about. 'These people are not working; why should anybody want to do anything but work.' " [1] He could not play happily nor gracefully nor understand others at play. Bradford is bolstered in his conclusions by Bascom Slemp who puts these words in Coolidge's mouth: "On a little church high on a Vermont hillside I saw this inscription: 'No man who lives a life of ease leaves a name worth remembering.' Industry pays because it is right." Hercules even in miniature could not for a gay moment cast himself as a little Falstaff.

In those first fair, fine days of his South Dakota residence, he seemed happy, though his smile under the sombrero looked silly and he knew it. In the meantime, as floods were sweeping down the Mississippi Valley, Secretary of Commerce Herbert Hoover was dramatizing himself in his best form as the hero of a relief crisis. The people rose to it. The price of the Hoover presidential stock stiffened. It was in those days that the President in his genial dried-apple wit began to refer to Mr. Hoover as "the wonder boy," or "the miracle worker," indicating a subconscious, unformed wish that the Secretary of Commerce would fall in the floods and choke. But outwardly the President was bland. Strolling railroad presidents and an occasional colonel, a brigadier and a captain of industry called upon him. Politicians wending their way to and from the Pacific coast were invited to the summer White House. The President was having, for himself, a gala time.

June 27, Mrs. Coolidge set out with James Haley, secret service man, as her guide for a morning's hike. It was her first long walk in the Black Hills. She was dressed in high, hiking laced shoes and a white sport skirt with a sweater. Secret Service Man Haley had been assigned to Mrs. Coolidge soon after she entered the White House and was familiar by sight to thousands who had seen him by her side walking about Washington. His departure with the Presidential party for the Black Hills was supposed by the reporters to be a promotion for him. That day Mrs. Coolidge took a downhill trail. She walked with a quick, long stride and swiftly put distance behind her. As was her habit in New England, she stopped to pick wild flowers, finding many new specimens in the Black Hills. Lovely gumbo lilies and shooting stars were her favorites. As she started on her hike, she

[1] "The Quick and the Dead."

stood and watched the secret service men kill a four foot gopher snake with a stick.

She had been gone more than three hours when the punctual President arrived at one o'clock sharp from the executive offices twelve miles away in Rapid City. It was a hot day. He took off his coat and sat down on the front porch in the cool shade. Mrs. Coolidge was late. Upon inquiry he found that she had left shortly after nine o'clock. He began to pace the porch nervously, and after he had been waiting fifteen minutes, called some of the secret service men and asked about her trail. Half an hour passed with the President pendulating from one end of the porch to the other. The secret service men standing nearby could detect a rising temperature. Another half hour passed and at 2:15, an hour and a quarter late, just as a searching party was being organized, Mrs. Coolidge, evidently fatigued from the long uphill walk, appeared cheerfully calling as she saw the pacing President:

"Hello, papa! Sorry to keep you waiting!"

She and the President went into the house, and as she has many times written in her memoirs of her husband "he is prone to visit his irritation upon the first person he meets." Very likely it was not entirely a happy homecoming. James Haley, the secret service man, explained to his fellows and Mrs. Coolidge corroborated to her friends, that he had repeatedly insisted that a mountain walk downhill would take much more time to retrace when they started uphill. He had several times urged her to turn back. It is inconceivable that as pleasant, just and kindly a person as Mrs. Coolidge would not have explained that in detail to the President. A few moments after her arrival she sat on the front porch sipping ginger ale with the President still in his shirt sleeves. Of course, the episode of the cold lunch at 2:15 and the President's rising anxiety and evident displeasure, made the headlines of the Boston papers. We read in the *Boston Herald* of June 28 and June 30:

WIFE'S LONG HIKE
VEXES COOLIDGE

———

PRESIDENT PACES PORCH AS
FIRST LADY HITS 15-
MILE TRAIL

———

MRS. COOLIDGE'S
GUARD RECALLED

In the *Boston Post*, of June 28, we read the following:

FIRST LADY
ALMOST LOST

PRESIDENT WORRIED, ON POINT OF
FORMING SEARCH PARTY JUST
AS MRS. COOLIDGE RETURNS

In the *Boston Globe*, of June 28, appeared these headlines:

WIFE'S DELAY TAXES
COOLIDGE'S PATIENCE
SHE GOES OFF ON LONG HIKE
AND LUNCHEON GETS COLD

PRESIDENT SITS ON PORCH AN HOUR
WAITING FOR HER TO EXPLAIN

The *Globe*, including the story, printed this: "If Mr. Coolidge was just a little vexed over the long wait while he sat anxiously waiting, Mrs. Coolidge must have assured him that her good time was worth the now cold luncheon for they soon entered the Game Lodge together, everything apparently adjusted."

Those who read newspapers with perspicacity will find much concealed in those few lines—much to nag the President.

A few days later in the *Boston Globe* of June 30 and July 1, we read these headlines:

HALEY OUT AS MRS.
COOLIDGE'S ESCORT

Veteran Secret Service Man to Go
to Washington—J. J. Fitzgerald
Named in His Place

HALEY SHIFTED TO
RETURN TO BRIDE

Recent Marriage Given as
Reason for Change

And then July 3rd, when letters from the South Dakota White House had apparently been received in the *Globe* Office, M. E. Hennessy, Coolidge's biographer, in his "Round About" Column printed these lines which also were significantly written:

Johnny Fitz, Mrs. Coolidge's new bodyguard, is not as striking a figure as Jim Haley, relieved and sent back to Washington to take up a different line of work. He is a fine horseman and familiar with the Black Hill trails. It will be hard to lose "Fitzy" in a hike, and he's always on time for his meals.

But all the Boston papers which used the first story emphasized the fact that Haley was blameless.[2] The papers also recalled that Haley was a veteran in the service, who had been employed in the Food Administration, under Administrator Hoover, and had done splendid police work there against violators of the food law. When the Food Administration was dissolved, Haley's work attracted the attention of the United States Secret Service, and he was taken on as a secret service man. He did good work running down counterfeiters and was assigned to the White House staff as being one of the best men in the service. He went ahead of President Harding to arrange for his Alaskan trip and was with Harding when he died. Obviously he was an exceptional man. His assignment as Mrs. Coolidge's bodyguard would seem to indicate Haley's unusual qualities.

Curiously the Boston papers gave the episode much more prominence than the New York papers devoted to it. The Boston papers with their annoying headlines did not find their way to Rapid City and the President's attention until after the episode was all but forgotten. That they renewed the irritation of the June morning is a reasonable assumption. That they would wound and deeply stir a man so sentimental, so sensitive, so keenly conscious of his own crusty temper, and of those qualities which he himself in his autobiography calls his "many palpable weaknesses," of course is evident. No one had a more acute sense of drama and his own relation to the spotlight of publicity than President Coolidge. He did not court the spotlight greedily. But he knew when its rays were falling and he could not have been insensitive, being what he was, to the disagreeable attitude in which the Boston newspapers placed him. It might have irked him pro-

---

[2] In his "Forty-two Years in the White House," Irwin Hoover insists that Officer Haley was immediately and incontinently discharged but we must remember that Hoover disliked President Coolidge and was sometimes unfair to him.

foundly. It might have sunk deeper under all the circumstances than the men who were most closely associated with him and most intimately near to him in his daily tasks knew or suspected. For the still waters of Calvin Coolidge's life ran deep.

The episode was entirely forgotten three weeks later when Senator Arthur Capper, of Kansas arrived in South Dakota as a personal guest of the President. Senator Capper and President Coolidge had many things in common. Both were quiet men, both were reserved, both were kindly, sentimental, courageous when the need came, but never given to advertising their courage. Both were soft spoken. Both sought the lines of least resistance. Naturally an affection had grown up between the President and the Kansas Senator. Capper did not agree with many of the Coolidge policies. He was the official head of the farm bloc which often opposed and greatly annoyed the President. In the years of Coolidge's official life Capper more often was found voting against the administration on vital matters than with it. Yet the two men held each other in that wholesome respect which is the basis of an enduring friendship. So the President welcomed Capper to Rapid City and then characteristically let him rather severely alone. Mrs. Coolidge was left to entertain him. It was she who prepared him a luncheon when he left on the train, fearing that the diner might miss rail connections. It was she who entertained him on the most memorable morning of that summer, August third, when the President drove off to the morning's work in the executive offices. Capper in his account of what followed then seemed to remember that he rode to Rapid City from the Game Lodge with the President. In a letter to Frank Buxton, Editor of the *Boston Herald*, Mrs. Coolidge explained that Capper did not go with the President. And she thought enough of making the correction to add that it was "significant" that the President rode alone that morning, down the long ten miles to his office. She wrote in *Good Housekeeping* ten years later that her husband never respected her education—probably meaning her feminine mind. That particular morning his wife sensed that the President had something upon his heart, that he was in one of his deeply meditative moods. For she knew him like a book. She understood later why it was "significant" that on that most fateful day he wished to be away from the gentle, kindly Kansas Senator. "Thinking, thinking, thinking," wrote Bascom Slemp of the President, describing certain of his moods. "He often sat silently for an entire evening," wrote his White House stenographer, describing the Coolidge throes of composition. In that ride down the mountain Calvin Coolidge was composing his renunciation. Mrs. Coolidge, with whom he would never discuss public questions, knew some kind of a pot was boiling in his brain—boiling down his decision into its lowest, simplest terms. She had no remote idea of the nature of the brew in the kettle.

It was a lovely morning, August third, as he sped down the broad mountain highway to his desk. He barked a casual, dry, hard begrudging Yankee good morning to the chauffeur—"Morning!" and spoke no other word. His boiling pot was simmering his language as he went down the hills. He was skimming off adjectives, dipping out participles, conjunctions, getting down to the bare, cryptic bones of his exact meaning. Probably debate in his heart had ended, and that, by instinct, his wife knew also. In the debate behind the closed portals of that hard, stubborn mouth, what had the affirmative pressed to a conclusion? Was it the headlines in the Boston papers? Did he inject remote implications between the lines of those stories of Jim Haley's discharge—the petulant, unfair thing he had done in removing Haley, a deplorable gesture? If he realized, did he repent and in Puritan shame did he find solace in some grand act of renunciation to appease a harrying, nagging conscience? Or perhaps the roar of the rising tide of the market speculation stirred some deep intuitive fear within him. He could not but know that in the undertow of the booming tide, nearly a million men in the heavy industries were jobless. He knew—what the nineteenth century economists had taught him at Amherst—that the first sign of a cyclic crash is idle men! Could the call of these men have weighed in the affirmative the syllogism which ground his logical decision to its conclusion? Then there was the warning of a tired heart, his slowly waning body—and the dust and ashes of his fame when he thought of the dead boy and the grave on the Vermont hillside. "The power and glory of the Presidency went with him!" And the bedspread, Mrs. Coolidge's bedspread with its forty-eight squares, one for each month of the passing four year term. Only a dozen squares, a year's quota and seven more squares—September, October, November, December, January, February, March! March fourth 1929 and she would be happy, she of whom he was so proud, she who in his heart he knew had been his strongest link with the people, his one link with happiness, she who had posed for "happy wife" pictures with him, who watched over his gaucheries, who loved him dearly enough to mock his Yankee words and ways, she who had always been—and then the Boston headlines, the peremptory discharge of Jim Haley and then— that swift gay thrust of Mike Hennessy's words, Johnny Fitz "not as striking a figure as Jim Haley" and "he's always on time for his meals!"

Calvin Coolidge looked neither to right nor left as the Presidential car sped down the hill, but the habitual, hard monosyllable he barked as he left the chauffeur may have carried in it the hot turmoil of self-impeachment, the stale weariness and disillusion of his high place, and the haunting, illogical fear of evil days ahead for his country.

CHAPTER XXXIII

And All the World Wonders

WHAT happened after Senator Capper arrived at Rapid City with one of the President's aids, the Senator can relate better than anyone. For first of all Senator Capper is a newspaperman, and most important of all as a newspaperman he is a reporter and a good one. He remembered for instance that the night before, sitting around the fire, when he had told the President that most of the Republican Congressmen and Senators felt that the President should take a second elective term, the President was interested but did not commit himself. Capper recalled also telling the President that while the Kansas farmers disagreed with him about his veto of the McNary-Haugen Bill, they would be for him in the election. Which obviously pleased him. But also Senator Capper noted in his story of the evening's visit that the remark did not draw any commitment from the President. But the President rarely committed himself unnecessarily upon anything.

When Senator Capper arrived at the executive offices in Rapid City, the President let him sit around for an hour, reading newspapers and amusing himself while the President ground through his mail.[1] At eleven o'clock, he called for Everett Sanders, his private secretary, and said:

"Bring in those newspaper fellers about noon. I have an announcement I want to make."

He gave no hint of its nature. Sanders later told Capper that he hadn't the slightest idea of what the President was up to. No friend of the President has ever intimated that he had the remotest hint of what was in the

[1] This part of the narrative has been written several times by Senator Capper and was related in a personal conversation a few months later and again checked up and set down shortly before the publication of this book.

President's mind at that time. After Sanders left, the President called one of his personal stenographers and in Capper's presence dictated without hesitation or correction these ten words:

"I do not choose to run for President in 1928." [2]

Capper said nothing. Coolidge said nothing. He added to the stenographer:

"Have twenty-five copies made."

Still without comment to Capper, who was in hearing of the dictation, the President went on with his work. In due time the stenographer returned with the slips on which were written the ten cryptic words. It was nearly noon. On the dot, at twelve, according to custom the newspapermen came crowding into the President's office. Without comment, without flicking a muscle, he handed each man a slip of paper with the strange phrase on it. Suddenly there was a tremendous buzzing. Charley Michelson, correspondent of the *New York World*, hurried over to Capper and asked in a whisper what it meant.

"You know all I know," answered Capper.

Someone else asked: "Well, Mr. President, can't you give us something more on this?"

The President clicked the steel slit of his Indian mouth with:

"There will be nothing more from this office today!"

He stood there, motionless, unblinking, like a man carved out of cheese. The men, anxious to make the mid-afternoon editions of their papers, rushed from the office to tell the world. When they had gone, he turned to the Kansas Senator and said:

"Well, let's go back to the Black Hills!"

Capper and the President rode back together. Still no word, no comment, or no explanation to clarify the announcement came to Capper from Coolidge. He began to talk of a fishing hole where they would go, bragged about some big trout he had left there—making talk. Finally Capper said, to show some interest in the performance of the morning:

"Well, that was a sudden announcement you made this morning, Mr. President. It took them by surprise."

That pleased the President. He squeezed the juice of his delight out of

---

[2] Probably the most intelligent interpretation of the oracular sentence came from William Howard Taft who wrote in a letter to his son, Robert, a few days after the Dakota cryptogram: "With reference to Coolidge, I think he really wishes to avoid running for the Presidency, that he has had enough . . . but he does not care to put himself in a place where, by the unanimous demand of the party he may feel under obligation to run and will not be confronted with a previous statement by him that he would not run." The same day Taft wrote to Dr. Henry Coe, at Pointe-au-Pic, Canada: "I think that the Republican party may find itself obliged to force upon Coolidge the acceptance of another candidacy." So felt all his intimates!

the way he had put it over on the newspapermen. Finally after a period of silence punctuated by Capper's mild queries and monosyllabic replies, the President said:

"It's four years ago today since I became President. If I take another term, I will be in the White House till 1933; a Vice President two years. Ten years in Washington is longer than any other man has had it—too long!"

Under the stress of his repressed emotions, he could say only that.

"In due time," Capper declares, "we came to the Lodge and had lunch with Mrs. Coolidge, talked about some Indians who came visiting bringing him gifts, still making talk to avoid a silence which might be used to revert to the morning episode. The President made quite a story of the Indian ceremony and seemed anxious to keep the talk going. He succeeded during lunch. After lunch we went to the living room and still the President chattered on about nothing. At last he said: 'Well, I guess I'll take my nap!' He disappeared, leaving me with Mrs. Coolidge. I knew that sooner or later she would get the story, and would realize that I sat there alone talking with her and did not tell her. So I began cautiously: 'That was a surprising announcement the President made this morning.' She asked: 'What announcement?' And I told her. She exclaimed: 'Isn't that like that man! He never gave me the slightest intimation of his intention. I had no idea.' "

A little while afterward, Senator Capper talked to Private Secretary Sanders and he said the same thing. A few weeks later, in Massachusetts, Capper saw Frank Stearns and he said it was an absolute surprise to him, as did Senator Butler, his political manager and adviser. Capper declared that in Boston they were going ahead at full speed with plans to nominate Coolidge in 1928 when they read the story in the afternoon papers.

How politics bubbled over the pot that summer after the ten ambiguous words from the Black Hills! In Boston, Senator Butler, who was busy with the plans for the President's renomination, had to stop his machine. In every state capital, where a setup for Coolidge had been forming, the astounded politicians who were building up Coolidge organizations in every county, felt the sudden grip of the brakes in Boston. This jolt sent consternation and anger thrilling down the line to the wards, precincts and county districts. A sudden scramble for the band wagon occurred, Lowden's, Curtis's, Hoover's, anything with moving wheels! And the grim-faced Yankee who had come up in politics from the precinct, knew the hullabaloo he had made! He enjoyed it, otherwise he would not have made it in exactly the way he did—leaving that small element of doubt that brings irritation and futile rage. Never did he turn the rack or the thumbscrew, but how he grinned when he saw the high gods fidgeting, when he tickled their toes with a straw!

The news reached the Wall Street tickers shortly after the close of the market for the week. That also the President must have realized. But Monday the market fell.

Instinctively Wall Street and all its brood and breed knew that it would soon lose its best friend. Wall Street loved him not for what he said and did, though what he said many a time had been a life saver to the market. The market in Wall Street delighted in him for what he was, for his attitude toward organized wealth, toward aggrandizing capital, toward the wide range of madness of his day and generation. The market shivered for an hour that Monday morning when it knew Coolidge's race was done. His background and his outlook were the Coolidge policies: They were the Coolidge policies though bulwarked with a masterly inaction that was itself *laissez faire* to the predatory minority of business men who paid their debts, who lived within the law, who burnished the outside of the cup and who whited the sepulchre wherein they decently entombed their victims' bones and where alas their own would soon be bleaching! Hard words! Not at all. For our millions of speculators in numberless things themselves represented in that day the dominant American spirit. And in so far as Coolidge was the embodiment of the big bull market of '26, '27, and '28, he was democracy in its perfect flower. He represented in his attitude toward the market the American mind, the controlling mind, the ruling class. Privately market gambling offended his sense of thrift and order. But as a public figure he was Heaven sent to the market. He breathed life into the nostrils of the raging beast that was the American gambler. In those days when he was whittling the phrase: "I do not choose to run" down to its peglike symmetry, the *Commercial and Financial Chronicle* of New York, a dignified periodical of Wall Street, was printing this three days after the Presidential cryptogram in an issue of August 6, 1927:

"However much or little Mr. Coolidge individually may have had to do with the prosperity of the past four years, he is credited with having been the backbone of that prosperity. The present Bull market is going down in history not only as one of record proportions, but as one labelled with the name of Coolidge."

Commenting on these words from the *Commercial* and *Financial Chronicle*, Ralph West Robey [3] wrote in an article in the *Atlantic Monthly*, September, 1928, a year before the crash:

[3] A lecturer on banking at Columbia who served with the National City Bank of New York and had performed research work for the Reserve Board and later was one of the Magi who served in Alfred Landon's court.

"If the bull market is to be labelled with the name of its chief sponsors, it would be much more courteous to recognize a second backbone and name it the Mellon-Coolidge market."

Within two weeks after the President had renounced an avowed candidacy for a second nomination, Herbert Hoover took a lead in the Presidential race which he held until the election, a year and three months later. Hoover came to visit the President in the Black Hills on a western trip, the fortnight following Capper's visit. But Mr. Hoover received no blessing save one of those conspicuous fanfares of silence with which Coolidge sometimes proclaimed his cryptic maledictions. There can be no doubt that as a Coolidge successor, Coolidge himself regarded Hoover at least as the lesser of several evils. The President was not happy in the latter days of his Black Hills vacation. The burden of a second nomination campaign having rolled away did not leave him in an exalted frame of mind. When he broke camp, the Presidential party with its royal entourage went to Yellowstone. The itinerary led the President into the Park by the entrance at Cody. In the canyon, when they reached the rather imposing Cody Dam, the caretaker showed Mrs. Coolidge and John all over the place and enjoyed their enthusiasm. The President stood with his back to the dam, and would not look at it.[4] He stood with his back to the dam, typically and symbolically—perhaps typically because with his cautious New England temperament he rejected automatically this newfangled western rigamajig. Possibly he rejected it symbolically, because out of the web and woof of his Yankee heritage, he and his kind were old sturdy individualists. They had conquered the continent single handed and they viewed with irritation, the inevitable changes in American life following with necessary scientific regimentation, which that dam, in a way, personified. To him it was a black devil, not merely harnessed water. He had seen water under leash along the Connecticut and its tributaries in Vermont, but that chained water knew its owner. There in the Connecticut Valley the harnessed water served a man, a family, corporation or at most a town, a local group of coöperating capitalists. But these western dams not only harnessed water; generically a western dam put capital into traces. The dam controlled the land below the dam and its fertility. The dam gave power to villages, to counties, to great valleys, to whole regions. This chained water channeled men in the new way they had to go. The Cody Dam probably represented to Calvin Coolidge as he stood there with his back turned to it, all the new social surrender that the new American must make to retain his liberty. Here was a new "x" entering into the equation of his life, demanding a new answer. So he turned inwards, gave the dam the contumely of his Presidential back, and let

[4] Interview with the crestfallen caretaker by the author ten days later.

Mrs. Coolidge and John represent the family at the miracle of the Cody Dam. Or maybe he was just grumpy! [5]

At the first overnight stop on his Yellowstone trip, the President went fishing on the lake. A game warden of high degree went with him, a person of rank and consequence in Montana. Under the broiling Montana sun, the President sat—silent. He made the high placed official bait his hook, said no word but every hour or so drew out a black cigar of an inexpensive brand, handed it in an angular gesture to his companion and said in laconic Vermontese something nasal that meant:

"Here!" Just that. And repeated it with clocklike regularity during the day. [6]

They took the President to the geysers. Mrs. Coolidge and John evidently were thrilled at Old Faithful and the others. The President said nothing, looked down his nose, seemed impatient, bestowed no word, passed on. At the Grande Hotel, he ate in his room. Harding and Taft, who had been there, mingled in the lobby, ate in the public places. The West did not care for the Coolidge isolation. Wherever the party went sight-seeing, forest rangers, quite outside their duty, rode ahead and after him on motorcycles, outriders clearing the way. They rode in the dust. They lined up in formation before his car when he came to the next hotel on the route. No word of recognition spoke he. They noticed it. [7]

The night when Massachusetts executed Sacco and Vanzetti, the chief ranger thought it wise to put a guard about the hotel grounds there in Yellowstone where the President slept. He had a bedroom not far above the foundation of the hotel in an exclusive suite. He changed his room that night. The Park authorities suggested it. [8] But forest rangers again quite outside their duty, patrolled the ground, and two men stood by his window through the long frosty night. He saw them there. They went with him sight-seeing the next morning. Their uniforms told him they were not soldiers. But never a word of greeting, never a smile of acknowledgment, never a kindly look he gave them. [9] What Gamaliel Bradford calls [10] Coolidge's "bitter self control" was upon him in those days. The great renunciation of August 3, had brought no solace to his heart.

When he left Yellowstone, heading a long parade of citizens, park

[5] In a letter to William Howard Taft, a few weeks later, the Superintendent of Hampton Institute who beguiled the President out to hear the singing of the colored students writes that the President sat glumly and conspicuously refusing to enjoy the singing and "looking as though his wife had made him come!" He was that way!

[6] Yellowstone myth born a week after Coolidge left, repeated at the time to author.

[7] Same Montana myth collected ten days later by author.

[8] Letter from Horace Albright, Superintendent of Yellowstone Park, in 1927.

[9] Story of the unhappy ranger who stood guard under Coolidge's window, to the author shortly after the episode.

[10] "The Quick and the Dead" by Gamaliel Bradford, p. 239.

officials, Montana dignitaries, led and heralded by the rangers on motor-cycles at the close of the two days' Yellowstone visit, the rangers lined up at the station platform and the President marched between the men who had been guiding him through the wonders of the Yellowstone. He marched stiff and straight with eyes to neither the right nor the left, shaking no hand, smiling no greeting, giving no sign. The Indian on his mother's side of the family tree going to torture could not have been more grim and unbending. Gamaliel Bradford,[11] a fellow Yankee who has known these hard spirits of the New England hills, writing of Calvin Coolidge, pictured the soul of the man as he turned his back deliberately upon his escort:

"That temperament was the inherited cumulative, aggravated temperament of New England in which the sense of duty is the overriding force and an uneasy conscience suggests that we are not in this world . . . to have a good time at all. . . . Always there is that New England face with its subtle implications, and the face seems peculiarly out of keeping with merrymaking or any sort of riot of set publicity."

Nor was he a man to be forced into the unexpected. His "bitter self control" was always well in hand. At a ceremony in Washington that autumn where there was a tree to plant, President Coolidge came properly attired in the formal garb of the Presidency. Perfunctorily he turned over a bit of earth and waited. The crowd also waited. The master of ceremonies waited. Finally: "Mr. President, won't you say a few words?"

The President looked down at the result of his spading: "That's a fine fishworm!" he said.[12] Washington could no more coax him out of his mood than the Yellowstone.

That year when the Coolidges returned to the White House, the gossips across the land began to buzz with the story that "Mrs. Coolidge would divorce the President as soon as he left office!"[13] It was a cruel, groundless story. The publicity that followed the Black Hills episode, of course annoyed the President. It must have cut him to the quick to realize how unjust he had been, which of course sharpened the edge of his devotion. Miss Mary Randolph, who was Mrs. Coolidge's White House secretary, may testify expertly. She writes:[14]

"Knowing as I did—and intimately—the tranquil tenor of their family life, the depth of understanding and affection that existed between them, this story hit me straight between the eyes. And if ever a tale was made out of whole cloth that one was. I was aghast at the stories which found their

[11] "The Quick and the Dead," by Gamaliel Bradford, p. 233.
[12] A story in the New York Times printed after his death.
[13] "Presidents and First Ladies," by Mary Randolph.
[14] "Presidents and First Ladies," by Mary Randolph, p. 71.

way to the drawing-rooms and dinner tables, and at last burst with a far-felt detonation."

Miss Randolph took the matter up with the close friends of the Coolidges. They didn't recognize the seriousness of the situation. Finally she went to the President with the story and all hands decided that in addition to his usual appearance with Mrs. Coolidge at church services and other public places, they would be seen walking together at "crowded hours through the most frequented thoroughfares, motoring through Washington's most popular parks." Then the story faded and was forgotten. Reference to it here seems to be necessary only because back of it possibly may have been the motive of an important decision in American history.

About that time stories began to appear in the newspapers concerning Mrs. Coolidge's health and someone in the White House—certainly not without the President's knowledge—gave out an embarrassing story which made women rage who read it in the newspapers of the nation. It was all most unpleasant.

At least, when large events were moving in the undertow of American financial history, by some subconscious coincidence these rumors about the Coolidges and the stories of her illness came after the market drop that hailed his announcement that he did not "choose to run." That they followed the unhappy display of one of his curiously inept moods in the Black Hills made a coincidence which was pure witchcraft. When it came to his attention, this also must have weighted his mind. Also the story of the Wall Street slump may have given him pain and pause,[15] yet he saw clearly how much he meant to the world of business. This knowledge must have brought some comfort. It may have satisfied some sadist sense that fed his dwarfish vanity.

[15] Dr. Claude M. Fuess, official biographer of Calvin Coolidge, writes that "one day President and Mrs. Coolidge were having guests, one of whom was endeavoring to pry out of the President what he meant when he said 'I do not choose to run in 1928.' Cal's answer was the same one that he used on many occasions—silence. But Mrs. Coolidge supplied the answer with: 'Poppa says there's a depression coming.'"

CHAPTER XXXIV

A Series of Plots, Visions and Illusions

THAT summer of 1927 when the President was in the Black Hills, stirring up American politics with his puzzling renunciation of a third presidential term, a small but significant gathering was meeting in early July in New York City. It came about in this way: That year a number of important developments in the United States and in other countries combined to create a demand for further liberalization of the Federal Reserve credit policy. The alarming business recession which occurred in the United States in the early part of that year and which Mellon and Coolidge checked by optimistic statements was part of an international slump. European nations had been losing gold to the United States. A number of these nations, notably Great Britain, were in danger of a collapse of their recently established gold standards unless the pressure on their gold reserves could be relieved. This condition intensified exchange difficulties. Also foreign purchases of American products were hampered, particularly agricultural commodities.

In an effort to resolve these difficulties a conference of four central international bank governors was held in New York early in July, 1927.[1] In attendance at the meetings were Governor Norman, of the Bank of England; Professor Charles Rist, deputy governor of the Bank of France; Dr. Hjalmar Schacht, president of the Reichsbank; and Governor Strong, of the Federal Reserve Bank of New York. Governor D. R. Crissinger, of the

[1] The authority for statements hereinbelow about this meeting comes from an American banker of the highest standards, who was in touch with the meeting while it progressed, and who talked at the time with one of the foreign conferees during their American visit, and who later discussed the matter with one of the other foreign conferees. This narrative also has been checked by others who were in a position to know definitely what happened at the Conference.

National Federal Reserve Board, also sat in at the meetings. But Strong of the Federal Reserve Bank was the dominant American at the conference.

Apparently, the conclusion of the conference was that the situation required a further easing of credit in the United States. Certainly, Strong held that belief. An effort was made to win over Dr. Miller, of the Federal Reserve Board, who was known to be stoutly opposed to such a policy. The party came to Washington. At the conclusion of a luncheon given in Washington while the President was in the Black Hills, on July 7, in honor of the visiting central bank governors, and attended by members of the Federal Reserve Board, Governor Norman followed Dr. Miller out of the luncheon room and pulled him into a side room, indicating that he wished to talk with him privately. Governor Strong and two of Governor Strong's junior officers, however, followed the two of them into the room, as did also Dr. Schacht and Professor Rist, who had consistently opposed Norman.

Governor Norman was in a jittery state and was obviously in no condition to weigh dispassionately the factors involved in a further easing of credit in the United States. He was plainly and of course properly apprehensive over the ability of Great Britain to maintain the gold standard without the aid of the United States. Governor Norman pleaded eloquently for the adoption of easier credit conditions in this country. Dr. Miller vigorously opposed the adoption of such a policy by the Federal Reserve system, pointing to the call money situation in this country and insisting that pumping additional Reserve credit into the money markets of the United States would give rise to a gigantic speculative boom, with dangerous consequences for America and for the rest of the world.

"Oh, hell, you've got that call money market on your mind and can't get it off," put in Governor Strong in reply to the argument of Dr. Miller, adding:

"Norman, what do you do when the call money rate takes a jump in England?"

"We pay no attention to it," replied Governor Norman.

"Can't we," persisted Governor Norman, "agree at least to one thing: that we have got to stand together to save the gold standard?"

"In my opinion," countered Dr. Miller, "the gold standard cannot be saved by any such device as that proposed here today."

And thus ended this clash between Governors Strong and Norman and Dr. Miller over Federal Reserve credit policy.[2] Incidentally, it may be safely said that Professor Rist and Dr. Schacht took little part in the discussions at this meeting. But their position was well known. France and

---

[2] The authority for this episode comes in a letter from one who was present whose name is withheld at his request.

Germany had entirely different and more or less opposite and conflicting situations from the British problems.

Rist, a distinguished economist, was gravely troubled [3] over the extent to which credit expansion had already gone in the world. He was troubled by the fact that central banks in various countries were using bank balances in other banks as their reserves instead of gold in their own vaults—which meant that the same gold could be used many times as the basis of bank expansion. The fact that the Bank of France was hampered by law, worried Rist. France could not receive gold in Paris at a fixed rate without assuming the risk of a later legal change in the rate, though it had a guarantee from the government against losses on the foreign exchange operations needed for stabilizing the franc. The result was that the Bank of France, in holding the value of the franc stable, operated chiefly in buying and selling foreign exchange, especially sterling, with some purchases of gold in foreign countries, which the law permitted. When sterling was weak and the franc was strong, this meant that the Bank of France had to accumulate vastly more sterling than it felt safe in holding. Rist wanted to stop the unsound credit expansion that was going on, and wanted England to stop financing speculation in the franc.

At the Washington conference both Rist and Schacht appeared anxious not to be placed in a position of opposing either Norman or Strong. Dr. Schacht, however, did make one statement that revealed a penetrating grasp of sound credit principles for central banks. Discussing the proposed reduction of the Federal Reserve rediscount rate, he said: "I am not interested in a low discount rate as such. What I am interested in is a true rate, and by that I mean a rate which correctly reflects the credit needs and requirements of the United States. While a low rediscount rate in the United States would make my problem much easier, it would be of no advantage to Germany for the United States to adopt a rediscount rate that did not truly reflect its credit needs and then be obliged to raise its rate later, perhaps to a higher level than would otherwise have been necessary."

Following this meeting the above-mentioned central bank governors and Governor Crissinger, political appointee of President Harding on the Reserve Board, returned to New York, spending the week-end together on a Long Island estate. On the 27th of July the open market investment committee of the Federal Reserve system, comprising the governors of the Federal Reserve Banks of New York, Boston, Chicago, Cleveland and Philadelphia, met and formally recommended the adoption of a more liberal credit policy by the Federal Reserve system. That recommendation

[3] See his published utterance and official reports.

was subsequently approved by all members of the Federal Reserve Board except Dr. Miller.[4]

The Chicago Federal Reserve Bank, better informed as to what was going on, opposed this inflationary expansion. But the Federal Reserve Board forced the Chicagoans to lower their rates after the conference of powers in New York City. Now this was in the Coolidge administration. He was responsible to the people. But as he told Raymond Clapper of the United Press [5] in an interview: " 'The only way to succeed is when there is a job to be done, to look around and find the best man to do it and then let him do it.' He drew reflectively at the white paper holder which held a long cigar with an inch of undisturbed ash at the end."

He had no cares in those days. The oracle he had issued that summer of 1927 in the Black Hills had the politicians guessing. The market was booming. If a group of international bankers cared to trust to Washington to settle the affairs of the world—all right. That was their business,

[4] The governors of country Federal Reserve banks were not fully informed as to what happened at the international bankers' conference. Nor were all of the regional governors given the real reason for the new credit policy. In particular, Governor Bailey, of the Kansas City Federal Reserve Bank, which was the first to lower its discount rate in 1927, was told that the new policy was instituted to help the farmers. So while President Coolidge was still in the Black Hills stirring up the political pot, Bailey went back to Kansas City and gave out a public interview that stirred the financial world. It was to this effect that at the time he lowered his discount rate, it was done to help the farmers, as it was—but only incidentally! Bailey was honest; a farm boy, once Governor of Kansas, a successful country banker. He led out in all sincerity to help the farmers.

In the late autumn of 1927, when the President had returned from his vacation, the governors of the twelve federal reserve banks met in conference for three days in Washington. He kept away from them with Coolidgean caution. At the conclusion of their conference, they dined together and a few informal speeches were made. Among these was one by Governor Bailey of Kansas City. Hardly had he begun to speak when one of his colleagues shouted:

"Tell us Governor, why your Federal Reserve Bank led off in the reduction of the discount rate?"

"I don't mind telling you," said the Governor, "I did it because Ben Strong wanted me to."

At this remark there was much laughter. When Governor Bailey resumed, he said with some show of feeling:

"I'm not afraid to follow the lead in financial matters, of such men as Ben Strong and Andrew Mellon."

Mr. Strong, Governor of the Federal Reserve Bank of New York, was present at the dinner. He was the author and instigator of the policy of forcing easy money conditions in the United States by open market operations of the Federal Reserve banks and low discount rates, in order to bolster the European monetary situation. When Governor Bailey made this frank avowal Governor Strong all but went under the table.*

(*Letter to the author from a diner; verified by letter from Mrs. W. J. Bailey, widow of Governor Bailey; also verified by letter from member of Federal Reserve Board present at the dinner.)

[5] Interview reprinted in the Boston Post, January 6, 1933. The interview was given two years earlier and the ex-President was recalling his White House days.

not his! When Will Rogers asked the President how he kept fit in a job that had broken Wilson and Harding, Calvin Coolidge replied: "By avoiding the big problems!" Here was a momentous decision, an international policy. It affected America as no other single decision made in America since Wilson had taken his country into war. Yet the President apparently knew nothing of it. Certainly he did nothing about it. He found the "right man," "avoided the big problems," and so became an attitude rather than an executive.

2

When the President returned to Washington from the Black Hills, he found a new issue waiting in the anteroom of history, demanding his attention. It was the treaty which afterwards became known as the Briand-Kellogg Peace Pact; Pact being used as a short necessary word for headline purposes. This treaty had been in gestation and birth for nearly a year. It was conceived and first brought to Coolidge's attention in 1924, when he was running for President, by Raymond Robins, a distinguished civic leader who was afire with the desire, specifically and in terms, to outlaw war. Two peace parties had been raging in the United States since the defeat in the Senate of the Covenant of the League of Nations. Raymond Robins and Salmon Levinson, a Chicago lawyer, Indiana born and educated at Yale, headed one party which maintained that war should be legislated out of international law. About the same time, James T. Shotwell, a Columbia professor, Canadian born and educated, director of the division of economics and history of the Carnegie Endowment, who was an adviser of President Wilson in Paris, led another group working upon a theory that the United States should take world leadership in the peace movement, which should set up the machinery of peace as far as the intelligence of the world would let men go in international matters. Early in the Coolidge administration both peace parties began to work together to support the World Court on the promise that a world conference would be called within two years to outlaw war. President Coolidge announced his support of the World Court, but not the outlawry of war. So peace went seething over America in two camps.[6] Senator Borah, the most militant American pacifist of distinction in his day, was inclined to fight with the Levinson-Robins group which was self-financed. Shotwell was financed somewhat by the Carnegie Endowment. Then from out of the clear sky, in mid-spring of 1927, Premier Briand, of France, made a proposal in a speech for a treaty between France and the United States, which would outlaw war one with the other—obviously a diplomatic trial balloon. Three

[6] The story herewith set down is condensed from "The American Diplomatic Game," by Drew Pearson and Constantine Brown, pp. 10–19.

weeks later Salmon Levinson started to Paris, and almost at the same time Professor James T. Shotwell left Europe for America, each burning with a desire to put the power of his organization behind the Briand proposal. Armed with a letter from Borah, of the Senate foreign relations committee, Levinson went to Briand. Shotwell appealed to the New York newspapers and to Nicholas Murray Butler. Butler wrote a letter to the New York Times, sounding a clarion note. Levinson, with Borah's letter to Ambassador Herrick, laid siege to ,Briand through his secretary, and cabled to Borah, sending Borah to Cleveland in late April making an eloquent plea for outlawing war, mentioning the Briand proposal and demanding that it be broadened, and instead of being restricted to France and the United States, be open to the wide world. Borah undoubtedly originated the wider scheme of multilateral treaties outlawing war. After that, agitation began. America became conscious of the Briand offer. The newspapers discussed it. France was not allowed to forget it. Raymond Robins tried futilely to persuade Coolidge to lead out in some sort of extra diplomatic reply to Briand. Robins "found resentment against the public appeal" in Coolidge's heart—too much Borah probably! Early in June of 1927, the American end of the Paris peace conspiracy persuaded M. Briand to send a formal diplomatic note to the American state department advocating a treaty between France and the United States, outlawing war. Secretary of State Kellogg discussed the matter with the President. Secretary Kellogg [7] and the President felt that such a treaty might be misunderstood in Europe for an alliance, or an entente with France. And when the President returned to Washington in the autumn of 1927, he and the Secretary of State talked the matter over. The Secretary did most of the talking. The President listened faithfully and at the end of the parley said:

"We can do that, can't we?"

The Secretary answered: "Yes, I think we can without any doubt."

The President stared at the Secretary a moment, and the interview was ended. Coolidge was committed. Secretary Kellogg went forward with the negotiations. The idea had become a world idea in 1927. Six months after the Briand note of June, Secretary Kellogg replying to the Briand suggestion, December 28, 1927, proposed a multilateral treaty to be signed by the principal powers and open to all the world—the first official suggestion of such a treaty.

But the particular tactical maneuver which unquestionably made the treaty possible, was the suggestion of Secretary Kellogg that the whole negotiations be conducted in the open.

"The President approved of this," writes Kellogg,[8] "and I know it was

[7] Letter to author, April 24, 1933.
[8] Letter to the author, April 24, 1933.

his opinion as well as mine that this was the only way ever to obtain the consent of the powers to such a treaty. Time demonstrated that we were right."

From that day in January until the treaty was ratified in Paris, in August, 1928, it was an American treaty. The Kellogg tactics, the Kellogg energy, and America's prestige framed it, took world leadership for whatever it was worth, convinced mankind that the outlawry of war at least was a realizable ideal. Perhaps as a subconscious recompense for the materialism of the hour humanity was considering peace, aspiring to a day when wars should be no more—the more ancient chimera of man!

3

During the mid-years of his elective term, President Coolidge tried with great earnestness to make the United States an advocate of peace on this hemisphere. America contributed definitely to the settlement of the Tacna-Arica dispute between Chile and Peru, and Ambassador Culbertson, of Chile, whom President Coolidge had promoted from the United States Tariff Commission into the diplomatic service (for cause!), was instrumental in achieving what other diplomats had tried to achieve for a generation. The President sent a special message to the Senate, outlining the unfortunate situation in Nicaragua; another message concerning the settlement of a boundary dispute between Peru, Colombia and Brazil; and other boundary disputes between Guatemala and Honduras, between Nicaragua and Colombia. His selection of Dwight Morrow as Ambassador to Mexico was one of those touches of genius which revealed Coolidge at his high point. Morrow's Mexican work, which essentially was a Coolidgean triumph, was done, nevertheless, with that free hand which Coolidge gave to others in important tasks.[9] In giving to Morrow a free hand, Coolidge displayed his wisdom. Probably the episode in American foreign relations in which Coolidge took the greatest interest was the sixth Pan-American Conference held in Havana during December, 1927, and January, 1928. The President went to Cuba and made the opening address. He sent a strong delegation headed by Charles Evans Hughes. Then he went off and let it alone—avoiding "the big problems!"

At the end of the sixth Pan-American Conference, a multilateral treaty for arbitration of all juridical questions was negotiated between all the sovereign nations of the western hemisphere, and another multilateral treaty providing for conciliation. In these treaties, America went further than she had ever gone. It was a Coolidgean triumph in that he had nothing to do with it save to choose his men and give them the reins. No brain-

[9] And alas, save as a memory of good will in Mexico, Morrow's work was all undone in ten years!

sweat of his, no heartache in his bosom went into these treaties. He had "found a good man to do the job and let him do it." Yet possibly in the run of the luck of the game of progress Coolidge may be known in some other day as the father of those treaties which if they are followed may hold the western hemisphere in the ways of peace.

4

As the year 1927 drew to its close and as Hoover's strength as a Presidential candidate began to gather, President Coolidge seems to have become conscious that the tide of his fate had turned. One gets from the Taft correspondence [10] something like a mirrored view of the President's attitude. For Taft continued to be that winter a White House familiar. He came on many errands. Oftenest he came on judicial errands following his vision for a non-partisan judiciary. Taft's letters after the renunciation of August and through the autumn until early 1928, reveal a slowly changing conviction among the President's political friends, notably Taft, Hilles, of New York, and Butler, of Massachusetts, that the President would only serve as a Presidential candidate in 1928 if he was drafted by the convention. In the autumn of '27, the Chief Justice sighs in his letters to his family that he and his political friends could get no positive word from Coolidge about his willingness to be drafted. As the winter of 1928 passed into spring, the Chief Justice's letters disclosed a fading hope of the conservatives and the regular Republicans that Coolidge would give some sign of his desire for the nomination. The plight of the Republican politicians grew tragic as it became more and more evident that they would have to take Hoover as their Presidential candidate. In the mid-winter of '27, the Chief Justice who was still fondling a dying hope that some miracle would produce a Coolidge tidal wave, wrote to C. S. Jobes, a Kansas City Republican stalwart, lamenting the fact that "Hoover does not speak the language of the politicians or of the delegates." Judge Taft reveals the political attitude of the cult in a letter to J. M. Ulman, of New Haven, dated a few days before the Jobes letter. To Mr. Ulman, Taft explains "Charley Hilles's position in respect to the Presidency," which was that he had always been for Coolidge, "but," gossips the Chief Justice, "in order to make somebody for them to vote for in the New York delegation," Hilles tried to tie up the delegation with instructions for Hughes. But Hughes as a stalking horse "would not stand hitched." "Alas for Hilles," writes the Chief Justice, who always could smile at a funny situation, "he is in a bad way!"

As for the men opposing Hoover, in the Republican convention, they were for the most part a shoddy lot. The Chief Justice peeping from be-

[10] In the Library of Congress.

hind the scales of blind justice writes to his son, Robert, that none of them could defeat [Governor Al] Smith in the election, and adds, "most of them are jokes." Later in a note to his son, Chief Justice Taft remarks that Hilles is pressing Coolidge but is getting no encouragement. The letter continues: "I don't quite understand Hilles's attitude." It was the typical attitude of regular high caste Republicans. Coolidge was their kind. He was the god of the market, their market. They were the prophets of prosperity. Hoover, they knew, had grumbled at the market. His challenge of Mellon's wisdom more or less publicly was Republican sacrilege. Hoover's convention opponents might be "jokes" but to the Republican bosses Hoover was a Judas to the type of mind and cast of thought that turned on Young Ted Roosevelt, as a gubernatorial candidate in New York four years before, because he did not stand by Harry Sinclair in time of trouble! [11]

In the spring of 1928, it was obvious that the political-minded magnates of Wall Street were uneasy about the nomination of Herbert Hoover. The Taft correspondence reveals several of those letters. From this correspondence it is obvious that both Hilles, the National Republican committeeman from New York, and Andrew Mellon, "deep in their hearts hoped to draft Coolidge just as do others in the financial world . . . and if he refuses at the last moment they want Dwight Morrow." Apparently the Wall Street group, but not Mellon or Hilles, were trying to create prejudice against Hoover by spreading the story in the West that Hoover was responsible for the veto of the McNary-Haugen Bill. The unfairness of this attitude roused the Chief Justice. Twice in his correspondence he glorified the veto by calling it a "sockdolager." After that he veered strongly to Hoover as a matter of sheer decency. In a letter to his son, Robert,[12] he protested against the wicked union of this Wall Street crowd with the "wild professional crowd of progressives" in trying to defeat Hoover. And in a letter to his daughter at Bryn Mawr, Taft is puzzled that "the New York capitalists and 'Wall Street' are not Hoover men." He adds: "My friend Hilles is determined not to go for Hoover if he can help it."

Conservative that he is, the Chief Justice, in this letter, writes that "if the Republican party is to win in this election, it has got to put a candidate who does not smell in any way of the old Republican organization which . . . in 1920 smeared the whole party." In another letter to his son Robert, he declares that this political financial group called "the organization" is trying to produce a deadlock and then demand the nomination of Coolidge. Finally in mid-spring, Taft was definitely persuaded—and no one was closer to the President than he—that Coolidge would run. But alas he

<hr>

[11] Revealed in the Taft correspondence.
[12] June 3, 1928.

had no word to substantiate his belief, and the statesmen from Wall Street and New England and the little Wall streets of all big towns kept drifting into Washington, trying to interest the Chief Justice in the conspiracy to balk Hoover's nomination by creating the deadlock and drafting the President. It was a time of petty futile intrigue and it does not reveal the participants in the backstairs strategy in that high nobility of purpose for which the wealth and brains of the Republican party were supposed to be famed. It was a day of decadence for wealth and brains in America.

As Hoover's strength began to consolidate in the spring primaries, it became evident to the Chief Justice that it would be impossible to renominate Coolidge even by a coup in the Republican convention, that such a course would challenge Coolidge's sincerity. His chief asset would be a liability. Those were sad days for the Sanhedrin. But for all that, probably the President did not relinquish a secret hope! Also probably he did not nourish his dying hope. It was in that season when Senator Spencer, of Missouri, joined him in a walk and as they approached the White House grounds, by way of persiflage, the Missouri Senator asked:

"I wonder who lives there?"

"Nobody," quacked the President, "they just come and go!" [13]

In the meantime, into the vortex of this Babylonian turmoil were drawn the interests and activities of men from the four corners of the earth. The pull of that current caught trappers in the Arctic circle and sent them hurrying to the great fur markets of the world. Moscow was crowded; Peking felt the throb of the Babylonian pulse. The great hotels there were thronged with American buyers who came scuttling across the world at top speed, bearing furs, silks and precious stones to heave in the swirling current. From the Andes came traders with strange wares, from Japan, silks and furs were sprinkled on the turbid waters of the maelstrom. Paris, Berlin, London, taking their profits, brightened up, and the night life of Europe glowed in reflected gaiety. From the Kimberley diamond mines out of the degradation of the blacks and of the bestial cruelty of the whites came jewels to bedizen the hands and shoulders of the women around the table where Calvin Coolidge sat daintily, competently, silently munching his food like a tomcat sphinx. Out of the deep workings of the gold veins on the Rand in a delirium of triumphant money lust came gold, well refined of the filth, fumes and dust that caked the blood in men's veins. That gold, swirling in the vortex of lower Broadway, released its hidden poison that drove men mad and slew its second quarry. From every corner of the earth where the unfettered industrial system was grinding out the raw materials for wealth, crushing men's bones, parching their blood, following them in a perpetual orgy of chicane and de-

[13] Story in the New York Times.

bauchery, came the onrushing flood of pennies pilfered from the poor, from the ignorant and from the savages, from indentured slaves who treaded the death mill to beat a rhythm to the saturnalia of America's unbridled profits. The Puritan in the White House saying Garman's litany gave his pious blessing to the devil's dance.

Gradually the personality of Calvin Coolidge was becoming the ectoplasmic substance of this vision of the American ideal of that day. Coolidge was business, big business, little business, the magnified horsetrade which is American commerce. The Chief Justice coming from a palaver with the President, wrote to his son, Robert, that he felt that Coolidge was "giving more body to the Republican party than it had had for a long time, the representative of prosperity." He was the substance of things hoped for in the hearts of them who held the old faith. Back of Coolidge in this portrait of privilege and plenty stood Andrew Mellon, always in the shadow keeping watch upon his own! Whatever the international understanding between the Bank of England and the American Reserve system was, to make expanded bank credit take the place of actual investors savings, or to postpone with cheap and abundant credit, the threatened British economic collapse or readjustment, the President seems to have been ignorant of the significance of the plan. For in truth Andrew Mellon was only a shadowy figure in the picture. Evidently he felt that the plot or plan or compact would finance the flow of goods to America over tariff walls—a one-sided flow of goods. The inevitable boom and the inexorable collapse—effect following cause—he did not foresee. Actually he seems to have written no official document or order that fundamentally affected the situation. Apparently he did not influence nor try to influence the Federal Reserve Board in those two years before the final up-rush to collapse. If he bolstered the New York Federal Reserve Governor and his official family in their expansive activities, no record has been published of Mellon's interference. Probably what Mellon did was to wear his aura: The outer circle shimmering "the richest man in the world," the inner lining gleaming, "the greatest Secretary of the Treasury since Hamilton." Under his halo he smiled a wan, detached, other-worldly fishy-eyed benediction on the raging bulls in the pit —and issued benefices in cheering interviews when the bulls needed stimulation. It was for him and his kind—and he typified the American mind, the very heart of our democracy in his day—that the founding fathers changed the phrase "life, liberty and the pursuit of happiness" in the Declaration of Independence to "life, liberty and property" in interpreting the Constitution. So in the days before the deluge, Calvin Coolidge and Andrew Mellon, two innocents smiling above the rosy clouds upon a happy world, "grew in beauty side by side"!

CHAPTER XXXV

"The Queen in the Parlor Eating Bread and Honey"

BUT even the Chief Justice, would-be keeper of the President's judicial conscience, with all his amiable optimism, was beginning to feel uneasy. In a Christmas letter in the middle twenties he wrote to his son, Robert, that he was afraid that we would be led into folly, counting on the continuance of good times. But later in the same letter he cheered up and expressed reassurance at having "Calvin at the head of the state with his frugal and careful economies derived from Vermont and Massachusetts."

In Coolidge's elective term, another inscrutable phenomenon began to appear in the business world. When a business man engaged in any commercial activity desired to borrow money, he began to find it easier to sell stock in his enterprise than to go to the note window of the bank. That was a new thing in the American business world, at least new in the American business world of the twentieth century. "Business," writes Edward Eyre Hunt, in his chapter on Credits and Money Movements in "An Audit of America" drawn from the "Recent Economic Changes in the United States," "has been financed less by borrowing from banks and more by borrowing in the capital market through security issues. Industrial requirements for bank loans also have been lessened by the prevailing custom of reducing inventories." This development, the substitution of securities issues for borrowings at commercial banks, really was a natural response to the immense flood of cheap money created by Federal Reserve policy and inflowing gold.[1]

[1] The year 1923 saw little if any bank expansion and the volume of new securities issues publicly placed was 4 billions, 300 millions. In 1927 bank credit expanded something over $3,000,000,000 and the volume of new securities issues was 7 billions, 791 millions. It is interesting to contrast the one year during which there was no bank expansion and the years of bank expansion with respect to the growth of commercial

Many seemingly unrelated trends appeared in the developments of this mad decade, but when the clue of cheap money and excessive bank expansion is followed, most of them fall readily into order.

We held vast liquid balances from other countries.[2] Commodity prices were not advancing. Wages were advancing only slightly. The farmer's case did not improve in the middle of Coolidge's elective term. But the agitation for the passage of the McNary-Haugen measure and the veto stirred politics deeply and agitated business. All the hullabaloo left the imperturbable President unruffled. In the midst of this hubbub, when the seeds of death and calamity were being sowed in the economic system, the President moved about his squirrel cage routine with no sign that he heeded the warning which economists and the minor prophets of politics were sowing in his path. A former Secretary of Commerce,[3] calling at the White House to get at the truth about Coolidge's political cryptogram, learned with amusement from an old friend of other days in the President's retinue that he had taken an hour or so off the day before, when things were growing too hot for him, to disappear into the cellar. A barrel of apples had come from a neighbor in Vermont. He was counting the apples. He was always, even in the midst of alarms, the unruffled, prim, parsimonious Yankee of the Green Mountains. He never shed his smock!

The sky was filled with ominous signs. While he was counting apples, the charts of all the government agencies and of all the economic researchers indicated that competition for some reason was not keen for labor and goods. Yet everywhere bank deposits were swelling. Time deposits grew miraculously. These bank deposits were not investors' savings. They were in fact, the product of bank expansion. And as the stream of deposits swelled, it swirled eastward into the banks of the seaboard metropolitan area from Boston to Baltimore. Yet curiously the commercial loans of American banks grew comparatively slowly, but the increased security holdings and collateral loans on bonds jumped to unbelievable heights. So the financial picture as well as the industrial picture began to change,[4] in

loans in banks. From March 1922 to March 1923 bank credit expanded rapidly under an easy Federal Reserve Bank policy. Stock and bond collateral loans in banks increased 13.0 per cent and bank holdings of bonds increased 27.6 per cent. Commercial loans increased 3.9 per cent, but commercial loans relative to the rest of bank loans and investments declined from 50.6 per cent to 46.9 per cent. In the following year, March 1923 to March 1924, the Federal Reserve banks were holding back. Collateral loans in banks did not increase. Bank investments in bonds declined approximately 5 per cent, but commercial loans increased around 3 per cent and rose to 48.4 per cent of commercial bank earning assets.

[2] These liquid balances from other countries are likewise to be explained as the product in large measure of our own bank expansion.

[3] Conversation with and letter to the author, January, 1937.

[4] See chapter III of "Modern Corporation and Private Property," by A. A. Berle, and "Liquid Claims and National Wealth," by the same author.

those middle Coolidge years, and the volume of trading in the New York stock exchange began to swell like an unhealthy tumor on the body of American commerce. The volume more than doubled from Harding's day to Hoover's in March, 1929. The increase was fairly healthy down to the late spring of 1924. The immense purchase by the Federal Reserve banks of government securities made the great cheap money episode of that year, followed by Mr. Coolidge's campaign speech in the autumn in which he said: "It has been the policy of this administration to reduce discount rates." Then the tumor began to bloat. Stocks commanded ridiculous prices when their earnings were considered. Toward the close of Coolidge's term the insatiable demand of the stock market was reflected in a gradually rising interest rate on all loans. At first the bankers took care of the brokers' loans, but in the last Coolidge days someone, perhaps the banker's alter ego, the investment banker's sales agent, began reaching into funds of investors and surpluses of business men and corporations.

No influence around the President cautioned him to veer away from the philosophy of a God-anointed plutocracy. Neither Mellon, its high priest, nor the Chief Justice who often ambled into the White House offered the President counsel against placing his trust in the priesthood of mammon. The type of counsel which the President was getting in the day of the bull market may be inferred from a letter which the Chief Justice wrote to his associate upon the Supreme Court, George Sutherland, in which the Chief Justice bewailed the fact that the Democrats and radicals were trying to avoid a reduction of taxes for those in the large brackets. He saw in that endeavor only an effort to "get even with the wealthy and the prosperous," and decried it to Justice Sutherland as he surely must have bewailed to the President as the tax program of the Progressives and declared that "when the whole business community is against the Democratic party, it cannot win." As the stock market began to swell, three months later, in a letter to Sidney A. Perkins, once secretary to Mark Hanna, the Chief Justice insisted that it was not "a tremendous boom" and contended that while business was so good, America was lucky to have a President who was "not disposed to interfere in any way with the even flow of events," who was indeed willing to let "the people work out for themselves the prosperity they deserved." The atmosphere created by men who held such views became a part of the Coolidge aura, an active force in American economic life. It was the soul that vitalized the bull market without any need of voice or deed.

Never before in our history were vast sums available from corporation lenders, individual investors, and others for loans to the stock market. In other days and times of boom men like Russell Sage, shrewd, hard-headed, scurried out of the market when it got too high. They lent when rates went dangerously high, but turned their call loans into purchases when stocks went low. Useful people were these hard-headed, shrewd men who could come into a crisis with cash and courage, and who, in making profits—sometimes unconscionable profits—none the less steadied the market and prevented disorganization. But in 1928–29 we had the extraordinary phenomenon of vast sums coming from non-bank lenders in the midst of a boom. Where did the money come from? The answer again—and there is where the President's advisers were confused—is to be found in the vast bank expansion which had preceded it. Business corporations, able to place stocks at high prices, financed themselves beyond their needs, held the excess funds in cash, and lent that cash at a high rate back to the stock market. Foreigners placed their bonds here in excess of their current needs, held part of the proceeds in cash and lent them to the stock market. Uncle Sam was a sort of cosmic Santa Claus. Investors selling old securities at fantastic prices to the stock market, unwilling to reinvest at prevailing prices, held cash and when the rates went to 10 per cent lent it to the stock market.[5]

This Santa Claus reputation of Uncle Sam gave the President one of the gorgeous moments of his life. The Roumanians needed cash. They sent their lovely Queen as a pilgrim to our cosmic Santa Claus. So from the consideration of drab problems, like the stock market, Vermont apples and Presidential politics, the President turned one evening to the dinner and reception given to Queen Marie. For days the White House was busy with preparations. The President was in the midst of it. Indeed he was in every domestic detail of the White House. A man who counts the hams before a dinner and goes up to look at his wife's dress on the bed before she dons it for a reception, and who occasionally ventures to buy her a dress, and most daring of all, a hat, is not going to let the fun of preparing a dinner for a reigning Queen go by. And what if the market is hurdling over the high places, and what if the financiers of Europe are huddling in secret conferences in Washington and New York, and it is all a puzzle and a mess to the President, the Queen's party—and he loved a party—he could understand. The Associated Press reports told of Queen Marie, of Roumania, coming to the White House that night "blazing with the crown jewels of her royal Russian mother," and she sat at the right hand of the little boy who came down the hill from Plymouth to Ludlow with the bull calf—the first reigning Queen ever entertained at a White House state banquet—she whom the reporters called "the most beautiful Queen in Europe." There, with the President's official family, the heads of foreign missions in Washington, and their wives, sat Prince Nicholas, Princess Ileana and other nobles of the Royal party. And how the quiet, saturnine, lean-faced,

[5] All this was explained at the time in the *Chase Economic Bulletin* called "Brokers' Loans and Bank Credit," published in the autumn of 1928.

little Yankee enjoyed it, glancing sideways at the Queen's "diamond tiara with great pear-shaped pearl drops." [6] For the first time in his life he got a square look at three strands of famous pearls and pearl earrings large as grapes.[7] That he could understand. There was something unlike the market, unlike Mr. Mellon's pronouncements which were bothering him. There was "the cash equivalent," stark and real. He marched with the party in stately tread down the long arched corridor to the east room where the reception was held. A funny thing happened there. The President held his place in line, took program and made no error. Mrs. Coolidge had been instructed by her social mentors to stay on her sofa with the Queen because on her sofa no one could come to the Queen uninvited and Mrs. Coolidge would be mistress of the prestige which was the dominant motive of the grand show. In an ardent, unguarded moment, Mrs. Coolidge left her sofa and quick as a flash the Queen beckoned to Alice Roosevelt and the Queen kept the sofa in her hand during the evening. No one could approach the Queen and most of the crowd who went to meet her, went away unhappy.[8] And then, after it was all over, they sat in the bedroom like good Vermonters talking it out, proud of themselves. The Black Hills episode had faded. Mrs. Coolidge's bedspread contained more than two dozen squares. The Coolidges were happy.

As a matter of fact, the White House under the Coolidges after 1924 was as gay as the times required. Mrs. Coolidge was particularly gracious. Her generation knew no lovelier First Lady. She dressed for it well, but not extravagantly. She took counsel from those who knew and walked the mazes of the social labyrinth without error, save now and then as when she left the sofa. But her errors were errors of a generous heart and they were so few they were not noticed and so kindly they were forgiven. Their table was not extravagant, but ample and decently spread. The President talked when he pleased and was glum when he pleased to be and she tried to make up for his deficiencies. He knew it. Rarely in the White House in the memory of living man had there been so romantic and devoted a husband despite his many curiously obtuse and blundering ways.

It was not the puzzle in the market which made him absent-minded. It was not the booming tide of a miraged prosperity that diverted his attention though that, if he could have envisioned its consequences, would have made him sad. He was what he was, and the times gorgeous and splendid as they were did not affect him. He was amazingly insensitive upon some matters and unhappily sensitive on others. To the roar and clamor of the

[6] Associated Press report.
[7] Associated Press report.
[8] "Presidents and First Ladies" by Mary Randolph, p. 58.

market and all it meant and all it might be, he was deaf. He was an economic fatalist. He had a God-given inertia. In a time when all other men in his place and situation were doing the wrong thing he did nothing. He knew nothing and refused to learn. Congressman James Strong, of Kansas, brought Norman Lombard, a Wall Street economist who believed in managed currency, to the White House one day to talk over the matter of banking credit which Congressman Strong's bill was aimed to accomplish. The President, after greeting his guests, treating himself to a cigar but not offering one to his guests, leaned back in his chair, put his feet in the second drawer of his table, and listened with close attention. After Lombard had spoken a moment, the President clipped in: "Who's behind you?" His guest indicated such men as Owen Young, Silas Strawn, Otto Kahn, Pierre DuPont, John Raskob. The President listened to the list, made no comment and after Lombard had talked fifteen minutes, getting on the subject of gold, the President remarked: "Oh, man found money—God made it!" apparently grounded in the belief that gold was the only money. Mr. Lombard tried to develop his thesis. Whereupon the President leaned back with a sigh: "All right, go ahead. See what you can do about it," and then turning to his secretary, "Who's next?" And in a few moments that particular attempt to educate the President upon the fundamental value of money passed into history.[9] Probably he felt that it was the task of Mr. Young, Mr. Strawn, Mr. Kahn, Mr. DuPont and John Raskob to educate the public and the public would in turn prod its Congressmen to pass the necessary laws and finally it would become the task of some President in another day to execute the necessary laws, provided the Supreme Court was willing. It was a long way off. The expansion of bank credit which was spinning its fine web over the land was merely an academic question.[10] Calvin Coolidge had no taste for academic questions. His mind was geared to grapple concrete things. Politics to him presented solid problems. The oracular message he proclaimed in the Black Hills that summer when the filmy threads of bank credit were enmeshing his country, was presenting a problem that the President could handle. He saw that the political uncertainty he had created made him an interesting figure—indeed gave him more power than if he were seeking power. For he was not an outgoing President—precisely. Nor was he seeking to hold his place. Politicians feared to ignore him, yet hesitated to court him. Politically he was out of focus, befogged—a mystery, which was the pose he loved.

[9] Letter from Norman Lombard, July, 1936.
[10] Again in all these statements of fact and conclusion the author is indebted to "Recent Economic Changes In the United States" and the "Audit" of it by Dr. Hunt. See also economic bulletins of the Chase National Bank (N.Y.O. for this period).

2

Now as he stands in the romantic gloaming of political mystery let us survey from a business viewpoint the latter days of Coolidge. It is fair to ask what was the net of it. What did the American people really get from 1927 to 1929 out of this turmoil of trading that rose above the hum of industry and the quiet business of farming? Apparently in 1928, the realized income of the people of the United States reached about eighty-nine billion dollars including many billions of capital gains. Equally divided it would have furnished $745.00 a year for each man, woman and child. Roughly speaking, hired workers received fifty-seven per cent of this realized income. And about two-thirds of this realized income went to wage earners as opposed to salaried men. The chief gain in income during Coolidge's days went to salaried employees which of course along with others includes government employees and banks. Rents were receiving a smaller portion of the national income in Coolidge's day than in the decade before. The proportionate total of dividends in the realized income suffered a slight decline. And roughly speaking again, all income paid to all kinds of labor, industrial, farming, white collared, totaled nearly sixty-five per cent of the realized income. The rest of the eighty-nine billions came from property.

As for the farmer's income, the allowance for declining real estate values showed that he had a deficit of nearly two billion dollars for his work in 1927, a bookkeeping deficit. Actually, even though he was running in debt, the rural standard of living generally, and particularly outside the cotton belt, was steadily rising. But even at that, at the end of the Coolidge administration, so much of American business was done by corporations or by the constantly expanding business activities of government, that corporation government, under Calvin Coolidge, disbursed, including wages of course, nearly half of the national income. A strange phenomenon! A study of these tables, again by Dr. Hunt,[11] forces one to "the conclusion that we Americans prefer to raise the economic level of the average life rather than to maximize national wealth." For Americans spent hundreds of millions in those Coolidge days for good roads, hundreds of millions for education, for radios, for automobiles, for telephones. Their use of power, including automobiles grew nearly four times faster than the growth of population.

And doing this they poured billions into foreign loans. Tenant farming was increasing and the submerged groups in both industry and agriculture had begun to realize the pinch of want. No one starved but many unemployed and irregularly employed in the bituminous coal mines, in some of the commodity industries like textiles, were undernourished. Two mil-

[11] See also A. A. Berle's books.

lion people in 1928 lived upon a family income of less than five hundred dollars, and two million more upon a family income lower than one thousand dollars. That income did not furnish much better than a subsistence diet, meager and scarcely decent homes with not enough medical and dental care. Statistics [12] seem to justify the statement that in 1929 one third of the homes in small towns and nearly eighty per cent of those in villages, had no sanitation or running water and nearly half of the houses were heated by stoves. Alas! America, under the millennium which seemed to be almost here in the Coolidge days, had its seamy side.

Consider for a moment another phase of the economic situation—the protective tariff. Here again Calvin Coolidge was the spirit of the times— an old oaken, an ironbound, a moss covered embodiment of high protection. The barons who once were moated behind the Connecticut River were tariff barons—his friends, even their new anonymous heirs and assigns. Their interest was his religion. To them, to touch the protective tariff was to violate the Ark of the Covenant. New England, for a hundred years had been nurtured on the tariff. There, the tariff was both politics and business.

That is how it happened—with our tariff policy what it was—that if we had not lent billions to buy our goods the surplus of our goods which we produced in excess of our domestic needs, notably wheat, tobacco, lard and cotton, would have backed up on our domestic markets.

So we changed from a debtor to a creditor position in our international relations and we still tried to keep up an export surplus or "a favorable balance of trade." A debtor country when it has payments to make abroad is like an individual producer when he has payments to make. It cannot afford to consume all it produces. But a creditor nation, like an individual creditor capitalist, can afford to consume more than it produces with its own labor. We were afraid of foreign goods. The Coolidge economic theory was grounded in his New England high tariff policy. We jogged along in that rut, on lending money to the outside world to get our own goods out and to keep foreign goods out. And the same cheap money that got the goods out made our fantastic boom at home. In that American economic megalomania, the Coolidge Bull Market was conceived in iniquity and born in sin.

Calvin Coolidge only typified his country, the ruling class, the middle class, honest, earnest, hard-working, high-minded optimists.

Who shall blame them for their blindness? For these people, the millions, lived well in their gorgeous prosperity during the Coolidge boom. They thought they could abolish poverty by speeding up the wheels of industry, and in agriculture by plowing deep where sluggards sleep. It is

[12] "Rich Land, Poor Land" by Stuart Chase, Harper & Brothers.

possible to read now what the prophets said in that day. We may marvel how exactly their prophecies came true. But when has man ever listened to the warning voices of sages until it was too late? It is neither honest nor intelligent to hold Calvin Coolidge up to the contumely of the perspective of another decade that his wisdom may be questioned.

Yet probably his mannerisms, his shyness for one thing, kept the truth from him many times when it might have led him to act. Yet he winced from the truth even when he seemed to seek it. Sometimes in other matters far from the market, he acted as he did when exposed to advice about the market. Occasionally he would pump his guests dry and make them talk about their favorite subjects. At other times officials summoned to his presence came ready to talk but he left no opening through which they might begin to ride their hobbies. Once in the midst of a quest for economy, he sent for one of the highest placed officers in the war department who might know about the army budget, asked the General to spend the week-end in the country with the Coolidges. His departmental colleagues gathered about the officer and stuffed him full of facts, gave him a valise filled with statistics, and the necessary data to defend the claim of the war department for appropriations. He rode out with the Coolidges, contributed to the small talk of the journey, met them at dinner Saturday night with a casual guest or so, spent the evening in pleasant conversation utterly removed from the subject nearest his heart. In the morning after breakfast, the President said:

"Well, we'll be going to church and your train goes about that time and we'll take you down. It has been good to see you." [13] Just that.

Why did he invite this casual acquaintance for this weekend visit? What quirk of his mind made him avoid discussion of a subject to which this General might have contributed? Many puzzles of this introverted heart never will be solved.

This incident is set down here, not as mere anecdotal trivia but to show how he blinded himself. Scores of honest, wise and willing economists, financiers and business statesmen were near him always, and yet he was either too cautious or too indifferent to probe their minds and to profit by their wisdom. Probably no statesman would have dared to act. But we must know, in appraising his public life in the White House, that Calvin Coolidge was never without access to the truth.

[13] Story to the author by General Briant H. Wells, U.S.A. retired.

CHAPTER XXXVI

A Gaudy Pageant Moves Gaily to Its Doom

It is impossible to make these stories of business and politics synchronize. We must for a moment return from the political story of the spring of 1928, to examine the business picture of that time:

In mid-November, of 1927, after a notable decline in business when railroad incomes were off eleven per cent, the President seemed to feel that a few encouraging words to the stock market were needed. So he told the Union League of Philadelphia that the nation was entering into a new era, that "the test which now confronts the nation is prosperity. Nothing is more likely to reveal the soul of a people. History is littered with stories of nations destroyed by their own wealth." [1] The echo of Professor Ripley's words about the "prestidigitation, double-shuffling, honey-fugling, hornswoggling and skulduggery" may have been reverberating in his heart. The next day after this Union League statement was printed, the market fluttered. But on the following day stocks began to climb again with a notable rise in sugar, corn and wheat. Secretary Mellon, returning from Europe, again had given his pontifical blessing to the market and again the market purred in response. The *New York Times*, commenting in its financial section in the autumn of 1927 on the President's attitude, fairly glowed with approval of the President's optimistic comments, which it declared "had the effect of converting an aimless, colorless stock market into a lively buoyant affair." And then this: "The President's statement . . . dispelled much of the pessimism that recent reports of reduced industrial and commercial activity had created." Wall Street brokerage houses made the most of the President's remark by emphasizing in their afternoon circular the "confidence that exists in high places." Referring to Mr. Mellon's au-

[1] Associated Press report; also *New York Times*, November 17.

tumnal beatification of the market, the New York Times declared that
Washington has become the source of most of the "bullish news of late.
Wall Street interests with a hand in the stock market seem to feel justified
in predicting encouraging developments in that quarter."

From the summer of June, 1927, and for practically a year later, certainly in the autumn of 1927, it was evident from the market trends and
from the effect of the Presidential outgivings upon the market that two
widely varying views were held in high places about the stock market boom.
Apparently certain members of the Federal Reserve Board were circumspectly but definitely trying to deflate credit. But apparently as fast as these
members prevailed upon the Board to withdraw its credit support from the
market, outside money came rushing in. Floods from the country banks all
swirled through the regional banks into Wall Street, and as fast as the Federal Reserve policy drained out credit, what might be called bootleg money
filled the power caissons and kept the tape machines going. After the President's market bolstering statement of mid-November, 1927, and until early
February, 1928—and despite these soothsayings—the Federal Reserve people
were reducing their holdings of government securities.[2] But much good it
did! Money remained easy. Bank expansion went on. The stock market
rose and reserves in member banks in that day of repression actually increased twenty-six million dollars.[3]

In the meantime, Herbert Hoover, Secretary of Commerce, who had no
official duty in the situation, was backing up his friend, Roy Young, of the
Federal Reserve Board, in an effort to tighten the money market, to slow
down speculation. Few powerful actors in the ironic tragedy kept their
heads. Friends who were familiar with Governor Young, of the Federal
Reserve Board, remember that sometimes at the end of the day they would
come in and find him laughing; laughing and reading the long ticker strips
strewn about his desk that told of another day's inexorable march to catastrophe. To a friend standing by the ticker Young said:

"It's funny to be laughing, but look here!"

And then he would point to the news tape which showed a sharp rise in
stocks that day and he would break out with something like the ironic

[2] They cut them down by one hundred and fourteen million and also raised their buying
rate for ninety day acceptances from three and a quarter late in December, to three
and a half toward the end of January. Toward the end of the period a movement upward of rediscount rates from three and a half to four was visible in Chicago, January
25, 1928, in Richmond two days later, and a week later in New York. In that period
there was an appreciable stiffening of rates in the general money market from November 30, 1927, to early February, 1928.

[3] The figures hereinafter quoted and which shall follow for the next paragraph, are
taken from "An Analysis of the Money Market," in the Chase Economic Bulletin by
Benjamin M. Anderson, Jr., economist of the Chase National Bank, published in
June, 1928.

cackle of a major god on Olympus looking down on the mad caperings of men:

"What I am laughing at is that I am sitting here trying to keep a hundred and twenty million people from doing what they want to do!"

Roy Young was a fine figure of a man in those days. He came into the Federal Reserve Board after the mischief was done. But he certainly was one of the motivating forces of a restraining policy. The President of the United States, who had built his career upon a policy of delegating his authority, probably knew as little about the tragedy which Young, and men of his kind, saw in the stock markets of the great cities as though he had been ruler of another land. The bullish statements given out by Coolidge and Mellon in October and November, 1927, were forceful enough to give a spurt to the stock market which carried prices at a fairly even level across the New Year's goal. And on January 28, 1928, the New York Stock Exchange announced that brokers' loans reported to that organization had increased more than three hundred forty-one million during December to an all-time record of four billion four hundred thirty-two million, a billion dollars above the figure for a year earlier. The following day the Federal Reserve Board announced that brokers' loans reported by member banks in New York City alone had risen to more than three billion eight hundred million. The stock exchange was uneasy. Brokers' loans were troubling conservative bankers. Stock prices were breaking badly. The New York Times for January 5 declares that the marketing community "sensed before the close of trading that a very large expansion in borrowing would be disclosed. Some of the swelling that swept over the market in the last few minutes of the trading was ascribed to this premonitory impression. When the figures came out at 2:30 brokerage interests made no effort to conceal their concern."

Nearly four billions in brokers' loans was enough to concern anyone. January 6, 1928, the New York Times saw a "general and in some stocks a violent reaction," though a reduction in official call money later in the day "had the effect of arresting the pace of the decline in stocks." It was evident to any man who had two thin dimes to rub together and another to throw into the bucketshop, that the country was teetering upon the brink of calamity. Now calamity, business calamity, in his administration, is exactly the thing which a President does not want whose wagon is hitched to the star of business. So certainly without consulting the Federal Reserve Board at Washington, which in the main was playing another game, which would like to have seen a gradual liquidation in this Bedlam, President Coolidge did an unprecedented thing. He used the weight and prestige of the Presidency to uphold prices on a tottering stock market. In a public statement he expressed the opinion that the volume of brokers' loans, approaching four billion dollars, was not large enough to cause apprehension. In the *New York Times*, of January 7, we read:

"Although loans to brokers and dealers held by the New York Federal Reserve member banks have reached the unprecedented heights of $3,810,-023,000, President Coolidge does not see any reason for unfavorable comment. . . . The President, it was said at the White House today, believes that the increase represents a natural expansion of business in the securities market and sees nothing unfavorable in it."

Of the effect of the President's remarks, the *New York Times* for January 8 stated:

"Stocks were turned over in huge volume on the New York Stock Exchange yesterday, largely as the result of enthusiasm aroused by President Coolidge's statement that he saw no reason for alarm in the large expansion of brokers' loans. . . . The reassuring statement by the President did not come until after the close of the market on Friday, and the stock market community, after reflecting on it overnight, concluded that it invited heavy buying of shares."

The financial section of the *New York Times* for the same issue contained the following comment:

"Old-timers in Wall Street tried without much success yesterday to recall any precedent for Mr. Coolidge's remark on brokers' loans, quoted in the morning newspaper dispatches. None of them could remember an instance in which the country's Chief Executive had made a public declaration on a controversy of just that character. . . . The Chief Executive has traditionally avoided expressing opinion on subjects purely technical or which are surrounded with problems of speculative activities."

Nevertheless, Calvin Coolidge always played safe. Here is a contemporary story from his cousin, H. Parker Willis,[4] former editor-in-chief of the *New York Journal of Commerce*, which indicates the doubt in the President's heart about the state of the nation, despite his public assurances:

"After the business which we had had in hand had been disposed of, Mr. Coolidge invited me to come back to luncheon. I did so, and after luncheon sat down with him in his study for a while, and he asked me a number of questions about pending financial affairs. It so happened that two or three days before that he had given out at the White House a statement that brokers' loans were not at all too large. On the occasion of this visit to Washington I had been testifying before the Senate committee that I thought they were very much too large. President Coolidge, referring to what I had said, remarked that my opinion had seemed to show a great difference from his, but he added:

_____
[4] Letter to the author.

" 'If I were to give my own personal opinion about it, I should say that any loan made for gambling in stocks was an "excessive loan." '

"I replied: 'I wish very much, Mr. President, that you had been willing to say that instead of making the public statement you did.'

" 'Why did you say that?' Mr. Coolidge queried.

" 'Simply because I think it would have had a tremendous effect in repressing an unwholesome speculation, with which, I now see, you have no sympathy.'

"Mr. Coolidge thought this over for a moment or so and then he said: 'Well, I regard myself as the representative of the government and not as an individual. When technical matters come up I feel called on to refer them to the proper department of the government which has some information about them and then, unless there is some good reason, I use this information as a basis for whatever I have to say; but that does not prevent me from thinking what I please as an individual.' "

Coolidge's position in all this was anomalous. He was torn between his natural Yankee contempt for prodigals and wasters, for gamblers and for men who treated lightly the holy dollar on one hand and on the other hand his desire to accelerate business, to keep things humming, to whirl the wheels of industry.

An apocryphal story, rather more Coolidgean than this one just quoted, was current in those days and still is repeated in New England to illustrate Coolidge's sour wit. This story declares that he greeted another old friend from Wall Street who was writing vigorously against the extension of brokers' loans, with the question:

"So you are against my policy of brokers' loans?"

The editorial visitor began to stammer, being awed in the presence of the ruler of a hundred and twenty million people. Then Coolidge stopped him and kindly explained his own position. The conversation drifted off and finally as the editor, who had been protesting against the brokers' loans, rose to go, Coolidge at the door, returned to the subject and said:

"Well, now about those pieces you are writing denouncing brokers' loans—" He paused a minute and added as he extended his hand, "just keep on writing 'em." And turned to his desk.[5]

The episode like this apocryphal story was true but not factual. The version of Mr. Willis is factual. But how strange a figure President Coolidge cuts there in that place of great power basically uninformed about either the source or the direction of the great tides that were washing around him, the resistless undertow that was dragging his country and the world

[5] In this connection it may be well to recall Governor Coolidge's letter to State Treasurer Jackson, pp. 175-176. Story above told to author by M. A. DeWolfe Howe, June 1935.

out of the old times into the new. There he stood, blinking at the tidal forces he could not fathom, uncertain whether to follow his prudent New England instinct against the day's prodigality or to take his hand off the helm and let the ship ride the tide to its strange harbor. In reality he was not quite the hero whom men saw in the White House in that hectic day, but a puzzled, befuddled smalltown New Englander torn with inner conflict while he turned to the world a smooth, flinty, unrevealing face. Courage, honesty, and common sense, the foundations of his character, were not enough in his great time of trial. For wealth in that terrible time was not brains; "the business of America" fundamentally was not business! And here is probably the true story of how the President stumbled into the blundering statement that caused the scandal inside market circles.

Judson Welliver recalls a visit at the White House about this time.[6] The President was puzzled at the hornets' nest he had disturbed by his statement about brokers' loans. It occurred this way: At a press conference someone had queried him about brokers' loans. He held the question over, took it to one of his secretarial advisors and asked: "What about brokers' loans?" The young man said: "I don't know, but I'll get you up an answer." The President nodded in acquiescence and went on. At the next press conference he used the amazing answer prepared for him, possibly in consultation with the Secretary of the Treasury. Mr. Welliver, probably at the President's suggestion, went down to Wall Street the next day to see if he could pick up Humpty Dumpty's eggs. He told the story in several important places and was told unanimously by those whom he questioned that nothing could be done about it. Welliver came back to the White House, reported that the damage was done, and the least said soonest mended, which was the Coolidge idea in general. So there the matter rested, and the President was out of sorts for several weeks after the episode. He had moved himself unconsciously to the wrong side of the checkerboard.

2

In those bewildered days he may have found some release in buying dresses for Mrs. Coolidge, dresses that he saw in the shop windows in Washington as he went about town with his secret service guard.[7] He was fond of dresses with vivid ribbons or a dash of color, and would order them sent to the White House on approval. They were expensive dresses. This was the "one extravagance of an otherwise over-conservative, somewhat inhibited and economical man." [8]

[6] Letter to author, July 1936.
[7] "Presidents and First Ladies," by Mary Randolph, pp. 94–100.
[8] Ibid.

But while the market was careening around like an untamed skyrocket, the President with his habitual avoidance of unnecessary responsibility went right on letting things go and took his delight in the furbelows and follies of his wife's millinery. Miss Randolph remembers that one day when Mrs. Coolidge was having a final fitting on a dress with a court train, for a most formal occasion, the President strolled into the room and after a long critical survey barked: "Very handsome dress!" And then to the horror of the dressmaker he calmly walked up the long white brocaded train which swept for yards upon the floor, deliberately pacing off the length of the material. Shrieking with the dressmakers, Mrs. Coolidge flung herself on all fours to protect the precious train. He had had his joke. It didn't tear the train. He sauntered out of the room. He had received his compensating thrill for the anxieties of the day, and if the market wanted to swish and swirl and flash its comet-like tail, little did he care; at least little did he do.

In the month of February, the restrictive policy of the Federal Reserve Board began to show itself. From February 28 to May 2, 1928, the Federal Reserve system was apparently getting the situation in hand. Its restraint was becoming effective. The holdings of government securities for the Federal Reserve system decreased by one hundred and fifteen million in this period, sending ninety-five million of gold out of the country, naturally reducing the reserve of member banks and tightening money. Money rates rose sharply.[9] But the infection of speculation still gave the country moments of delirium. And despite the quiet pressure of the Federal system, expansion went on and the stock market rose. In the month of May, 1928, the measures that the Federal Reserve authorities had been taking actually succeeded in pulling down reserves in member banks. Call money moved up to six and a half on May 28.

Bullish statements by Coolidge and Mellon in 1928 were infrequent. Such statements were no longer necessary. Their previous statements, in conjunction with other favorable factors, had given rise to an upswerve on the stock market which, with few setbacks, rolled on and on, gathering momentum and increasing in size to the end, which was still a year away.

This in brief is the history of the causes that produced the business depression when it came in 1929. The speculative debauch being so widespread, the business situation got out of hand. This detailed account of the part played by Coolidge and Mellon in stimulating and encouraging speculation in the stock market is a necessary part of a story of Calvin Coolidge and his times. For certainly these two Pied Pipers did call the Wall Street community and the great body of stock speculators into the cave. Defenders of Coolidge and Mellon naturally counter with the claim that the White

[9] Chase Economic Bulletin, June, 1928.

House statements should have been taken only at their face value. The defenders contend that what these men said was often exaggerated or mis-construed by newspaper reporters. This probably is true. But a great body of investors and stock speculators were taking these statements at more than their face value and on the basis of such statements were making commitments in the stock market that were fraught with consequences of tremendous importance for their future welfare; Coolidge and Mellon were under heavy moral obligation to disavow any statements attributed to them that had been exaggerated or misconstrued by newspaper reporters, or to cease making statements altogether if they were being misinterpreted or were being used in ways other than those intended. Never once did they deny, qualify or restate their published opinions. The historical conse-quences of the statements are properly on their doorsteps.

Yet Coolidge and Mellon deserve only part of the blame. Some part—and a considerable part—of the responsibility must be laid at the door of the Federal Reserve authorities, regional and at Washington. The Federal Reserve system, by its easy money policy initiated in 1924 and extended in the summer of 1927,[10] when Coolidge was in the Black Hills, although the aims sought to be attained by that policy were praiseworthy, made credit available for stock speculation at low rates and thus played into the hands of the Wall Street community. Credit was made too plentiful and too cheap. That was bad enough. But large amounts of reserve credit were find-ing their way into the stock market. There winked the red lights.

Judge Cooper of the Farm Board, finding himself in disagreement, like many members of the administration, with the treasury department poli-cies, asked for the President's time to go to the White House and reason with him. Coolidge listened patiently to the Chairman of the Farm Board, then said:

"Well, Governor, I don't know enough about the subject to advise you one way or another, but if I were in your position, I think I would yield my views to that of the Treasurer and make the Treasurer assume the responsibility!" [11] And so Mr. Mellon's policy continued.

Speculation alone did not create our economic structure in the Coolidge years. The whole business of blame, credit, responsibility or careless im-punity in the journey down the rapids to the economic holocaust lies not in any man, not in any group or institution. Given the American ideals of 1921–29, indeed for two or three previous generations, and the resultant development of those ideals in the American mind alone—with or without Calvin Coolidge or the Federal Reserve System or Andrew Mellon—and

---

[10] That was the date of the conference of the British and European bankers previously described.

[11] Letter from Judge R. A. Cooper, to the author.

the New York bankers would have sped inevitably, sooner or later, to the abysm of catastrophe. For we had what all Christendom had—a Chamber of Commerce complex. It was and is fundamentally creative, instinctively geared to grapple economic problems of production, in a day and a time when invention should have been contriving improvements with distributive systems.

For a generation before Coolidge came to the White House, minority groups representing the various trades, callings, and crafts, had been establishing propaganda bureaus in Washington. The jewelers, the lawyers, the doctors, the bankers, the merchants, the grocers, the preachers, the teachers, the farmers, the railroad brotherhoods, the A.F. of L., the Council of Churches, the prohibitionists, and the anti-prohibitionists, the National Democrats, the National Republicans, oil, textiles, coal, copper, lead and zinc, wool and cotton, all focused their propaganda machines through paid secretaries and groups of lawyer lobbyists in Washington. In Coolidge's administration, the national Chamber of Commerce, organizing all these pressure groups, became the super-lobbyist of the land. In a nation whose head declared "the business of America is business," it was something more than a passing coincidence that the United States Chamber of Commerce housed itself during that administration in an imposing marble palace across Jackson Park from the White House. There the nation's business was transacted. That phrase "the Nation's Business" became the title of a magazine which represented the commercial interests of the United States. During the Coolidge administration, the national Chamber of Commerce, through its sixteen hundred member organizations scattered across the country, representing nearly a million firms and corporations, campaigned actively for a return to a peace time taxation basis. Playing a game of echo with Calvin Coolidge in the White House, across Jackson Park, the Chamber was politically impossible. For Coolidge had his Progressives to satisfy, who often longed to use taxation as a weapon of social progress. Also Coolidge in those days was faced with a surplus. Congress had set up statutory provisions for the retirement of the war debt. Coolidge desired to retire the war debt faster even than Congress had contemplated. He cut it down four billion in four years with the excise taxes left over from the war period flooding the Treasury with funds, even though Secretary Mellon's estimates of tax collections fell far short of actual receipts. Indeed Mellon would have been discredited as a financial seer if the awesome sense of his vast wealth had not hallowed his outgivings.

Andrew W. Mellon was President Coolidge's bad angel. The President could cope with Daugherty or with Denby, two men of vastly different political minds—and each a wide variant from Coolidge. He and Fall could get to terms—but they would be Coolidge's terms—in short order. Will

Hays, Coolidge might have used with more profit to Hays and the country than Harding used Hays. But Mellon was too much for Coolidge. For Mellon's millions spoke gospel to Coolidge. He represented in those years by his social attitudes, by his official policy, a strictly plutocratic philosophy of life. He was the guardian angel of all that the Chamber of Commerce held sacred in its white marble palace. In another day when democracy had changed its mood and manner, Mellon was to sit in the witness chair in an investigation of his income tax returns, shy, rattled, and bedevilled, while certain unpleasant facts were shaken through the meshes of the corporate organizations which he established to ward off taxes.

3

In the first years of the Coolidge regime, as the scores of special groups with their paid secretaries in Washington saw expanding United States surpluses with greedy eyes and went after them, it was the business of the Chamber of Commerce to back up President Coolidge in saying "no" to these groups and to encourage him to pay off the national debt. The Treasury Department estimated that in the last fiscal year of the Coolidge administration, taxes could be cut by nearly a quarter of a billion and still leave a balanced budget. The Chamber recommended that the taxes be cut four hundred million. The Chamber's position for a four hundred million tax reduction advanced in 1927, was based upon a referendum vote of its member organizations. They demanded a reduction of the corporate income tax to ten per cent, a repeal of the war excise tax, and a repeal of the estate tax or inheritance tax. The President echoed back favoring the ultimate repeal of the inheritance tax, because men with large fortunes gave lavishly to national benevolences.

A scholar interested in coincidences might set down the projects favored in the Coolidge years by the United States Chamber of Commerce and then look at the achievements and endeavors of the Coolidge administration. Each list matches the other. The Chamber of Commerce was Coolidge's alter ego. The Chamber's activities and desires, generally based upon referendum votes of its member organizations, lifted the voice of hundreds of thousands of business men, big and little who followed the Coolidge program. In 1927, both the Chamber and the President demanded a reduction of the income tax and curtailing the inheritance tax. The Chamber and the President backed the American Federal Reserve Act promoting legislation to help ship-building interests, advocated easing down on the anti-trust laws to promote combinations in export trade, stood for a flexible tariff, asked for aid to ocean mail subvention and in restricting government aid to American built vessels, called for aid for private operation for the merchant marine, opposed government construction, desired excise

·taxes repealed, asked that the police duties of the Bureau of Internal Revenue be transferred to the Treasury Department. The rate of corporate income applicable to the net income of '27 was reduced to twelve per cent, the President and the Chamber smiling in a common grin of the cat that swallowed the canary! They were both for flood control in '28 and both stood for controlling aeronautics. When the Chamber asked for a foreign commerce service for commercial attachés and trade commissioners, the President acted with mesmeric alacrity. The adoption of the McFadden-Pepper Act extending Federal Reserve bank charters came as an answer to prayer but who prayed first, the President or the Chamber, it would be hard to tell. Both were for the international highways conference. Both were for the revision of postal rates. Once the President got out of line. He asked for the recognition of Soviet Russia. But when the Chamber gasped, the President changed his mind. The radio act of 1927 was one of those cases of thought transference between the White House and the Chamber of Commerce building. It took Congress six years to achieve the return of alien properties asked by the Chamber in 1922. But Congress created the court of tax appeals quickly after the Chamber asked for it. And when the Chamber and the President agreed upon the question of Japanese immigration, Congress overrode them both as it did when the Chamber and the President declared against the soldiers' bonus. But he and the Chamber had their way when he vetoed the McNary-Haugen Bill. In the booming stock market the President and the United States Chamber of Commerce were making one big noise in the same rain-barrel!

CHAPTER XXXVII

"The Clock in the Tower Struck Twelve and All Her Fine Clothes Turned to Rags"

But Calvin Coolidge had other things on his mind than the whimseys of the stock market and the demands of the Chamber of Commerce.

As 1928 opened, Herbert Hoover was in full cry after the Republican presidential nomination. The President's renunciation in the Black Hills had given Mr. Hoover the opening he needed. It was evident that he would be the next Republican Presidential nominee. But it was also obvious that President Coolidge was not happy at the turn the political game was taking. For he was gradually being eased out of control of the Republican party.

In the spring of 1928 the battle of the stock market roared on. The official activities of the Federal Reserve Board were rather ineffectually seeking to liquidate the market, to stop the upward rush of stock prices that could only end in destruction. And on the other hand, all that the President represented, all that the Republican party stood for, was lending the force of a tremendous prestige to the Wall Street boomers. But this contest did not deeply interest President Calvin Coolidge—if he knew much about it. His heart was in another struggle, the great game of politics which culminates every four years in our national conventions. The phrase, "I do not choose to run" seems to have been purposely oracular. It left the President free to accept a deadlocked Republican nomination at the hands of the convention without going into a contest for delegates in any state. Over and over Chief Justice Taft, in his letters to his family in those days, explains this phase of the President's position. Yet Taft, who was frequently the President's guest, was never able to squeeze one word out of his host illuminating that theory of the cryptic message. Yet in decency after that pronouncement the President could not ask for delegates. Even as late as the autumn of 1927, Senator Butler and National Committeeman Hilles, of New York, had gone to the President. They came away without encouragement. Later they were gossiping with Chief Justice Taft, each trying to find out what the other two knew or felt about the South Dakota letter, and all were baffled. Taft wrote to his son Robert that they were unable to say whether or not Coolidge could be driven into accepting another nomination! He was too much for the masters of politics, old hands as they were, at the great American indoor game! Two days after the futile conference when the Chief Justice failed to persuade the President to go to Yale for a formal speech at an academic dinner, Taft wrote to President Angell that the President was "hard-boiled" about it. He was too cunning to be coaxed. But often cunning men yield to commands. Murray Crane found this out when the crisis in the Boston strike reached its climax. It was common knowledge in the Massachusetts assembly that a distinguished lobbyist would get Coolidge's vote in extreme unction. But in those days after the return from Dakota, who could bark at a President—not the amiable Taft, nor the velvet voiced Hilles—certainly not Senator Butler to whom the President had given orders for years. Ten days later the temple priests were trying to find someone who could voice a command. The Chief Justice was still hoping. He wrote to his son, Robert: "It really looks as if Coolidge was determined to defeat any movement for his being drafted, but I doubt if he can prevent it!" This doubt was a basis for hope among the palace favorites which lasted through the autumn and winter well into the spring of 1928. In the winter of that year, the Chief Justice wrote to Dr. Coe that Coolidge would not be able to withstand the pressure, and to his son, Robert, that while "Hilles has come out for Hughes" it is only to "keep his delegates together" to plump them for Coolidge when "the matter becomes so serious that they must resort to him!"

As the spring of 1928 deepened, it seemed certain that Herbert Hoover would control the National Republican Convention. And reluctantly, even sadly, the Taft letters in June reveal an acceptance of the truth.

Favorite sons of various states were entering the Republican Presidential contest, bringing their delegations with them. But like most delegations picked in states by favorite sons, the delegations were human beings, eager for the band wagon, and few delegations following the banner of a favorite son were free from undercover disloyalty. In June it was evident that with a few delegates from the weaker states who were espousing favorite sons, the Hoover forces could control the convention. Two conflicting facts arose from the situation; first, the President was unhappy about it; second, he told no human being, least of all Hoover, what he thought. During May, day after day, as the Republican Presidential convention in June drew nearer the President's friends, Secretary Mellon, Charles Hilles the national committeeman from New York, the Chief Justice, Senator Butler, of Massachusetts, Everett Sanders, the President's Secretary, all had gone to the White House, hoping to have some hint, some vague indication that the President would accept a nomination which they assumed he desired. But he would not give them authority to whisper even to their intimates any remote notion of his expectation. They felt forlorn. He must have known how they felt. But his jaw was locked. His lips set.

Many of those who were near the President felt sure, with no evidence to prove it, that he was expecting a deadlock to arise in the convention, and that out of the deadlock would come his nomination.[1] William Jardine, his Secretary of Agriculture, went to him in May, before the June convention, at Kansas City, and suggested a special message to Congress advocating a farm bill which was being urged by the Secretary of Agriculture and by Mr. Hoover, who, of course, had every political reason to desire the farm question removed from the campaign by a law which would settle it, at least temporarily. It was not Coolidge's measure. But he saw that the measure carried an advantage to Hoover if he was nominated. He talked for a time with Secretary Jardine, parried the farm question and then when the Secretary innocently urged the special message to pass the farm bill as a political favor to Hoover, inferentially assuming his nomination, the President cut in:

"That man has offered me unsolicited advice for six years, all of it bad!"

After which, the interview drew abruptly to a close. Parenthetically considering that much of Hoover's advice had been in disagreement with the advice of Mellon, in the matter of curbing the stock market, it is easy to see how, in that day, the President, viewing the stock boom, considered Hoover's advice "most of it bad." In the conversation with Jardine that May day, the President evidently was counting strongly on a deadlock.[2]

He saw the weeks before the convention dribble away into days. Friend after friend came to the White House, left an opening into which he could drop a wish, and went away sadly disappointed. Yet they knew he was unhappy. Herbert Hoover knew that the President was unhappy. No word had the President said or sent of congratulation, as state after state wheeled into line for Mr. Hoover. So June came and while the President's friends were on the trains speeding to Kansas City, and while the newspaper reporters there and in Washington were spinning stories to the effect that Mr. Hilles was coming with a word from Mr. Coolidge, that Mr. Mellon had

[1] Letter from W. M. Jardine, Secretary of Agriculture; also conversation with W. M. Butler, Chairman Republican National Committee; also conversations with Herbert Hoover, James M. Beck, and Irwin Hoover. Taft gave up hope.
[2] Letter from W. M. Jardine, August, 1936.

the open sesame which would start the Coolidge fight, or that Mr. Sanders would bring a message that would create the deadlock out of which the Coolidge nomination would spring, like Minerva, from the head of Jupiter, Herbert Hoover, in Washington, reading these stories was puzzled. He sent for James M. Beck,[3] member of Congress, constitutional lawyer to whom Mr. Hoover had turned when in need of a dignified eloquent speech presenting his name to the coming convention. Beck, answering the summons, saw the Secretary of Commerce pacing his room, hands in pockets, head down after his manner. He turned on Beck and exclaimed in answer to an inquiry:

"Everything is all right. The Kansas City situation seems to hold no element of doubt. But the thing that gets me is the man there in the White House. What is his game? What cards will he play and how?"

And the upshot of it was that Mr. Beck did not know either and they let it go at that. Five members of the President's Cabinet were in Kansas City. Senator Butler, of Massachusetts, Chairman of the Republican National Committee, named by Coolidge for the place in 1924, was in Kansas City also. The President's secretary, Everett Sanders was there. Among the politicians, and particularly among those in charge of the candidacy of Mr. Hoover, it was assumed that these Cabinet members, with Mr. Butler and Mr. Sanders, were in Kansas City looking after the political interests of their Chief. Three days before the convention met, all day Saturday, all day Sunday, and until Monday noon, reporters and politicians generally felt that some kind of coalition among the so-called allies against Hoover might be made to deadlock the convention and throw the nomination to Coolidge. During these preliminary days, every hour the pre-convention crowd felt that "the word" from the White House would come thrilling down which would authorize the coalition. But no word came. The President's friends were assuring him Saturday and Sunday before the convention met, of his strength there. For they hoped against hope.

But Sunday, the newspapers in New York and Washington carried the story that Mellon and Butler had surrendered to the Hoover forces.[4] From

---

[3] Mr. Beck's conversation with the author, June, 1935.

[4] I have a curious story told to me by an employee of the White House who is still living and cannot allow me to use his name, confirming Ike Hoover's story. The employee said that when President Coolidge read Butler's interview of surrender, the President sent his old friend and political manager a furious telegram. A few minutes later he recalled it and sent to all relaying junction offices of the Western Union, messages to recall and destroy copies of the President's telegram. Another evidential circumstance: When I heard this story, I wrote immediately to William M. Butler asking him to verify it. He had been usually prompt in answering his letters and always had given me a yes or no answer. He did not reply to this letter. He died a few months afterward. I believe that the White House employee who told me the story is telling the truth. I cannot vouch for it.

now on let the story be told by Irwin H. Hoover, head usher of the White House. His notes unchanged, as he wrote them not as they were published in his book, carry an illuminating picture. After the President read the news stories that Mellon and Butler had declared for Hoover, Irwin Hoover writes:

"There was dismay at the White House. Sadness, disappointment, regrets. Among the stories that came was one that carried word of Butler inferring that the President favored the candidacy of his Secy. of Commerce to that of any other. A telegram to Butler was promptly dispatched to the effect that he had no right to so announce. A short snappy message inspired only by the word that had come thru the sources before mentioned. A denial was promptly received, in effect, that no such effort was being made. Thus with this state of affairs, still so indefinite from a White House standpoint but now apparently settled in so far as the Convention was concerned that Mr. Hoover was to be the nominee.

"The President was not long in vacating the Executive Office. He came to the White House visibly distressed. He was a changed man. It was evident to all.

"On the same floor where the President's room is located and to which he retired immediately, there was a radio machine in the course of full operation. There was also one in Mrs. Coolidge's room pealing out its strains of all the preliminary proceedings of the Convention. The President took no notice of either of them, forgetting even to visit in Mrs. Coolidge's room (She was ill at the time.) which was invariably his custom when he came over to the house from the office. He threw himself across the bed continuing on indefinitely to lay there. He took no lunch and only that the physician came out a couple of times to inquire, at the suggestion of the President, for word of the Convention doings, was it known, the drift of his thoughts. In this room he continued on to remain thru the rest of the day and night, not emerging therefrom until nearly eleven o'clock the next (Monday) morning. Even then it was a different President we knew. There came up immediately the question whether he would or would not attend the luncheon that was to be given in honor of the Mexican flyers at the Pan-American Building. So much was this in doubt that word was passed along to be prepared for his absence. In the end he attended but it did not alter the situation in so far as the evident keenness of disappointment was concerned. That night he left for Wisconsin."

2

And in Kansas City Sunday followed Saturday and Monday followed Sunday, and all the wise men from the White House—Sanders, Wilbur, Mellon, Hilles, who was ill in bed, came, looked sagacious, spoke in cryptic language. Then late Monday afternoon, a little pudgy, rheumy-eyed, thin-haired, squeaky-voiced Pennsylvania boss named Bill Vare, sizing up the cryptic silence of Mellon accurately, took the ball, ran around the end, announced the Pennsylvania delegation was for Hoover and closed the contest. Mellon, the richest man in the world, was left suspended on a limb, his political feet dangling in the air. Vare, his man servant in Pennsylvania politics, had outgeneraled Croesus. Tuesday morning's papers announced that Vare had exploded the possibility of a deadlock. The favorite sons could not unite without Pennsylvania. "I do not choose to run" had done its deadly work. The Republican party had a new master. After that, the nomination in the Convention was only a matter of form.

Mr. Irwin Hoover explained in another part of his notes what was possibly [5] the Coolidge theory of his defeat. We read:

"One of these close friends brought back from the Convention what is considered the real story of the downfall of their idol. It narrates how, even with the chances apparently against them, with no word from the President to help, they yet had hopes of stampeding the Convention. One of their leaders (meaning Charles Hilles of New York) who had been faithful until the last, lay ill in bed, having been so stricken since coming to Kansas City. How, as he lay there still scheming and figuring how it could be done, there appeared in his room, one, very close to the President, supposedly, one who was in daily contact with him and shared his burdens of office, if such there be.[6] This person boldly advanced the suggestion that the President would not accept the nomination if tendered him and that for the good of the party it must not be done; picturing how humiliating it would be to have to offer it to another, especially the foremost candidate in the person of the Secretary of Commerce. This suggestion, coming from this source, threw consternation into the camp of those who had the intent of nominating the President in the foreground, regardless of 'I do not choose'.

"It was told of how this word was passed around. First the man sick in bed sent for certain ones who had been affiliated with him and who stood close to the President. They in turn passed the word along until it became a living understanding of what was to be expected. It became more definite in its course of passage and accepted, even by the most optimistic that it would be out of place to continue to hope for the naming of the President in the atmosphere that had grown up. Thus it came to be a scramble to get on the band wagon of the Secretary of Commerce and the story of the Convention was told.

"Every sign and indication during and after the Convention, pointed to the fact that the President's friends who wished to nominate him regard-

---

[5] And possibly not. For Ike Hoover loved gossip—as who does not—but if he sometimes did not get things straight, he was never deliberately malicious.

[6] The head usher apparently is gossiping here about Everett Sanders. But evidently he has mistaken the man.

less, were right in their conclusion that he would accept. They sounded him out as best they could and above all, learned that no one was authorized to say that he would not accept. Then to be showed by every act that he was a disappointed man when another was named.[7] The word coming just before the convention met, which he recognized, meant the nomination of Secretary Hoover, took all the interest out of the convention. He was a sick man and neither listened to the proceedings, in any manner or form but lost all interest in what they were doing or what was done. When told of Senator Curtis being nominated for the Vice Presidency he remarked he did not care who was nominated and said it with a show of anger as if he did not wish to be bothered. At another time when told of the proceedings he said he didn't wish to hear anything about it. So the Convention went on and adjourned with as little interest for him as could be possibly imagined.

"Upon his arrival in the West [8] where he was enroute during part of the Convention he became a nervous wreck. He could neither eat nor sleep with satisfaction and as compared with his former self was helpless for at least ten or twelve days after his arrival. The employees who came in contact with him were concerned lest his condition prove serious. But time was the great healer and fishing overcame the distress until by the time of the return to Washington, he was to all outward appearance, quite his normal self once again.

"But he never recovered fully from the shock, and a shock it was. He has never been able to enter into the spirit of a successful campaign for the nominee who was named at Kansas City. It is different around the White House from what previous situations of this kind have brought to light. The candidate's name is seldom mentioned and when so done it is more with an indifference than with any word or thought for his success. The wonder is that such can be and can only be measured in the light of the failure of 'I do not choose' to have brought about the nomination of the President in a way, just different from what it had ever been done before. It serves for a lot of gossip, a lot of talk and to those of us about who are in a position to observe, the President would not shed a tear if Mr. Hoover was defeated." [9]

In due form the convention at Kansas City assembled Tuesday after Vare's coup and Mellon's surrender. It met without great enthusiasm, but still was fizzing with the ginger of victory. The fight of the convention came not upon the nominee. That issue was all but settled. The fight occurred over the platform. Two issues demanded settlement—prohibition and the farm problem—each equally impossible of adequate solution. The Hoover

---

[7] This is transcribed verbatim as Mr. Hoover wrote it.

[8] His summer camp in Wisconsin.

[9] Remember always that "Ike" Hoover disliked the President and was liable to misinterpret his actions and to misunderstand the meaning of his sometimes Delphic utterances!

forces favored prohibition, but were also set against any solution of the
farm problem by price fixing. A night and two days the resolutions committee wrestled. The platform is always adopted before the candidate is
named. In those two days hope still burned in the breasts of Coolidge
followers that some platform row or some bolt from Heaven would break
the Hoover control of the convention.

The day came when the platform was adopted endorsing prohibition
and refusing to bind the party to a policy of price fixing for farm products.
The next order of the day was the nomination of a candidate for President
of the United States. The roll of states was called. A California delegate
stepped forth and started to recite his speech which Mr. Hoover's friends
under the guidance of Mr. Beck had written so meticulously.[10] It was a
good speech. But rather too early in the text the name of Herbert Hoover
incidentally appeared. The crowd was tense after two days' fight on the
platform. At the mention of Hoover's name, in a tremendous snap of relief,
the crowd burst into cheers, kept cheering. A quarter of an hour passed.
Twenty minutes, and still the crowd cheered. The ovation lasted twenty-
four minutes.[11] The orator with his set speech again and again tried to
launch into his oratory. The crowd would have none of it. In his speech
he had not scheduled the applause for that moment. But when the crowd
roared its enthusiasm, the orator let himself loose with it and when the
chairman's gavel had hammered silence out of the mob, the orator went
on with another speech which Mr. Beck did not recognize. It was a good
speech and like all true oratory came "from the man, the subject and the
occasion."

Later in the evening, at a little before nine o'clock Congressman J.
Napoleon Tincher, of Medicine Lodge, Kansas, appeared on the speaker's
platform to nominate Senator Charles Curtis for President. Before Tincher
had spoken a word, and because Tincher, an old guardsman of the pre-
Coolidge days was scheduled to nominate Senator Curtis, who dated his
congressional service back to 1893, the last of the old guard of Hanna's
day, of Aldrich's regime, and of Penrose's rule, the convention spon-
taneously broke into a wild demonstration. It lasted twenty-two minutes,
two minutes less than the applause which greeted Hoover. But applause
for Mr. Hoover, the nominee, was in order. It was cordial but was on the
whole something less than ecstatic. The Curtis demonstration was spontaneous, sincere, entirely out of order. In it was the substitute for pathos
for the past. For these old regime Republicans, with all their congenial
Puritanical virtues of thrift, diligence and repression, in their high moments do blubber. And the Curtis demonstration in which the galleries

[10] Mr. Beck's conversation with the author, June, 1935.
[11] Official Proceedings of the Republican National Convention, 1928.

and the delegates mingled their emotion was genuine of its kind. Calvin Coolidge, enroute to Wisconsin that night, might easily have squeezed out a pebbly tear for all that was behind the tribute to Curtis. For an hour, his hope that Curtis would create a deadlock, might have been fanned into life by the convention's show of ardor for Curtis. Here was the moment he had hoped for. He was baffled and confused that a man of Curtis' popularity and political acumen—virtues for which Calvin Coolidge had great respect—could not gather the fruits of victory from that hour.[12] When Tincher had finished his short speech, again the convention burst into disorderly applause, a long, waning demonstration like the echo from Blaine's day, or McKinley's in the days of giants. Possibly after Tincher sat down, the speech of Delegate Lilly, nominating Senator Guy D. Goff, of West Virginia, may have given some comfort to the President when he read it the next morning. For the peak of Delegate Lilly's oratorical flight was this left-handed swipe at Hoover:

Our candidate is a Simon Pure Republican, and stands four-square and double-breasted as the peer of any man, born of woman, in service to his party and service to his country. To him the Dragon of China, the Tricolor of France, the Union Jack of Great Britain have no patriotic significance. He is not a citizen of the world. He loves with a love that is stronger than love, the Stars and Stripes.

The applause that greeted this sneer at the cosmopolitan candidate who was about to be named by the party, was significant. These words certainly were spoken for the man in the White House, the man who for a year had been referring to his successor as the "wonduh boy" and the "miracle man." The evening dragged on. Other candidates were named. Governor Lowden, of Illinois, withdrew his name. A Nebraska delegate presented the name of Senator George Norris. And an Indianan made an eloquent speech nominating Senator James Watson. Then an odd thing happened: Ralph D. Cole, delegate from Ohio, was recognized by the chairman. Mr. Cole explained that he was elected to present the name of Senator Frank B. Willis, of Ohio, but "in the midst of the battle in Ohio this magnificent young man was called from earthly scenes of action." He continued: "My second choice for the Presidency was Frank O. Lowden, of Illinois." But Lowden had withdrawn. Whereupon Delegate Cole continued: "I wish now to choose my third candidate. . . . My really first choice and the first choice of America in the beginning of this campaign."

A delegate cried across the Hall: "Coolidge!" And then a burst of cheering came pouring out of the radio to America. Possibly by that time the President had recovered from the shock he had when Mellon left him to

[12] Letter from W. M. Jardine, August, 1936.

tag after Vare into Hoover's camp. If so, it must have pleased him, for this applause was the last note of applause he would ever have from a Republican National Convention in his life time. After the demonstration calmed Mr. Cole continued:

"Some delegate in this Convention has the power of reading the human mind." He continued: "My candidate is a Republican." Again applause hailed out of the radio. "This is a Republican convention." [More applause] "We have adopted today a Republican platform. We must nominate a Republican for President of the United States." [Great applause. During the applause there were several cries of "Hurrah for Hoover!" and "Hoover, Hoover!"] "My candidate cast his first vote in 1896 for Ohio's illustrious son, William McKinley, [Applause] the great champion of protection. He did not fail to vote in that campaign. [Applause] He next followed the leadership of that greatest American of his day and generation, Theodore Roosevelt. [Great applause] But he did not forget to vote. He did not forget his name when he came to vote. [Applause and shouts for Hoover] My candidate voted the ticket in 1916. [Applause] . . . My candidate was chosen Vice President in 1920, elected President in 1924— an exalted specimen of American manhood, better than wealth, better than all power, better than all position, to have the courage, character and conscience of Calvin Coolidge, my candidate." [13] [Great applause]

But the clock of fate had begun to strike twelve. As an answer to the wave of applause that swept from the convention into the radios of the land, the unemotional voice of George Moses, Chairman of the convention, named to nominate the new hero of the party, droned on drowning the final ripple of applause:

"The Secretary will continue the roll of the states." And after the reading clerk had mouthed the call from Arizona to Porto Rico, the Chairman's voice again rang across the Hall: "Seconding speeches are now in order. They will be limited to five minutes each."

And the last phrase evoked the cackle of approval in the great Hall. Even Coolidge, famed for brevity, could have smiled at that obvious economy of time. It was eleven o'clock P.M. For twenty minutes the roll-call ground on, a dull procedure. The drama had been extracted from the scene. Eleven twenty and the Permanent Chairman was saying:

"And Herbert Hoover, having received the majority of all votes cast, I declare him to be the candidate of the Republican party for President of the United States."

Seven minutes more were required to hear and receive the motion to make it unanimous. Senator Moses read the Permanent Chairman's telegram to Herbert Hoover. It was all but midnight when the Chair recog-

[13] Official proceedings of the National Republican Convention, pp. 202–203.

nized "the delegate from New Jersey, Congressman Franklin W. Fort," [14]
and he made a motion to adjourn. And so came midnight. And Cinderella,
hurrying from the Hall in rags and tatters, stepped into her pumpkin
coach and scampered away from the grand ball and from the potentates
and the powers; that day was done.

[14] Official Proceedings of the Convention.

CHAPTER XXXVIII

The Long Day Closes

How the market in Wall Street boomed in those summer days of 1928! The boom must have soothed Calvin Coolidge. It was his boom. Whatever of bitterness he drained from the cup of politics when Herbert Hoover was nominated was sweetened by the big bull market—the Coolidge market, men called it. Any man with his perspicacity would have felt sure and proud that his name would be forever linked with this period of his country's prosperity. With all his heart he believed in it. He must have known of the quiet struggle of Roy Young,[1] Chairman of the Federal Reserve Board, and Dr. Adolph Miller against Ben Strong, of the New York Federal Reserve Bank, to control the market, to keep down broker's loans. Coolidge's head might have been, probably was, with Young—but not his heart! Anyway Strong was sick and passing out of the picture. Time and again callers at the White House and visitors who saw the President at public functions, noted that he always sought out Roy Young, of the Federal Reserve Board. Sometimes he drew a chair alongside Young. Often he began:

"Well, Roy, tell me how is the market?"[2]

But the President could not bring himself to give Young encouragement in his contest. His heart was with Benjamin Strong of New York, wherever his Vermont head may have led him. So the Coolidge stock market boom rose with the summer. But the spotlight had moved from the White House to the headquarters of the Republican national committee where Herbert Hoover, the Republican Presidential nominee was in charge.

Once the spotlight returned to the summer White House in Wisconsin

[1] Mr. Young became governor of the Federal Reserve Board October 4, 1927.
[2] Letter from W. H. Grimes, Wall Street Journal.

when diplomats of the world gathered in Paris to sign the Kellogg-Briand Treaty to outlaw war. While, of course, it was the Kellogg Treaty for America, Kellogg was in truth only the Secretary of State. Calvin Coolidge was the President. If he was the business man's President, then certainly he was entitled to the spotlight, for essentially this was a business man's treaty.

In Paris it was as Calvin Coolidge would have arranged it, an occasion without pomp and ceremony. For the Peace Conference of 1919, was a politician's holiday. But on August 27, 1928, after a year of negotiations for peace, in which Coolidge had backed his Secretary of State with whole-hearted ardor, the business governments of the world, organizing for peace as an economic necessity and not as a moral ideal, came to that conclave.

At the check stand where the world statesmen and their guests left their hats that Monday afternoon before entering the Salon de l'Horloge, where the Kellogg-Briand Treaty was signed, only twenty-three silk hats were among the three hundred headpieces adorning the pegs and pigeonholes when the great clock struck three and the parade of peace filed into the historic room.[3]

In the gathering were a half hundred gray business suits. It was the middle class of Calvin Coolidge speaking through this treaty, the middle class caparisoned in its working clothes, that gave the tone to the picture.

There, in the ante-room of the Salon de l'Horloge, where less than a decade before Wilson came with the politician's high vision, behold a hat-rack filled with gray soft hats as business reached—even if futilely—for the sceptre. Business conjured with an ancient wand to witch away the war monster. So the false dawn of a new epoch, the pale peace pageant of the Coolidge bull market came to mankind. If the real dawn still lagged, Calvin Coolidge looking at his handiwork might well have called it good.

In the campaign that followed the nomination of Mr. Hoover at Kansas City, President Coolidge took but a perfunctory part. Certainly there was little evidence that he was unhappy, but he gave no sign of enthusiasm. His detachment was uniquely Coolidgean.

The summer slipped away. It was evident in the early autumn that Alfred Emanuel Smith, the Democratic candidate for President, would be defeated. The only question in men's minds was how badly. The President who had met Governor Smith in his gubernatorial days rather liked him. In many ways they were kindred spirits—the man from the back woods, the man from the back street; the last of the old rural Presidents, the herald of the new urban day. Smith and Coolidge were exhorters for economy. They lived simple lives according to the traditions of their widely different environments. They were family men. They had deep

[3] Was there, counted them myself. W. A. W.

loyalties. Both were sentimental. The story of each was the old American story of the poor boy who rises, the one to be President, the other to be the idol of his party and of his class. So even though Smith and Coolidge were separated by the abyss of party politics, Coolidge had no unkind words, public or private, for the opponent of Mr. Hoover, and precious few words of any import for Mr. Hoover; only enough to satisfy the amenities of a hot and crowded hour. The campaign closed without an episode so far as President Coolidge was concerned. He heard with but casual comment the returns which gave Herbert Hoover the Presidency. His enthusiasm was restrained. But it always was, and no one remarked at his lack of ardor.

2

A drop in stock prices came in the early summer of 1928, which did not continue. By the late summer of that year stock prices were again mounting. The upward movement had gained impetus from the credit-easing operations of the Federal Reserve Bank. This may be explained as a reluctant support of the acceptance market. Stock prices continued to leap upward at an alarming rate all that campaign year. No action was taken by the Federal Reserve authorities to curb the use of credit for speculative purposes until February 2, 1929. It was obvious then that the booming tide was a devastating wave. The Hoover administration was coming into power. Herbert Hoover knew the truth. The very influence of his knowledge availed somewhat in the situation. But for whatever cause the Federal Reserve system began to modify its inflationary policy. By February the restraining influence of the President-elect and the demands of the more intelligent bankers produced a change of policy in the Federal Reserve system. The Federal Reserve Board, by a division of five to four initiated its "direct action" policy of bringing pressure to bear on member banks of the Reserve system which were using Federal Reserve credit to make loans for speculative purposes. Calvin Coolidge knew this. He issued no statement. He was definitely committed publicly to another policy.

Shortly thereafter the demand for restricting the use of Federal Reserve credit in the stock market became general, and from that time on the question was no longer one of whether the use of Federal Reserve credit for stock gambling operations was to be restricted, but of how such restriction was to be effected. The question of method led to a disagreement between the Federal Reserve Board and the Federal Reserve Bank of New York and also between the Board and the Advisory Council of the Federal Reserve system—the pilgrims to Mellon. But alas by the time that controversy had developed, the speculative situation was already beyond control. All informed financial leaders knew in Washington—and Calvin

Coolidge even then must have realized it was only a question of the day when the crash would occur. But this episode runs ahead of our story. It has little to do with Calvin Coolidge who in January, 1929, was the bedraggled, neglected stepchild in the ashes. We must go back now to the riotous days of 1929—when the year was at the spring, and the lark on the wing.[4]

The Congress convened for its short session in December with the spotlight divided between the President and his successor-elect. It is the way of politics. It was always an embarrassing and sometimes tragic three months for an outgoing President. But no tragedy visited President Coolidge in those three months before March 1. He followed the even tenor of his way. He was not dramatic enough for tragedy. It was evident to the newspapermen whom he saw twice a week, and to his personal associates in the White House, if not to casual callers, that these were dreary days.

In that winter of his discontent James Lucey, his old shoemaker friend, dropped in on the President as casually as the President used to drop into Lucey's shop. Lucey told the reporters that the President greeted him cordially and that the President meant it. And Lucey ought to know. They chatted for a while about the old Northampton that had changed in the thirty years of their acquaintance. Then the President invited Lucey who

---

[4] There remains to be cleared up the question why in view of the alarming growth of speculation in the stock market, no action was taken by the Federal Reserve authorities to restrict the use of Reserve credit for such speculation between the late summer of 1928 and the early part of February, 1929.

The answer to that question seems to be that those who dominated Federal Reserve policy believed that the credit restriction measures adopted by the Federal Reserve system during the first half of 1928 would eventually prove inadequate to bring the speculative situation under control. Governor Roy A. Young, of the Federal Reserve Board, in an address before the Indiana Bankers Association on September 20, 1928, declared: "Many people in America seem to be more concerned about the present situation than the Federal Reserve system is. If unsound credit practices have developed, these practices will in time correct themselves. . . ." Yet quietly he was trying to correct these "unsound credit practices" by moral suasion!

Secretary Mellon, as late as December 31, 1928, had given out a New Year's statement saying, according to the New York Times, the following day: "In the industrial world conditions seem to be on an even keel" and "I look forward with confidence to continued progress in the year ahead."

Such statements as these from the head of the Federal Reserve Board and from Mellon under his holy aura, lulled the country into a false sense of security and created an atmosphere which made it impossible for those in the Federal Reserve system who recognized the danger of the stock market boom to secure the adoption of correct measures before the situation got entirely out of control. For after all, as Secretary of the Treasury, Mellon was ex-officio chairman of the Federal Reserve Board. He was then accepted as Hoover's choice for the Treasury portfolio, and Mellon's known opposition to the adoption of measures to restrict the use of credit for speculative purposes encouraged large stock market operators to believe that no restriction would be imposed on the employment of Federal Reserve credit in the stock market and thus led Wall Street and its stock speculators to plunge ever-more deeply into stock market ventures.

was pretty well dressed up in his Sunday best, to luncheon in the White House. Lucey had a son employed in one of the Departments in Washington. He had sent three girls through college, read their texts at odd times, and had become a man of more than ordinary information and mental acumen. The President and Mrs. Coolidge held him after lunch for a few moments and gave him a White House car for the afternoon. The Lucey episode in the life of Calvin Coolidge illustrates one strong Coolidge trait. He never lost a friend. The men whom he knew in his early days were with him at the last.

Coolidge continued to enjoy the pomp of the White House. When the band played "Hail to the Chief," which Taft and Cleveland forbade in their presence,[5] the pleasure upon President Coolidge's face was evident. But formal ceremonies irked him. He hurried them through as quickly as possible. The army and navy reception, which is the big tin sword and rooster feathers show of the winter, that last winter President Coolidge made an all-time low record of cordiality, receiving more than two thousand guests at the rate of thirty-two per minute. The reception began at 9:05 and ended at 10:10. Fifteen minutes later the dancing was over and the lights began to burn low. The thrill of the Presidency had departed.[6] In his family letters during that period, the faithful Taft declares that the dinner and reception to the Supreme Court was a dreary affair; no special music, no entertainment and as the Court wore no uniforms the occasion was without grace. Nine aging lawyers and their middle-aged wives sat around the table politely waiting for a decent time to go home and go to bed. Senator William Butler, of Massachusetts, calling at the White House that winter, found the President far from happy in his work. To Butler he complained about the terrific responsibility of the Presidency, and its loneliness.[7] The letters of the Chief Justice at the end of the Coolidge administration are hopeless about the improvement of the federal judiciary. Taft's five year mission to the White House had accomplished little. He bemoans the fact that while Coolidge was "a member of the Bar" he had no "practice or experience" to give him the "professional sense of responsibility" needed. As a President he played politics with the judiciary. "My dear friend, the President," wrote Taft to C. D. Williams, his law clerk, "has not the right conception about the selection of judges. They [Coolidge and Hilles playing politics] lower the dignity of the Senate by selecting a man for purely political purposes without the slightest regard

[5] Notes of Irwin Hoover.

[6] *Good Housekeeping*, April, 1935.

[7] To understand the gloom of those latter White House days one should know that Mrs. Coolidge's mother lay in Northampton, dying of a cancer. Letter from William Howard Taft to his son Robert, December, 1929. Also letter from Thomas Cocoran, in Boston papers after Butler's death.

for his ability." These were hard words but true. Later in a letter to Judge
Learned Hand, the Chief Justice wrote: "His one weakness is his lack of
judgment in the selection of competent persons for office." Senators needed
the federal receiverships and the pickings of receiverships as campaign
provender; the President needed the Senators in supporting his legislative
problem. So the President let the Senators name the judges, the judges let
the Senators have a hand in naming federal receivers whose clerks, attorneys and hangers-on were Senatorial claquers, and the President had his
way with the more regular and subservient Republicans in Congress. Chief
Justice Taft who had returned to earth from the Heaven of judicial isolation had to see his ideal of a hightoned judiciary wither and fade! His
letters betray his disappointment. About all he got out of his six year campaign was the association with a taciturn, sometimes crabbed and always
provoking Yankee who played politics greedily like a gambler at the wheel!
The Chief Justice, writing in this letter to Judge Hand, recalled in the
hour of his failure what Murray Crane had said of Coolidge's limitations—
his inability to judge men.[8]

It was evident that the outgoing President was low in his mind. He
spent more and more time alone in the Executive office on Sundays and
holidays. He was no hand—not in Boston, in Northampton, nor in Plymouth—to loaf about the house on Sundays. The home was for women.
They dominated it. He avoided it. It was interesting to read in one of
Irwin Hoover's unpublished notes, this observation:

"Ladies, Coolidge—nothing doing; not even his wife."

Apparently he was not a demonstrative man, for Hoover adds: "No one
ever saw a manifestation of affection demonstrated."

He was even that shy. So in those last months he hid out in his leisure
days in his office, sometimes went there at night. There he read his papers
and occasionally in daytime lay down and took a nap. He had always
liked to walk over the offices and then down through the kitchen and
pantry storerooms. And in those later days when ceremonies irked him,
the interest he had in pantry prowling did not desert him. He could be
alone, silent, thinking.[9]

3

In February before President Coolidge left the White House, he was
beset by editors and representatives of various magazines and publishing

[8] Letter of W. H. Taft to Judge Learned Hand, May 3, 1927.
[9] The data for the material and conclusions on these last pages have been gathered
from conversations with associates of the President in those days and from the unpublished notes of Irwin H. Hoover, head usher of the White House; also from conversations with newspaper correspondents stationed in Washington. All agree that the
winter of 1929 was not a happy one in the White House.

houses anxious to secure his memoirs or other articles. The *Cosmopolitan Magazine* and Cosmopolitan Book Corporation contracted with him at five dollars a word for his memoirs: [10] probably something like the top price that had been paid for a series of articles by any publisher. The contract included both book and magazine rights. One advertising firm offered him seventy-five thousand dollars a year to become director and advisor. Banks came after him and bond houses, railroads, manufacturers, newspaper syndicates, public utilities; every line of business that had big resources hoped to employ him. Toward the last of this siege Mr. and Mrs. James Derieux, representing the Crowell Publishing Company, came to the White House and the President talked with the two of them in his office for the whole afternoon. He told them why he could not consider any of these offers. He declared that if he took a job with a financial house that put out securities, his mail would be crowded with letters in varying forms complaining, "I bought these securities because of your connection with this house. I had confidence in you so did not question anything done by the firm. A man who has been President of the United States should not lead those who trust him into bad investments." [11] He told Mr. Derieux that he was afraid to acknowledge the offer lest the man or firm making it should print his reply and get even that pinch of advertising. This penury about publicity covered him when he said: "Everybody knows I smoke and I don't intend for any advertiser to make use of that fact." So he never was photographed with a cigar about him. Yet he was prodigal with his kindness. Mr. Derieux was deaf in one ear and recalled [12] as a man with one-sided hearing always remembers with gratitude that in the years of their association, Calvin Coolidge never forgot which side was deaf.

In the conversation with Mr. Derieux, the President reminisced: [13]

"I'd like to go into some kind of business, but I can't do it with propriety. A man who has been President is not free, not for a time anyway. I wish he were. Whatever influence I might have, came to me because of the position I have held, and to use that influence in any competitive field would be unfair. Some of the offers that have come to me would never have come if I had not been President. That means these people are trying to hire not Calvin Coolidge, but a former President of the United States. I can't make that kind of use of the office. I've had banking offers, but banks sell securities, and securities sometimes go bad. I can't do any-

[10] Letter from James Derieux, September, 1936.
[11] *Ibid.*
[12] *Ibid.*
[13] *Ibid.*

thing that might take away from the Presidency any of its dignity, or any of the faith people have in it."

Mr. Derieux remembered that his letters were super-cautious. He rarely in his replies revealed the subject under discussion. Answering a letter he would put it something like this:

"The matter you mention in your letter," or "the subject that you bring up," so that the careful Coolidgean biographer will have to obtain both ends of his correspondence to understand it.[14] He was proud to say that no confusing or compromising letters such as Theodore Roosevelt often left to his biographers ever would turn up. He told Mr. Derieux that in early life he had learned to be cautious in letters. He said: "I doubt whether a search of my correspondence would reveal any conflicting statements made by me even if the search included a long period of years." [15] His caution was constitutional. All his life he had a miserly habit of consciously hoarding the gold of consistency—choosing it rather than wisdom.

He liked to talk about himself to those he trusted. Rationalizing his conduct he pointed out that he was not a silent man. To Mr. Derieux he explained: "I don't know why people say I am silent unless it is because I have no dinner talk. I have made more speeches than any other President. If I tried, I suppose I could learn to talk at the table, but I always have good company around me and could have a better time listening than talking." [16]

Looking forward into the Hoover administration that day, he said: "The best thing I can do for the Hoover administration is to keep my mouth shut." He considered Everett Sanders for a moment and said: "He was the best secretary any President ever had." Then he added reflectively: "The trouble with newspapermen as secretaries is that newspapermen will tell all they know, and a secretary should know everything and tell nothing." He grinned and went on: "I guess I am not naturally energetic. I like to sit around and talk." Mrs. Derieux asked him that day what had worried him most of all. He hesitated a few seconds and then gave her a housekeeper's answer. With a straight face, looking down his nose he replied, remembering his conflict with Elizabeth Jaffray, the White House housekeeper:

"The White House hams; they would always bring a big one to the table. Mrs. Coolidge would always have a slice and I would have one. The butler would take it away and what happened to it afterward I never could find out." Almost wistfully he added: "I like ham that comes from near the bone." [17]

[14] Letter from James Derieux, September, 1936.
[15] *Ibid.*
[16] *Ibid.*
[17] *Ibid.*

As the day approached for the inauguration of President-elect Hoover, President Coolidge began the rather painful business of cleaning up his office. He put his correspondence in order, gathered it carefully, catalogued it meticulously, for he was a methodical man.[18] He sublimated sex as completely as a New England spinster. All his notes of sentimental affection were addressed to men. He wrote many of them. But in cleaning his desk no haunting feeling of inadvertent indiscretion could have annoyed him as he went through the tedious business of assembling and docketing his personal and public correspondence, his last letters to posterity, the abstract of his title to fame.

4

He arose early the morning of March 4, 1929, and for the last time before the stenographers and office assistants had come down, he wandered through his vacant executive offices. The janitors saw him flipping open books, looking under blotters, peering into pigeon holes, searching for any scrap that might be left behind. The servants at their early breakfast saw him in the White House kitchen and the storeroom, roaming aimlessly, grimly, merely nodding to the help to acknowledge that they were human, speaking to no one, saying nothing even to himself. At breakfast, he was the Calvin Coolidge he had been for six years in the White House, to whom one meal was more or less like all meals. He ate when served, waited for no guest or family member, made feeding a serious necessary business which he did effectively and then rose and again went over to his office. Again, casually but rather conspicuously in the presence of the clerks and stenographers that they might see him do it, he put his house in order. He left abruptly with few farewells to the men and women who had been working with him in that office for nearly six years. The best he could do was to wag a head to them in passing. As he was about to leave, he sat down at his desk and with his own hand wrote this note to Mrs. Derieux, the last line he wrote on White House stationery. She was a casual acquaintance. But he knew a White House letter would thrill her.[19]

My dear Mrs. Derieux:
   Before I leave I wanted to express my pleasure at having you call on me with Mr. Derieux when you were in Washington. I was sorry to learn that you have been ill since I saw you and trust this may find you much improved. Mrs. Coolidge hopes to see you sometime and would wish to add her good wishes to my own.

                                        Cordially,
                                        Calvin Coolidge

[18] Irwin Hoover, "Forty Years In the White House."
[19] Letter from James Derieux, September, 1936.

When Irwin Hoover wrote: "Ladies, nothing doing; not even his wife," he could not suspect how this repression of the romantic impulse sublimated itself and appeared in odd little tendernesses to charming women rather remotely outside his life. Noting here that the last letter he signed in the White House was directed to Mrs. Derieux, we should recall that the first letter he wrote was sent to the charming Mrs. Marie Burroughs Currier, regretting that he could not be the guest of the Curriers to witness the performance of "Romeo and Juliet" on their estate. Sentiment, though sternly strangled, nevertheless was always a moving force in his heart.

What romantic frustration was it that haunted this man, enshrined his mother's portrait on his desk for forty years, made him thoughtful but never amorous with many casual women? Was it the death of his sister? Was it the passing through his life of some childhood sweetheart like Dante's "Beatrice"? Did he turn instinctively to Dante, the frustrated lover who built a nation's philosophy on his lost love? Was it significant or just casual that in the days of his youth and young manhood this frigid, repressed Yankee spent long hours translating Dante from the original? Did he find in the cry of Dante's lost "Beatrice" an echo that he cherished in his own chilled heart?

When he finished his work in the White House offices, he came over to the family part of the White House, again wandered through all the rooms, upstairs, downstairs, in milady's chamber. Only a sentimental man, deeply moved, could have bid so silent and so carefully measured a farewell to all his greatness. Eight o'clock, nine o'clock, nine-thirty passed. On the Avenue the stir of inauguration, the rising hum of a busy day was penetrating the White House. He was not yet dressed for the occasion. At ten o'clock carriages were beginning to appear in the White House driveway. At ten-fifteen, on the second floor of the White House, he still was not formally dressed. Mrs. Coolidge had been busy during the morning saying gracious things to members of the White House organization, in the executive offices, in the kitchen, all over the place. Most of these people had lived with the Coolidge family for six years, intimately, and had given the best service they knew. Mrs. Coolidge, realizing her husband's vinegary mood, was exceptionally felicitous that morning, considerate, cordial to the people about her who had served the family. She maneuvered the President into his room where he dressed tardily, and while he was dressing she lined up the White House personnel before the door of his bedroom. He came out in his striped trousers, his morning coat with his high hat in his hand, and stalked down the line which he realized had been formed in conspiracy to break the ice of his mood.

Mrs. Coolidge said: "Why Calvin, here's Wilkins and the boys!"

They were hoping he would shake hands. He nodded curtly, looked down his nose, rasped out a little harsh bark which may have been goodbye or good morning, or how are you, but which was all the group had from him as a farewell. So he appeared to old Ike Hoover who hated the Presidential shadow. But Mary Randolph, who loved the Coolidges, felt that he did not bid her a formal goodbye because he could not trust himself. He was afraid he would slop over [20]—a crime worse than manslaughter in New England. She felt his heart was so full he did not dare to stop to shake hands with "Wilkins and the boys." That he was deeply moved, no one can doubt. How, no one can say! Thus he walked through the line to the elevator and so came down to the first floor of the White House. The head usher [21] escorted the President to the Blue Room where already the official guests, the President-elect, the Vice President-elect, the outgoing Vice President and their families were assembled. He nodded to the dignitaries, stood stodgily, then quacked: "Time to go!" They noticed that he was in one of his dried-ice moods, a state of some sort of suspended social animation. He managed to give a cold flipper to the incoming President and to one or two others; neglected most of the women; while Mrs. Coolidge went the rounds with an exuberance which made up for the President's shrivelled courtesy.

And so when the hour arrived, he walked beside the President-elect like a gyved prisoner to his doom. For three hours he was on exhibition to the public, a spiritual mummy of a man, dried and dead. Later in the day he cheered up. In the Senate chamber, waiting for the inaugural ceremony, reporters noted his smile. The pain of the wrench of the White House farewell wore off and he took pleasantly the plaudits of the multitude. When it was all over, he rode solemnly, through the sad drizzling afternoon, to the station. There a large crowd had gathered to greet the Coolidges and bid the ex-President farewell. A great roar of applause rang through the cavernous shed of the station as the train pulled out. He was to have but one other such greeting in the few short years stretching before him.

The forty-eight squares in Mrs. Coolidge's bedspread were finished. As the crowd turned away and the train pulled out of the station at Washington, it was journey's end for Calvin Coolidge: the top of the rise—a felicitous, remarkable, lucky, yet inevitable rise—of a competent, intelligent, hard-working politician; honest as his times would permit, courageous as the prod of circumstances and a political habit of mind could make him, endowed with such common sense and such high purpose as were the people whom he represented. Probably he believed that the muse of history

[20] Who wrote "Presidents and First Ladies." Letter to the author, September, 1936.
[21] The story of the morning written above came from a conversation in February, 1933 with Irwin Hoover—immediately transcribed.

would some day find him and give him his well-earned crown, but the belief would fill a man of his temperament with no joy as he rode away through the gloomy twilight. Hurrying down the long ladder of fame which he had climbed so steadily and so far in two score years, did he mark the rungs where his friends had given him a hand? Here McCamant, of Oregon, had boosted him; there William Butler and Murray Crane had guided his feet; further down the glamorous way Guy Currier, grand vizier of all New England, had given him a leg. Here he felt the cordial pull of Frank Stearns's handclasp. There Jake Wardell, the lobbyist, had put the Coolidge foot upon the round that made him President of the Massachusetts Senate. All the way up, Tom White had steadied the Northampton man's grip on the ladder. At the very bottom was former Senator Richard W. Irwin, of Northampton, who sent the young legislator to Boston with the letter which described him as "a singed cat!"

As Calvin Coolidge, in that March twilight, sat silently gazing through the Pullman window into the steam clouds that blurred the gray landscape, what strange figures took shrouded shape in the mist! The lovely Roumanian Queen in her jewels, the smiling heir apparent of England, the Morgan bankers with all their power, Lodge in his pride, Rockefeller meekly inheriting the earth—all these and hundreds of their kind had stood before him, hat in hand, curtsying for favors. And now they were departed from him forever. So he went to bed a prince whose glory was fading. He awoke a country town lawyer without a practice, wondering how he would spend his time and save his money. It was the old miracle of democracy in reverse. The ruler returns to the soil—politically earth to earth, ashes to ashes. How much went with him to oblivion! The old order was beginning to vanish in the mist of that gray March day. The pride of Henry Cabot Lodge, the power of Penrose, and the dynasty that fell with him—Aldrich, Platt, Hanna, Quay, back to Thurlow Weed, the bosses of the plutocracy, the arrogance of the political nobility, lingered only as a shadow. The rule of the party of wealth and brains in his own day, Mellon and Coolidge personifying the well born, "the wise and the good," was beginning to weaken. How impotent they all had been in the three generations since the war between the states, impotent to change the inexorable current of events. So at the end, by their futility they proved again the proverb: How little wit it takes to run the world!

CHAPTER XXXIX

Twilight and Evening Bell

A DARK day, with sleet and snow and rain, greeted the Coolidges when they returned to Northampton from Washington. They stepped from the train into a great crowd, which the reporters estimated at six thousand neighbors, gathered to greet their returning heroes. The former President shook hands with his old friends. Mrs. Coolidge greeted them with her accustomed cordiality. The returning townsfolk hurried through the station into their waiting automobile and through the congregation which still lined the sidewalks, standing curiously to see their great neighbors come home. The Coolidges went to their duplex apartment on Massasoit Street. They were two small town Americans, living again in the little house in which they began housekeeping. They were paying thirty-six dollars a month rent for it. The rent was increased from twenty-seven in the husband's gubernatorial term.

As small town Americans they fell quickly into their accustomed places.

Mrs. Coolidge resumed her church work at the Edwards Congregational Church. She associated with the Red Cross and with the women's activities of the town. Being a social creature, she enjoyed the intimacies of small town life, the various drives, campaigns, organized community enterprises. But her husband, also in character, kept away from these things. He returned to his law office. To his partner, Ralph Hemenway, he complained [1] that he could not walk around town freely. Some strangers would rush up and grab him by the hand and say:

"Mr. Coolidge, I have always wanted to shake your hand!"

Or some such expression. So his favorite pastime of window-shopping

[1] Says Bruce Barton.

421

was curtailed. To James Derieux who came to Northampton looking for a magazine article, Mr. Coolidge asked sadly: [2]

"Did you ever stop to think what a task it is to speak to every person you see on the streets? It is nice to say good morning to several persons, and to shake hands with them; but it is hard to say good morning to several hundred, and to shake hands with each one of them."

After the interview Mr. Derieux wrote: "When he is riding, motorists often drive past his car, then slow down so he will pass them, and so get a second look. Maps are distributed by the hotels showing how to reach the Coolidge home. Often at night, when Mr. and Mrs. Coolidge are sitting in their living room, they see faces peering through the windows at them." One reporter who adopted this method of news hunting discovered a piece of putty missing from a window pane and made a story of that.[3]

Another reporter tried to get into Mr. Coolidge's bathroom while he was taking a shower! A third reporter discovered that he wore old-fashioned night shirts.[4] Relating these sad tales Mr. Coolidge said to Derieux [5] that when he tried to tell any of his adventures or any quips and stories about himself, his auditor would say: "Oh yes, I read that in the paper." He added mournfully: "It is discouraging to have so many nice stories about yourself and then discover they are all second hand." That was sorrow's crown of sorrows!

The ex-President did not try to practice law. His law office furnished him merely a downtown loafing place. There his stenographer came, a pleasant young woman who after his death confessed that it took a tremendous amount of courage to ask him for a raise and when she got it he gave her a lecture on saving it. He could go into James Lucey's little shoeshop and loaf without being annoyed and this he did sometimes. He had a left-over knowledge of the intimate lives of the older settlers. He returned to the same barber he had patronized when he was a student, always for a haircut. He shaved himself and honed his own razor. When a stranger from the city came to see him and the two went downtown to have their shoes shined, he stopped for a moment before a closed door and said:

"Let's wait for him. He has two small children."

And so they waited. He knew Northampton that way. He began working on his "Autobiography" soon after his return from the White House and kept at it diligently, faithfully to the end. He sent it off, not unmindful of the tremendous price that it brought from the publisher.[6] It was not

---

[2] Letter from James Derieux, September, 1936.
[3] This was before the removal of the Coolidges to their new home.
[4] Letter from James Derieux to author, September, 1936.
[5] Ibid.
[6] Something over $75,000.

an intimate biography. One would hardly expect ribald details from Calvin Coolidge. But it was revealing in its implications. Between many lines he disclosed himself, his deep sentimentality, his sense of his own short-comings. He knew that he was sometimes brusque and mean and unwittingly cruel, and let that knowledge shine forth almost unconsciously in what he wrote.

When he could find the place that suited him, he bought a large house on the outskirts of Northampton, a pretentious place in well kept grounds, a place known as "The Beeches." In it was much more floor space than the family quarters of the White House. It was the largest house in which Calvin Coolidge had ever lived. Northampton tradition says that he was tremendously proud of it. He took there his old habits. He prowled about "The Beeches" as he used to rummage through the White House and as he moused around prisons and asylums as Lieutenant Governor, devoting considerable time to the kitchen, storehouses and pantries. He liked to walk about the grounds. The house faced the hills with a beautiful prospect. But too often he was chased indoors by tourists who seeing him in the grounds pulled their cars to the curb and rushed to greet him. Their effusiveness wearied him. He counted the cars that passed as he did at White House receptions and told a visitor [7] how many had gone by the day of the visit. Facts and figures still enthralled him.

He purposely cut off all antennae to politics. For the first time since 1896 he was not connected with the Republican organization as its national leader, the Governor of the state, member of the legislature, chairman of the organization, or precinct committeeman. No one consulted him about local appointments. State and County Republican chairmen came and went with no aid or comfort from Calvin Coolidge. He was out of it all. He took membership on a national committee to study the transportation problem and there renewed acquaintanceship with his old friend, Alfred Emanuel Smith whom he had known as Governor of New York in the Coolidge Massachusetts days. He applied himself faithfully to the study of transportation and contributed much to it. He was elected president of the American Antiquarian Society and enjoyed the work. He appeared as faithfully at its meetings as he appeared in 1896 at the Republican caucus in Precinct One, Northampton. When he went away from home reporters and photographers greeted him. A *Boston Globe* reporter [8] remembers seeing Coolidge pacing up and down in front of a hall where he was to preside patiently waiting for the photographers after the assemblage had gathered. When he was snapped, he smiled a reminis-

[7] Boston papers in his obituary.
[8] Louis Lyons.

cent smile of the old days and trotted in to open the meeting. He liked to sniff the old fumes of the lamp of fame.

At home one of his pleasant diversions was to autograph his picture or his book for a fee and turn the fee over to his favorite church or charity. A commercial autograph collector in Boston sent a bale of books down to Mr. Coolidge with the fee and had the books inscribed to Erskine Caldwell, Theodore Dreiser, D. H. Lawrence, Ernest Hemingway, Sherwood Anderson, a group of the young Rabelaisian literary skeesickses, who stood in gorgeous contrast with the hardmouthed Puritan of the Connecticut Valley. These books will turn up some day as literary curiosities worth ten times their cost with the fees added. He may have known, but probably not, the significance of his inscriptions when he wrote them. If he did, the fee seemed sufficient and he let it go at that. Early in this last phase of his Northampton years, he was made a director of the New York Life Insurance Company and went on quarterly pilgrimages to New York in state and style. He always put up at a good hotel and seemed to enjoy the Byzantine magnificence of the metropolis. But he avoided reporters. He liked to quip and quibble and do a bit of dry Yankee clowning with Al Smith and the financial bigwigs who met about the board. He was an earnest, honest board member with a surprisingly good memory for details, with comprehensive judgment in interpreting them and yet with becoming modesty in the presence of his financial betters. Probably he always remembered that just before he got to the bottom of the steps which landed him in the White House he had refused to be the president of a New England life insurance company at a prospective salary in excess of his White House salary. He was too smart to be caught by money then, but had the decent respect for it that one has who renounces it. He had the perfect heart of a small town man. Often he summoned to his New York hotel friends from the city and gabbled with them or sat more or less silent, according to his mood.

The city pleased him, gave him a sense of perspective on Northampton and Washington. But it was not his habitat. He wrote daily some little note of affection to Grace Coolidge, if he left her at home as frequently he did. And when he came home to Northampton he was happy. During the twenty years between the day when he went to Boston, a "singed cat," and the day when he came home from Washington in the sleet and snow and rain, Northampton was home. It had changed little in those years. The red soft brick mercantile buildings two and three stories high, built in the fifties, sixties and seventies when it was a mill town, still adorned its business streets. Wide streets they were, spacious, hospitable, kindly streets, bedecked with an architectural story that spanned the days from Van Buren to Hoover, an interesting narrative. Being an observant man, he could not fail to find some significance in the architectural story of the town's main street where his office had remained unused for so many years.

It was a second story office when he came back to it, in a shabby old building with plumbing only one cut above the primitive Vermont standard. Wooden floors of such offices in country towns still were scantily carpeted with runners from desk to door and desk again. His was a small inner office with a desk, two chairs and a bookcase filled with old, little used law books. Here was a safe refuge from the jostlers on Main Street. When he signed a contract to write a short Ben Franklinesque daily comment on passing things, here he installed his force; the stenographer, the representative of the Syndicate who paid him, who kept and filed his little daily grist of rather dull homilies (and he knew they were dull, probably purposely kept them dull—and uncompromising!), and the small machinery necessary to turn the little mill that ground these grains of wisdom. Toward the end of the contract the job must have irked him for he did not renew it. He explained that he had run out of ideas, which was true. Newspaper paragraphs were not his medium. He could not be wise for pay. Soon after his return from Washington he joined a man's club for dinner and discussion and seemed to enjoy the fellowship there, though he contributed for the most part only the charm of his silence and the distinction of his prestige.[9]

His investments were made in substantial things, bonds, gilt-edged preferred stocks, real estate. When in June, 1929, after he left the White House, a market slump occurred, it held no personal heartbreak for Calvin Coolidge. He was glad when the market recovered. If that little cave-in interested him even slightly, no letter or anecdote is preserved to indicate his passing interest. He thought probably that he had left the market March 4 of that year and with the market all the other trappings of his power and glory.[10] He had his big house in the wide grounds, his faithful,

[9] June, *Good Housekeeping*.

[10] When Coolidge's name appeared on a Morgan preferred stock list, after he had left the Presidency, it aroused little comment in his home state. The disclosure came at a time when few people were concerned with ethics or financial morals. His small favors from Wall Street could be defended under the ethics of the day. Withal, it is summed up when one says that the administration of Calvin Coolidge, and indeed his whole career were financial and economic, rather than social and political.

To many, Coolidge's remark on the loans to the Allies was indicative of the character and temperament of the man. "They hired the money, didn't they?" Thus spoke the honorable Yankee. Money was hired, and should be repaid with adequate rental. To competent economists of the times, the man who thus spoke was unversed in economics. Repayment at any time in the 1920's was an impossibility, either in goods or gold, in the view of unbiased monetary experts. To them the ridiculous whole was the failure to make adequate provisions at the time the credit was extended. To them the war loans and the burden placed on Germany at Versailles both seemed to be ridiculous.

Extending Coolidge full credit for a native shrewdness and acumen that could hurdle

loving dog who did not mind an occasional felicitous sadistic amenity, a pulled ear, a slyly twisted leg.[11] The market recovered. The summer deepened. The former President rode about in his big car, one he had used in Washington which he bought second hand from the White House garage. He went to shore and mountain but generally turned up of nights at "The Beeches," the home of his pride. And always Mrs. Coolidge paid her devotions to the formal altruistic institutions of Northampton, served, and served beautifully, her fellow citizens. Occasionally the former President dropped into the office of the newspaper [12] and gossiped a bit, for he loved to gossip, with the woman publisher and with the reporters. But on the whole he kept himself strictly to the character he had assumed at Cinderella's hearthstone, very much the drab step-sister in the cold ashes of private life. He brought home several thousand books from the White House, mostly gift books. Some of these he turned over to the Forbes Library. He was a patron of that library, read widely, mostly old books, standard authors. He was no hand to ride the frothy crest of advanced waves of thought or learning. Sometimes, being lonesome, in the afternoons he would call a friend on the phone:

"What yuh dewin tonight?"

And if the answer was encouraging, he would add:

"Can't yuh come over to suppuh with us?" [13] Then he would add: "We eat at six!"

And so the little procession of his old friends moved through the new house. Grace Coolidge served good food, made good talk at the table, did her full stint to make her mate comfortable and happy. The autumn deepened.

One fine day in late October, when the coloring came superbly upon the hills around about "The Beeches," and New England was aglow with fire in the maples and oaks and gold leaf on the willows, Calvin Coolidge opened the *Northampton Gazette* and read disturbing news. It told of the market crash in Wall Street. And then for ten terrible days the catastrophe fell which sent all of Calvin Coolidge's world, all of his high vision for

the obstacle of involved economics, it might be that he felt all along that the approach to settlement of the loans was all wrong. He said little about them, but what little he did say left an indelible print on the memory of America. Few remarks about the European monetary walk-out have ever been more widely quoted. The dull disquisitions of economists and professors of finance became duller when contrasted with that typical Plymouth Notch phrase. Its acceptance by the people as gospel frightened politicians away from any reasonable and possible settlement of the international debts. Like many another mere smart phrase this quip about the international debts left his country crippled and hobbled in the international financial arena.

[11] The unpublished notes of Irwin H. Hoover give a passing description of Coolidge's gay acerbities with dumb creatures whom he really loved.
[12] The *Northampton Gazette*.
[13] Conversation with Judge Henry Field, 1934.

justice through prosperity, crashing about his ears. No one can doubt that he was deeply stirred in mind and heart by the news that came in those autumn days of 1929.

Like all of his fellow citizens, the highest and lowest, he felt that the crash, however serious it was, could not be so terrible as it seemed. It was preposterous that a mere shattered market for paper property should break down the foundations of a world. The winter came and went with its woe. Millions of idle men walked the streets, homeless, hopeless job-seekers. The highways filled with hitch-hikers shuffling along in the snow, or standing wistfully signaling at the corners. In New England Calvin Coolidge could not go out in his big car without seeing evidences of the burst bubble of his shattered world. Northampton rallied to rescue the unemployed victims of the Wall Street slump. Grace Coolidge did her part, a good wholesome part. Her husband was not equipped by temperament or talent to step into any breach, however bitterly he felt the need of some strong man. The day came when all men were saying, "Times will be better in the spring." Then in the spring they chanted, "Prosperity is just around the corner," and chased the will-o'-the-wisp of their hopes while the slow subsidence of wealth drew new millions into the slough of despond as the foundations of an old era sank and sank.

Those were not happy days for Calvin Coolidge. He was not vocal. In all his daily newspaper paragraphs he gave no sign that he feared for the ultimate recovery of his country, or felt the press of the harrow in the hearts of the constantly widening circle of the poor. Yet he must have felt deeply. His sentimental nature sublimated his pain and bewilderment in many a quip and quirk that went dancing even if awkwardly through his daily offering of desiccated New England persiflage in the daily press. But Charles Andrews [14] of Amherst College distinctly had the feeling that the depression worried the former President. Mr. Andrews wrote that so keen a man certainly realized "what he had done in pumping wind into the Wall Street balloon during the two or three years before he left the White House." Mr. Andrews recalls that Coolidge said when they were talking of the national calamity:

"In other periods of depression it has always been possible to see some things which were solid and upon which you could base hope, but as I look back I can see nothing to give ground for hope, nothing of man. But there is still religion which is the same yesterday, today and forever. That continues as a solid base for hope and courage."

His stenographer, writing for *Good Housekeeping* [15] recalled how he used to sit thinking, looking out of the window a long time at a stretch.

[14] *Good Housekeeping*, June, 1935.
[15] May, 1935.

His old Boston friend and biographer, M. E. Hennessy of the *Boston Globe* [16] says that Coolidge worried about the depression, as his words to Professor Andrews seem to imply. A taciturn creature like Coolidge does not say that much even to an old friend like Professor Andrews unless he has brooded a long time. And Hennessy adds that the realization that Coolidge was on Morgan's preferred stock list while so many of his humble friends in Northampton were on the list of the Welfare Association, saddened him. When he left the White House he saw no turpitude in the fact that the House of Morgan made him one of its insiders [17] but when the crash came, when the price of Standard Brands which had been allotted him by the Morgans began to drop like a plummet, it is not improbable that he had a certain disturbing sense of the realities. Other bankers' lists in those days were being exposed to publicity; incidentally to public contumely. And there he sat looking out of the window, knowing he was on the Morgan list! When a man like Calvin Coolidge broods over his mistakes his heart feels the pressure of his disquietude.

Here was another Coolidge; not the man who, when he was waked up in the early morning of August, 1923, facing the awful responsibility of the Presidency, cried to himself: "I think I can swing it!" Nine years had passed. He had gone out of his forties into his fifties. The impact of his Presidential responsibility had weighted him down. Mr. Andrews, writing in the article just referred to, declares that even then Coolidge was a sicker man than he knew, that "his utter discouragement as to the state of our society, unlike the boy and man I had known for forty years, was the evidence of the toll his work had exacted from him."

Other visitors who had known him in other years saw the striking contrast that had come. Frank W. Buxton, the Editor of the *Boston Herald*, who had known Calvin Coolidge in his Boston days, visited him twice after he left the White House. Both times the Coolidges were in Plymouth. On the first visit, the former President trotted Buxton all over the homestead, showed him the walks that he and his father laid, took pride in a graceful elm tree planted three generations before, [18] traced the

[16] Letter to the author, September 3, 1935.

[17] May 25, 1935—A list of those who, in July, 1929, received from J. P. Morgan & Co., an opportunity to buy at 32, a new stock issue of Standard Brands, Inc., which was quoted at over 40 when it went on the open market, was given by the firm to the Senate subcommittee at Washington. Among the names on the list was that of Calvin Coolidge, whose term as president had ended on March 4, 1929. He was down for 3,000 shares. Many of the names on this list also were on the Alleghany Corp. list. The price 32 was the cost to the Morgan firm. That firm, at the committee's further request submitted a statement to the effect that, since January 1, 1919, it had marketed $6,024,-444,200 of foreign and domestic stocks and bonds of which $2,098,953,400 had been redeemed and retired; and had established revolving credits of nearly $300,000,000 for banks in England ($200,000,000); and in Italy, Spain and Japan.

[18] Conversation with F. W. Buxton, June 1935, Boston.

boundary of the neighborhood for years back, pointed out the old lime kiln, the fields best suited for corn, the good ground and bad pastures, the sugar orchard, the lot where they used to get firewood, all with a loving reminiscence and with a zest that was unmistakable. Buxton felt his host's deep essential love of the placid lonely little town with its white cottages, its stone walls and hills and streams, a sort of wistful, eager, robust joy in the place as a material link between his boyhood and his maturity. Another time, two years later, Buxton saw the former President. The depression was bogging into an economic horror—a permanent nightmare. Millions were falling into the pit of enforced idleness and riches were taking unto themselves wings. Northampton widows and orphans were puzzled and then aghast at the disappearance of their dependencies. England went off the gold standard. Panic shivered through Christendom. Here was a catastrophe which certainly brought heartbreak to Calvin Coolidge as he sat looking out of his office window into the familiar street. He was not the man to forget what his cousin said to him about the market—about brokers' loans.[19] Would his New England conscience let him forget what Roy Young of the Federal Reserve Board had said about the market at those odd times when the President had sidled his chair up to that of the young governor of the Reserve Board? For Young told him the truth. Moreover, the fallacies of the Mellon pronouncements were becoming a handicap to President Hoover.[20] The ex-President knew that Mellon was on his way out of the Treasury—easily, kindly but surely. Men in high places were blaming Mellon's optimism for the crash. And Calvin Coolidge was too sensitive, too perspicacious, to miss the inference.

It was in the gloomy days of 1931, that Guy Currier died at his home in Peterborough. He died a comparatively rich man worth something over a million, something less than five. Calvin Coolidge did not attend his funeral. He sent no flowers. He wrote no note of condolence to the widow, probably because he was low in his mind. In those depressed times, was he questioning all that Guy Currier stood for so valiantly, with such earnest conviction as a liaison officer between big business and high politics which seemed so fair and so fine ten years before?

In Calvin Coolidge's lifetime he had seen a change in the type and character of the men who were fulfilling Guy Currier's mission. Their influence began to wane when Currier and Coolidge reached their pinnacle of power. The President in the White House had seen younger men supplant his Boston friends and heroes; younger men, harder, cruder men whom he disliked and distrusted were tending the flame in his temple. These young altar boys of the priests of plutocracy boasted no erudition.

[19] Letter from Parker Willis.
[20] Hoover was easing Mellon out of the Cabinet.

College gave them sports, liquor, girls, business opportunities. These leaders bought and sold in politics in their own right, too honest to be squeamish about necessary bribery, too cynical to brag about it, too naive to lie about it. They crowded into oblivion elder statesmen who during the nineteenth century held regal court in the various capitals and ruled with a royal pomp. When Calvin Coolidge came back to Northampton, no small part of the bewilderment and pain of his declining years was the plight of his Boston comrades. They sat among the leather cushions of old clubs, eager for companionship. They toddled through the lobbies of hotels where the American Bar Association met, stuffy old ghosts. They attended college alumni dinners sitting among faded moth-eaten old classmates greedily clutching for the crumbs of their old fame. Sometimes they moved to Washington where they appeared at formal occasions at the White House meticulously clad three years behind the vogue, panting under the lard of forgotten feasts, covering up their blotching hands with white gloves, wearing little purple crisscrosses on their faces that marked the graves of past revels. Sometimes at an embassy soiree in a distant land one of these ancients of days materialized as a sort of ghost of his earthly incarnation, babbling to grizzled dames from pensions, with their wistful daughters, old tags from Tennyson, or Shakespeare, or the King James Version, or from long forgotten songs that echoed only in their own brave hearts. The gilt and tinsel of young embassy attachés was gossamer, and the light in youthful eyes a mockery. So passed the Samurai that ruled the land when Calvin Coolidge first came out of Northampton rejoicing as a bridegroom and Guy Currier was the benevolent despot, defender of the faith and the protector of the rich for all New England. Small wonder that Calvin Coolidge had neither the strength nor the heart to stand as a mourner at his benefactor's grave.

During the year 1931, Dwight Morrow died—Coolidge's boyhood friend. McCall, Weeks, Currier were gone. Death was calling out the members of the Knockers' Club. A generation was marching on, an era closing. He felt himself alone and lagging as his old friends were passing and as old customs staled.

CHAPTER XL

And After That the Dark

In 1932, the former President and his friend Frank Buxton went over something of the same journey in Plymouth that they had enjoyed the year before, and Mr. Buxton reported that Coolidge was a different being from the one with whom he had walked and ridden over the Plymouth farm so buoyantly a year before. "His face was drawn and pale. He moved slowly and wearily. He ate sparingly," and when his visitor drove away Coolidge "stood in the doorway and waved his arm in a farewell salute,. but it fell heavily as if it was an effort." Those were the days of the calamity which had overtaken his country. Then to cap it all, a bank in Northampton closed its doors. His former partner, Ralph Hemenway, recalls a characteristic incident of that sad day in Northampton. He wrote: [1]

"Once I was in need of funds owing to the closing of a local bank. I was seated at my desk deeply buried in thoughts that were not particularly cheerful when he came through the connecting doorway from his office, walked over to me, and placed a slip of paper on my blotter. As he turned away and went back to his room, he said quietly:

"'And as much more as you want.'

"It was a check for $5000."

Christmas, 1931, closed a prosperous year for Calvin Coolidge. His writings and his savings had made him a well-to-do man according to the standards of the town and time. And so as a sort of thank offering to his benevolent gods of friendship, he bought a small cedar chest designed ornately and in it dropped five shiny twenty dollar gold pieces for his old friend "Jim" Lucey. The sap of sentiment still was running in his veins.

But it never dimmed his microscopic eye for the main chance. It was

[1] *Good Housekeeping*, April, 1935—Ralph Hemenway.

the week after this generous gesture, according to Herman Beatty, his secretary, that Coolidge's wrath rose when he learned that some poor devil had been cutting wood on his sugar lot in Plymouth, Vermont. Shortly after the sugar lot episode the ex-President took Beatty out to "The Beeches," and led him to his bedroom, where he pulled down several old suits and overcoats saying not a word. "He handed me a pair of small scissors and took his own pocket knife. 'Take out any tailor marks or name tags. These are still too good to throw away but Mrs. Coolidge says I must get rid of them. I cannot let anyone resell them as mine.' I put them in my car, drove to a small town a considerable distance away and sold them to a secondhand dealer. 'Pretty good trading,' was my reward when I handed the twenty odd dollars for the clothes over to Mr. Coolidge."

Mr. Beatty also remembers and records that about this time he found Coolidge making up his own income tax report. He kept a series of small pocket memorandum books wherein he entered every check received, every bill paid. "He showed me one day," writes Mr. Beatty, "the first of a series of these books started when he first came to Northampton as a law clerk. His income was under $250 and there was a net balance at the end of the year!"

His secretary noticed the ex-President's growing weariness. Sometimes he took two naps a day instead of one. He liked to be alone. He had "recurring attacks of what was diagnosed as asthma" [2] but may have been induced by stomach trouble making it necessary for him to keep a supply of digestive tablets always with him on his travels. He was what doctors call "self-medicated." He took what he thought was good for him.

By the failure of the Northampton Savings Bank thousands of his friends, workers and merchants, lawyers and teachers, saw their life's hoardings threatened. The town was in gloom. It did not relieve the worry in Coolidge's heart to know that Northampton's fate was only the common lot of all his countrymen. The Reconstruction Finance Corporation, a socialistic financial device but made necessary by the calamitous times, was organized to save the banks great and small, the railroads, the insurance companies; all were facing ruin. The whole world of Calvin Coolidge and his pride in the power of brains and wealth was toppling. He hoped against hope that the collapse was not so serious as it seemed. Until the end of his life, he was reluctant to believe that the break was real. He cited our car loadings and the price of steel and other solid stocks to indicate that the situation was not as bad as was thought.[3] He read the financial pages of the newspapers carefully and tried to figure it all out.

Judson Welliver, who had been literary secretary of the White House,

---

[2] Notes from Herman Beatty, his secretary.
[3] Letter from James C. Derieux to author, August 9, 1936.

spent an evening in Northampton and the two men wandered in retrospect over old battlefields. The host fell to talking about the Boston police strike.[4] He went into details and for an hour he was proud of his prowess. Walking up and down the floor, the old Fox chuckled about the Sunday when he disappeared from Boston and everyone was looking for him. He was as pleased with himself as he used to be when he rang all the bells on his desk at the White House and brought the servants running to his office, or when he sent for Professor Ripley who had written about the stock exchange and by the mere gesture of feeding the Harvard Professor had given the stock market a congestive chill. Never did the mischief in his heart fail to delight him. But he was modest about his part in the police strike and recalled the fact that a do-nothing policy paid in the end. A few store windows were broken, a little looting followed, some men were cuffed about by the police, but he felt that it was worth it to have the lesson dramatized nationally that public servants have no right to strike. And then his message to Gompers comforted him and he was proud of that.

Referring to some matter in which he had been urged to take a position he said to Judson Welliver, in November, 1932: "I wouldn't take it. The situation had not developed. Theodore Roosevelt was always getting himself in hot water by talking before he had to commit himself upon issues not well-defined. It seems to me," and his eyes met Welliver's as they recalled in fond recollection a number of instances where this wise saw would cut into the truth, "public administrators would get along better if they would restrain the impulse to butt in or to be dragged into trouble. They should remain silent until an issue is reduced to its lowest terms, until it boils down into something like a moral issue."

Which was, in another mood, what he had said to Will Rogers, that he "avoided the big problems." He enjoyed looking at himself in retrospect, an enjoyment that is symptomatic of declining years.

But too often the times rode him. The Garmanian philosophy could not help him. His lifelong faith in working and saving was not enough! He was without anchor.

Yet he did not lose his love for simple living. William Z. Ripley remembers [5] that an artist who had come to Northampton to paint an official Coolidge portrait, arrived in town at half past six of a summer evening, appeared by appointment at "The Beeches" at six forty-five and was greeted by the former President, who said:

---

[4] The facts following in this paragraph are contained in a letter from Judson Welliver, July 28, 1936.
[5] Conversation of Mr. Ripley, former professor of political economy, Harvard University, with the author, June, 1935.

"We have supper at six o'clock, so Grace and I have eaten. But she can set you out something in the lib'ry, so's you won't go to bed hungry."

Mr. Charles Hopkinson, who painted his portrait for the White House about this time, remembers that when Coolidge looked at it, he discerned an ugly mouth indicating irritability, caution to the verge of timidity and temper. Gazing at it for a moment, Coolidge declared that it was the only honest mouth of his he had ever seen painted. He never blinked realities. His was a world of Yankee sentimentality wherein no fairies played.

The partisan spirit, his almost canine loyalty to party organization, apparently was dying in him. The Republican National Convention at Chicago which nominated Herbert Hoover for the second time, unanimously and with no dramatic show or play, interested the former President not at all. He was the only living ex-President. He did not greet the convention. The convention sent no greetings to him. Politics had passed out of his life; it went down in the crash. His subconscious distaste for President Hoover manifested itself in a careful avoidance of President Hoover's name. Where he should have used it, he spoke of Washington or the administration.[6]

That summer of 1932, he went to Plymouth. He had a severe attack of hay fever which his doctor felt might have left a lasting weakness upon his heart. That autumn, Mrs. Coolidge recalls [7] that scarcely a night passed when he was not compelled to use a spray. He suspected many foods of contributing to his discomfort and was insufficiently nourished. "He lost weight and seemed very tired, for he was not of a rugged constitution." Then she adds, sadly: "The death of our younger son was a severe shock and the zest of living never was the same to him afterward."

That summer the financial crisis in the United States was nearing its bottom. The industrial midwest, that region west of Pittsburgh, north of the Ohio and east of Chicago, was falling into the abyss. Banks in Cleveland, Detroit, Indiana, and the Chicago area were collapsing under the strain of financial tension and economic upheaval. The Reconstruction Finance Corporation was lending money to railroads to save insurance companies holding railroad bonds about to be defaulted. Bankruptcies were erupting the ashes of bad and sometimes scandalous assets all over the land. Shame in high places was too common for the comfort of a sensitive soul like Calvin Coolidge whose faith rested on the thesis that "the rich" are "wise and good!" Men whom he had invited to the White House board, major gods in his cosmos, were shrivelling in public esteem. His keen perceptions saw his world cracking. The face which Gamaliel Bradford felt was the "pinched drawn face of a man perpetually confronted by problems too big

[6] Notes from Herman Beatty, his secretary.
[7] Good Housekeeping, June, 1935.

for him" in these summer days of 1932 became haggard. Age was chiselling out the curves from his jaws, gaunting his neck, straightening the soft lines of his face. The man of fifty-nine looked as his father looked at seventy. Was it the toll that the White House takes which makes victims of its masters? Had the "big problems" which Will Rogers reported he always avoided come flocking back to corrode his heart? That new world which he never dreamed lay back of his orderly universe, a world governed by wealth and brains, everlastingly synonymous, was hurrying its new education too rapidly; "education," as he wrote a decade before, "is the process by which each individual recreates his own universe and determines its dimensions." What birth pangs were crashing through his soul as his own new universe was gestating! The embers of his spirit were burning low and the pulse of his life was slowing down. Thus men who brood their troubles in silence, "thinking, thinking, always thinking," sometimes die of a broken heart!

When he returned from Plymouth that fall, he dropped into George Dragon's barber shop for his monthly haircut. Dragon had been working on that head for thirty years and more. They had gossiped back and forth about town affairs rather freely. When he came back from Washington, Coolidge wanted to make it clear to George that he had been true to him and he said:

"George, I had a barber at the White House, but I didn't go into any shops."

So that their relations were cosy. Someone had been priming the barber to ask the former President about the depression which that autumn was nearing its nadir. The closed bank at Northampton had brought it home. So as Coolidge lolled into the long cushioned chair while Dragon was fixing the towels, he swallowed and began: [8]

"Mr. Coolidge, how about this depression? When is it going to end?"

He answered: "Well, George, the big men of the country have got to get together and do something about it. It isn't going to end itself. We all hope it will end, but we don't see it yet." This from the high priest of laissez-faire, was the bitter heresy of disillusion!

Coolidge was a great hand to pass jokes around at the barber shop. He and his doctor, who was in the next chair, got into a jollying contest and the barber remembers that Coolidge got one on the doctor and chuckled and chuckled while the scissors snipped. There was the small town man, the old-fashioned American at his happiest, and alas in that day at his saddest.

2

During the campaign of 1932, the Republicans felt keenly the need of Coolidge. John Q. Tilson, former majority leader of the House of Repre-

[8] *Boston Globe*, January 6, 1933.

sentatives in charge of the Speakers' Bureau during the Hoover campaign of 1932, writes in the *Good Housekeeping* [9] that he asked Coolidge to "help with some speeches," but that Coolidge indicated that his health would not let him speak. Others persuaded him. He came down from Northampton to Madison Square Garden and Mrs. Coolidge, who heard the speech on the radio, sitting in the little back room of the Plymouth store wrote [10] that she "realized that he was using his voice with care that it might hold out to the end." For some reason, the former President failed to rise to the response of the audience to his unconscious humor. He began a phrase modestly, "When I was in Washington," being a euphemism for "When I was President" and the audience burst into laughter. Afterwards, he said sadly to Mrs. Coolidge:

"They seemed to be in a strange mood. I never spoke to an audience which laughed before." [11]

Yet a few weeks later when an enthusiastic woman Republican gurgled at him:

"Oh, Mr. Coolidge, I enjoyed your speech so much that I stood up during the whole speech. I couldn't get a seat."

Quipped Coolidge: "So did I!"

Which was humor from the old trickling spring. But after the peak of the depression, when he realized the force of its devastation, when he knew that it had shattered for a decade, possibly for a generation or forever, the world he helped to build and loved, his little spring of humor began to go dry. Being sharp as he was, Calvin Coolidge must have anticipated that Herbert Hoover would lose the Presidency in 1932. But he made the New York speech at Madison Square Garden under protest, protest— not because he wished to stand outside the collapse of his party, but because he was profoundly tired. When the collapse came, when the November election indicated such a complete debacle for the Republican party, no degree of aloofness from the situation could keep Calvin Coolidge from being low in his mind. For even if he kept away from politics after 1929, he was a Republican of Republicans, regular to the core. And for the sentimentalist he was, the blow of the November election must have had a physical repercussion.

Almost four years to the day, almost four years from the minute when Calvin Coolidge walked out of the White House, the incarnation of the American spirit, the Puritan ideal of thrift and industry upon which the Babylonian debauch of plutocratic waste and splendor was founded, another man walked past the white pillars and across the portals of the

[9] June, 1935.
[10] *Good Housekeeping*, June 1935.
[11] *Ibid.*

colonial mansion, who embodied as Coolidge did in his own day, the spirit of the times. What a contrast between these two American idols! Franklin Roosevelt was everything that Calvin Coolidge was not. One was light and gay, suave and facile, where the other was sour and dour and harsh and heavy. If Coolidge was the Puritan in Babylon, Roosevelt was the cavalier in the seventh hell. Yet the American people chose to worship each in his day because each had what the people had not who set him on their altar. As Calvin Coolidge left the White House in the rain that March 4, 1929, how little could he dream that the whirling wheel of time would so completely change the people who roared their farewell to him on his way out of the Washington station into the portentous mists of that day! What dire and terrible events hovered in the offing to make his countrymen turn from a man like Coolidge and with all sincerity and all honor and all integrity follow a man so different and so strange. Never before in our history has the fickle mob of Washington which roared its farewell in the station changed so quickly, so completely as it turned in four fleeting terrible years from Calvin Coolidge to Franklin Roosevelt.

A month after the election, he went to New York on his last trip. Outwardly he appeared to be in his usual health. A month before, when his chronic weariness worried him, his home doctor in Northampton had gone over him, particularly examined his heart and found it sound. But the heart is a tricky thing. It does not always yield its secret to a cursory examination.[12] At least possibly the New York journey increased his lassitude. Evidently he was keen to get home. The day before he left New York for home he wrote this, his last love letter, to the woman who had been his refuge and strength for a generation—she, whom he wrote "has borne with my infirmities and I have rejoiced in her graces. We thought we were made for each other." His letter reads:

[12] In a letter to the author, a doctor of national standing who had discussed Calvin Coolidge's health with other doctors who had examined him, writes:

"In my experience I have rarely known a man to have the type of heart condition such as that from which President Coolidge suffered without himself having had some of the danger signs. Often the pain and discomfort and difficulty with breathing are ascribed to some other condition, sometimes deliberately by the physician to remove fear of worry. My interpretation is that such an introspective man as Calvin Coolidge would have formed his own judgment as to the physical difficulties from which he suffered. . . . The mere examination of a man's chest and heart with the usual physical methods often does not elicit the pathology that is present in angina cases or in the type of patient dying with sudden heart failure from disease of the blood vessels of the heart or the walls of the heart. It would take an electro-cardiograph and other refined methods, and even with these the picture may be by no means clear.

"I got the impression from some of those who were in immediate attendance on Mr. Coolidge that some of his symptoms went further than digestive disturbance and had to do with circulation. I rather interpreted Mr. Coolidge's decision and his stubbornness in maintaining it to his own attitude toward his own condition."

The Vanderbilt Hotel
Park Avenue at Thirty-fourth Street
New York   Thursday

My dear Grace:

Tomorrow I shall go home. Unless you hear send the car to Springfield
at 8:40 Friday.

I have thought of you all the time since I left home.

With much love,
Calvin Coolidge

Here is a flash into the heart of Calvin Coolidge, irritable probably,
peevish sometimes, not above the petty cruelties of a repressed and sublimated habit of inner wrath, but still always loyal, always deeply affectionate and through it all profoundly devoted. In the various loyalties of Calvin
Coolidge, and he had many and they were strong, none was more beautiful
than his loyalty to this woman.

The next day at the Vanderbilt Hotel, after Coolidge had finished his
business downtown at the New York Life Insurance Company's meeting,
he sat sprawled, after his fashion, slumping on his spine, talking to Henry
L. Stoddard,[13] who had been prominent in New York politics for twenty
years. Stoddard once published the *Evening Mail*. He was a friend of
Theodore Roosevelt the elder. Stoddard came to the hotel at Coolidge's
invitation. Stoddard remembers that as Coolidge grew restless he began to
move about the room. He took short, rather timid steps. His complexion
plainly showed slow heart action. Stoddard realized that his host had to be
careful of himself. But the former President finally sprawled down and with
the familiar smile on his face began his monologue:

"The election went against us much more heavily than I had anticipated.
I suppose that, since it had to be, it is just as well that the Democrats have
it lock, stock and barrel; but somehow I feel it is a mistake to break down
the Hoover administration just as it is making progress toward national
recovery. The Democrats probably will set aside the Hoover measures and
try some of their own. That only means more experimenting with legislation. The big thing this country stands most in need of just now is economy. Unless Congress can bring down expenditures drastically all other
measures will not count for much. Probably the people will have to find
a way out themselves."

"The Democrats don't propose to pay much attention to Hoover in this
session," Stoddard interrupted.

"That is not unusual," replied the ex-President. "It is no reflection on
Mr. Hoover. A President on his way out is never given much consideration.

[13] We have the same picture of him at the Union Club during the Boston strike.
The dialogue following is reported in Mr. Stoddard's own words.

That's politics. I remember that after the 1922 election, when the tide went only slightly against the Republicans, I could see a difference in the number and manner of visitors who came to the Vice President's office the next winter. It seems to be human nature to want to be with the winner."

"Well," his guest said, "they're surely crowding around the winner now—even some Republicans or presumed Republicans."

"I haven't followed matters closely enough to understand just what they're aiming at," said Mr. Coolidge. "I am out of it and have kept out. I made up my mind not to embarrass President Hoover by comments on his policies, one way or the other, and I have never made any. The surest way to avoid it was to put my mind on other subjects, and I have done so. It was hard work for me to do that *Saturday Evening Post* article on President Hoover last summer just because I had not kept posted. Up in Plymouth, you know, you are pretty well out of the current.

"I have been out of touch so long with political activities that I feel I no longer fit in with these times," continued Mr. Coolidge. "Great changes can come in four years. These socialistic notions of government are not of my day. When I was in office, tax reduction, debt reduction, tariff stability and economy were the things to which I gave attention. We succeeded on those lines. It has always seemed to me that common sense is the real solvent for the nation's problems at all times—common sense and hard work. When I read of the new-fangled things that are now so popular I realize that my time in public affairs is past. I wouldn't know how to handle them if I were called upon to do so.

"That is why I am through with public life forever. I shall never again hold public office. I shall always do my part to help elect Republican candidates, for I am a party man, but in no other way shall I have anything to do with political matters.

"I hear talk of nominating me for President in 1936. That cannot be. There is no way I can decline something not yet offered, but I am embarrassed by the discussion of my name. I cannot answer letters or give interviews about it, but I want to stop it before it gets too far. I authorize you now to say publicly, in your own way and in your own time, that I am no longer to be considered for any public office. I do not care to have you quote me directly, but you will know how to state it so that it will be accepted as authoritative. I do not think anything should be said until after the holidays; people will not be paying much attention to politics the next few weeks.

"We are in a new era to which I do not belong, and it would not be possible for me to adjust myself to it.

"These new ideas call for new men to develop them. That task is not for men who believe in the only kind of government I know anything about. We cannot put everything up to the Government without over-burdening it. However, I do not care to be criticizing those in power. I've never been much good attacking men in public office. If they succeed, the criticism fails; if they fail, the people find it out as quickly as you can tell them."

"But, Mr. Coolidge, when this so-called new deal fails to accomplish all that the people expect of it," Stoddard replied, "will they not turn to conservatism overwhelmingly and to you as its most conspicuous leader? Will it not be impossible for you to resist such a demand?"

"It was not in 1928," replied Mr. Coolidge, "and it will not be again. I am through with public life. You cannot state it too positively. Nothing would induce me to take office again."

On New Year's day his old friend, Charles A. Andrews, Treasurer of Amherst College, saw him, and in reply to an inquiry about his health Coolidge said:

"I am very comfortable because I am not doing anything of real account. But any effort to accomplish something goes hard with me. I am too old for my years. I suppose carrying responsibility takes its toll. I am afraid I am all burned out. But I am very comfortable."

He regretted missing the class dinner in New York and added:

"It is difficult for me to go anywhere; I have to be attended. A police squad goes with me. You know," and here came a grin and a chuckle, "it's my past life that makes all of this trouble. If I could only get rid of my past, but that always stays with one!"

It was a fine day, an open winter day, January 5, 1933, in Northampton. Calvin Coolidge rose, went about his daily chores, neglected to shave before breakfast. Breakfast usually punctilious—was on the tick of the tock. And at nine o'clock he went downtown to his office. He stayed there for a time, perhaps an hour, doing odd jobs, attending to the routine of his office work and business duties, then he rose and said casually to his associate that he was not feeling very well and that he thought he would go home. Just that. At home he sat down for a while, apparently reading. Mrs. Coolidge had gone into Main Street for her morning's shopping. Some casual errand attracted him to the basement. He went down, passed the man of all work there with a brusque "Good morning, Robert," and climbed the two flights of stairs to his bedroom. At noon he remembered that he had not shaved and went upstairs, took off his coat, got out his shaving tools and then—no one knows exactly what happened. When Mrs. Coolidge came in she called cheerfully to him as was her wont, but when there was no answer she went to the second floor to put away her wraps and there, face downward on the floor she found his lifeless body. He who had lived aloof, died alone.

CHAPTER XLI

The "Sadness of Farewell"

CALVIN COOLIDGE died the first Thursday in January and they buried him Saturday. It was a cold, dour, rainy day in Northampton, but the largest crowd assembled that Saturday that the town had ever seen. People from all over the wide Connecticut Valley motored in, and parking space, half a mile from the Edwards Congregational Church, was taken in the early morning. The Mayor announced the day before that the stores would not be closed.

"Every nickel counts," said he in a public statement. "If the business places close they might lose some sales and that is exactly what Calvin would not want."

So with Puritan thrift and to give the place a solemn air while saving the nickels, the Mayor asked the merchants to draw down the window shades. Nor did he drape the town in mourning, being frugal: "Calvin was a simple man. He would not want the people to go to all that expense," the Mayor said! So they tied crape on the City Hall.

From all over the land visitors who came thronging in, that cloudy day, saw there the bedraggled black cotton limp and listless, and a wilted flag hanging at half mast. Before nine o'clock the sidewalks on Main Street and in front of the church were packed. The windows of the stores and the high school across the street showed the buff of curious faces. Most of the seats in the church were reserved.[1] James Lucey sat in one of these—exalted in his friendship to the last. Only the galleries were free. When the doors of the church swung open it took less than a minute to jam the galleries. The crowd outside reflected symbolically the disorganization and disorder of the times. The President and members of the Supreme Court

[1] The Coolidge pew well back in the Auditorium was marked and remained vacant.

441

were jostled, elbowed, buffeted as the state police wangled a wedged opening for the great men to enter the church. Silk hats were knocked askew, clothing was twisted awry, and the procession of dignitaries—President Hoover, the members of the Supreme Court, the United States Representatives and Senators, the governors of adjoining states—ran a gauntlet to the church doors. As Mrs. Coolidge came in, with the Stearns family and Senator Butler, the crowd respectfully made way.

A young Congregational preacher, a few months out of college, less than half a year in Northampton, took charge of the services. He was a handsome curlyheaded boy with a voice like a bishop and he read the simple service of the democratic church.

The service opened with Handel's Largo from "Xerxes," and the church choir which Calvin Coolidge had heard so often sang "Lead Kindly Light." No eulogy was spoken. The prayer was formal and simple. Outside in the rain, which changed to snow at times, the throng stood silent when it heard the organ mourn. As Mrs. Coolidge rose to go, the organ lifted its voice in the solemn theme from Dvořák's New World Symphony, the Largo which has been given the words popularly "I'm Going Home, Going Home." The throng, hearing the dirge swell as the church door opened, made way for the Coolidges when they appeared in the street. As they entered their cars, a rift of cold New England sunlight flooded the place for a moment. But as the funeral cortege left the town, the clouds closed down. The rain began to fall listlessly.

The President and the dignitaries of the Senate and Court turned back to Washington from the church. Mrs. Coolidge and a few faithful friends, the Stearnses, Senator and Mrs. Butler, and the Coolidge family, motored behind the hearse up through Massachusetts into Vermont in a pelting rain. Yet at every farm site, village and town thousands of his New England neighbors stood uncovered, curiously, earnestly, sadly gazing at the cortege. It was afternoon when the hundred miles was spanned which separates the place where Calvin Coolidge lived for thirty years and died from the place of his birth. The funeral party was late. The afternoon was waning. Up the hill from Plymouth village the mourners came. Gathered outside the cemetery were a hundred old friends and neighbors standing under umbrellas in the sleet that was whitening the mountains.

The Plymouth graveyard rests on a terraced hillside. On the Coolidge terrace a canvas had been erected over the open grave to protect the family. With the family standing there were Mr. and Mrs. Frank Stearns and the Butlers with Young John and his wife, and a few more intimate friends. From a vantage point just below the retaining wall of the bottom terrace a homely group of villagers watched the mourners. The weird and somber contour of a score of covering umbrellas was varied by a bright splash where a man wrapped his little boy in a yellow horse blanket. Six sturdy United States marshals in their dress uniform from the states surrounding, and of course Republicans, bore the heavy coffin to the grave. The bronze coffin carried few flowers. President Hoover's Japanese Leothe leaves lay above his predecessor's breast. The young preacher's prayer was short. Dust was consigned to dust, ashes to ashes. No band wailed, and the village watchers heard no song or psalm. There was a pause and then the clear young voice of the preacher at Mrs. Coolidge's request intoned these lines by Robert Richardson:

> Warm summer sun,
> Shine kindly here;
> Warm southern wind,
> Blow softly here;
> Green sod above
> Lie light, lie light.
> Good-night, dear heart,
> Good-night, good-night.

For a moment the silent crowd stood, unmoving. Then the clear blast of a soldier's bugle cried taps to the lonely mountains. When the echoes had died, a strange and lovely thing happened. For a moment, scarcely more than a minute, the slanting western sun broke through the clouds, cast its horizontal light upon the hillside, then closed quickly down, leaving a lemon colored glow through the mist, brilliant like a shimmering halo. The mourners turned and went about the street and man to his long home. When they were gone the sleet fell in little gusts upon the frozen grass like grudging tears of some dejected god. Thus the earth of the hills that he loved closed over Calvin Coolidge's mortal clay. With this pale pageantry his day was done.

What of him—this rather drab, colorless figure who purposely kept drama from his life and staged himself as a primitive and solitary figure walking his appointed way to a place of power? What will be the verdict of his countrymen and of the world when the record finally closes? Of course it will be many years before the record ends. A cloud of witnesses writing their stories will appear and things unseen will become plain in another day and decade. It is too early to put a final estimate on Calvin Coolidge. But surely this much may be written—he was honest; he was cautious, but he never lacked at last for courage. He walked through the politics of his time touching elbows with the worst of his contemporaries as he met and passed them on his pilgrimage through American government from the bottom to the top. Yet he went unsmirched. He knew the obvious realities of his environment. He never blinked the sordid facts of any problem that came to his hands. Neither did he advertise those sordid facts, nor ever bewail them. He took American politics as it was, not perhaps as he would have liked it. He played the game under the established rules, with men about the table good and bad—as they came under the spotlight. He played his winning hand. He was wise according to his day and generation. His instinctive common sense was illumined by rather more intelligence than most men had with whom he gambled. He knew that politics was a cult in America, and he held no nonsense in his heart about its purification. To him that remained always "an iridescent dream." But he did, with all his heart, try to get out of the politics in which he worked, bad as it might have been, what he thought was the best for all the people. Yet he never played the demagogue. To his friends he was loyal, crabbedly cordial, and in the end always kind, whatever petty pouting he did to ease his tired nerves. To his country he gave unstinted devotion. In the terrible decade when he was in his place of greatest power, he lacked the vision to exercise the highest judgment. He was handicapped there by his background, circumscribed by his own life's pattern, his temperament, his experience as the pampered adopted child of what he felt was a benevolent plutocracy. He did not question its authority, nor its beneficence. Another wiser man coming to the helm from another ship, as for instance from business or perhaps from the academic cloister, if he had held the nation's wheel, possibly might have steered the boat from the rapids. Probably not. Such a man would have been a freak in our politics, an accident. His wisdom might have wrecked that wiser man and with him wrecked his country a few years before its debacle. Calvin Coolidge was democracy functioning at its best, which sometimes is its worst. Being what he was, he was forced, by the destiny of his own qualities, his own ideals, his high calling, into the way he took. Back of him were the urgent purposes of the American democracy, the lust for prosperity, the Hamiltonian faith that "the rich" are indeed the "wise and good," the Republican creed which identifies wealth with brains. His short, sage, oracular sentences polished like New England granite, reflected the American heart in the days when he guided the American people. But whither they would go he led them, not as a weakling, not as a demagogue, but cherishing the noblest purpose he knew, following the democratic vision of the America of his age.

Was it not well that when he laid him down to sleep his stern, scorned dying gods, the Puritan gods of things that were, rained their icy tears upon his hillside bed!

Index

choose," 399–400; effect on C. C. of surrender to Hoover, 401–402

Cabinet, C. C. at meetings by invitation of Harding, 224, 230–231; Harding, taken over by C. C., 250–251

Cal, use of, as a name for C. C., 82, 83, 89

Calvinism, C. C. ever a believer in, 72

Cape Cod Canal, 310

Capper, Arthur, in Senate Progressive group, 261, 262; friendly with President but head of Farm Bloc, 262; on good terms with C. C., 318; visits C. C. in Black Hills, 357; his story of C. C.'s "I do not choose," 359–361

Carpenter, Ernest C., teacher, memories of the boy C. C., 19, 20

Central America, dollar diplomacy in, 323–324

China, Boxer indemnity relinquished to, 312

Christy, Howard Chandler, portrait of President Coolidge, 254

Christy, Mrs. Howard Chandler, pies praised by C. C., 254

Cicero, orations of, 28, 47

Cinderella, C. C. as a, 68, 74, 215, 314

Cleveland (Ohio), convention hall in, 298

Cleveland, Grover, Presidency of, 25; Vermont attitude toward, 26–27; deserts party on gold dollar, 50

Cody Dam, the Coolidges at, 363–364

Cole, John R. (Speaker of Mass. House), reception of "singed cat" letter, 73; C. C.'s investment in entertainment of, 74

Commercial and Financial Chronicle, on Coolidge bull market, 362

Congress, Pres. Coolidge's first message to, 260, 262–263, 264, 265; C. C.'s 1925 message to, 309–310; warned by C. C. against further army and navy reduction, 323; regular Repub. organization of, conservative to reactionary, 348

Congressional Library, books borrowed from, by C. C., 234

Connecticut Valley, backbone of Mass. Repub. party, 76–77

Constitution, defended by C. C., 147

Continental Trading Co., history of, 225, 226–227, 228

Coolidge, Abbie, sister of Calvin, 11, 15, 23; effect of her death on Calvin, 29

Coolidge, Mrs. Calvin, begins married life, 64–66; in Jonathan Edwards Congrega-

tional Church, 75, 83, 114; as mayor's wife, 82, 86–87; as state senator's wife, 89; C. C.'s opinion of, 102; activities of, 126; at home, 127; visits to Boston, 138; at C. C.'s inauguration as governor, 139, 140; hears from C. C. of his nomination for Vice President, 214; looking back, calls C. C. "always sentimental," 215; C. C.'s table manners needed her care, 222; triumph in Washington official society as Vice President's wife, 231–232, 235; C. C. buys Gainsborough hat for, 238; at Plymouth swearing in of C. C. as President, 243; with C. C. on trip to Washington, 244; with C. C. on Harding funeral train to Marion, 244–245; unexcelled in charm and training for task as White House mistress, 255–256; never part of President's political family, 256; object of C. C.'s pride while he was her problem child, 256–257; tea to Supreme Court Justices' wives, 285; effect on, of son's death, 309; in Inauguration Day parade, 315; late return from walk with secret service man upsets husband, making news that irks him, 353–357; surprised by President's "I do not choose," 361; rumors as to health of, 366; at White House reception to Queen Marie, 382; C. C.'s interest in her dress, 392–393; last day in White House, 418–419; resumes old life in Northampton, 421; notes from C. C. to, 424, 438; at "The Beeches," 426

Coolidge, Calvin, Jr., home life of, 127; death of, 308–309

Coolidge, Calvin Galusha, grandfather of C. C., 12–13, 28, 29; entails farm to C. C., 13, 45; Bible of, used in administering Presidential oath to C. C., 315

Coolidge, Mrs. Calvin Galusha, 12; influences C. C. against dancing, 18, 36

Coolidge, Carrie A., stepmother of C. C., 31

Coolidge, John (son), birth of, 65; home training of, 127; reveals C. C. as a disciplinarian and a tease, 256, 257

Coolidge, John Calvin, public service of, 10, 13, 81; father of C. C., 10, 11; training of C. C., 14, 16; inventory of his possessions, 16; pay for superintending schools, 20; enrolls son in academy, 22–23; runs postoffice, 27; remarries, 31; ambition for C. C., 45; continued help of C. C., 47; his influence over C. C., 74; visited by C. C. as mayor of North-